Asia Unplugged

The Wireless and Mobile Media Boom in the Asia-Pacific

Editors

Madanmohan Rao

Lunita Mendoza

Response Books

A division of Sage Publications

New Delhi/Thousand Oaks/London

First published in 2005 by

Response Books
A division of Sage Publications India Pvt Ltd
B-42, Panchsheel Enclave
New Delhi - 110 017

Sage Publications Inc
2455 Teller Road
Thousand Oaks
California 91320

Sage Publications Ltd
1 Oliver's Yard
55 City Road
London EC1Y 1SP

Published by Tejeshwar Singh for Response Books, typeset in 10 pt Garamond by InoSoft Systems and printed at Chaman Enterprises, New Delhi.

Library of Congress Cataloging-in-Publication Data

Asia unplugged: the wireless and mobile media boom in the Asia-Pacific/ editors, Madanmohan Rao, Lunita Mendoza.

p. cm.

Includes bibliographical references and index.

1. Cellular telephone services industry—Asia. 2. Cellular telephone services industry—Pacific Area. 3. Mobile communication systems—Asia. 4. Mobile communication systems—Pacific Area. 5. Wireless Internet—Asia. 6. Wireless Internet—Pacific Area. I. Rao, Madanmohan. II. Mendoza Lunita.

HE9715.A78A85 384.5′34′095—dc22 2004 2004021779

ISBN: 0-7619-3272-0 (HB) 81-7829-456-7 (India-HB)

Production Team: Gargi Dasgupta, R.A.M. Brown, Mathew P.J. and Santosh Rawat

Asia Unplugged

To the endless quest
for the work-life balance

Contents

Part III: Book Reviews

All Book Reviews are by Madanmohan Rao

About the Editors and Contributors

Editors

Madanmohan Rao, a consultant and writer from Bangalore, is research director at the Asian Media Information and Communication Centre (AMIC). Madan is the editor of two book series: *The Asia-Pacific Internet Handbook* and *The Knowledge Management Chronicles*. He serves on the board of a number of new media firms, and tracks wireless and Internet developments via a blog for the Poynter Institute. Madan graduated from IIT Bombay and the University of Massachusetts at Amherst, with an MS in computer science and a PhD in communications. Madan has spoken at IT conferences in over 50 countries. He is also actively involved with the INSEAD Innovasia programme for tracking emerging technologies in Asia. His hobbies include swimming, hiking, world music and beer appreciation.

Lunita Mendoza is the editor of *Wireless World* magazine in Singapore. She has travelled extensively to wireless forums and summits across Asia, Europe and the Americas, and covers trends ranging from handheld devices and mobile payment to technology evolution and venture capital dynamics. She formerly ran three other publications in the IT world. Lunita has a degree from Australia's Royal Melbourne Institute of Technology, with a minor in broadcast journalism and a major in print journalism. She has been actively engaged in professional and competitive sport, enjoys diving, rock climbing and wakeboarding.

Contributors

T. H. Chowdary is the information technology advisor to the Government of Andhra Pradesh in India.

Paul Gardner is an independent consultant providing strategic management services to high-growth opportunities within early stage and mature companies.

June Gassin is associate dean for IT and multimedia at the University of Melbourne, Australia.
Laina Raveendran Greene is managing director of Getit Multimedia in Singapore and an IT consultant in Silicon Valley.
Ashok Jhunjhunwala is professor of electrical engineering at the Indian Institute of Technology, Madras.
John Lee is principal, Softbank Venture Capital, based in Seoul, South Korea.
Srikanya Mongkonsiri is the managing director of Chuo Senko in Bangkok, Thailand's leading advertising agency.
Som Naidu is associate professor in the Multimedia Education Unit at the University of Melbourne, Australia.
Loudon Owen is co-founder and managing partner of McLean Watson Capital Inc., based in Toronto, Canada.
Vicente Rafael is professor in the department of history, University of Washington at Seattle.
Sayeedur Rahman is IT administrator in the Governor's Office for Technology, Commonwealth of Kentucky, and the founder of Bangladesh Information Technology group Bangla-IT.
Subha Rama is an industry analyst in Enterprise Communications at Frost & Sullivan's Technology Practice division in Singapore.
Michael Sainsbury is the telecommunications writer at *The Australian* newspaper.
Shinichi Saito is an associate professor in the department of communication at Tokyo Women's Christian University.
Sunanda Sangwan is associate professor of marketing and electronic commerce at Nanyang Business School in Singapore.
Lara Srivastava is a policy analyst in the strategy and policy unit of the International Telecommunications Union in Geneva.
Chas Sweeting is consultancy group director at the technology practice of GCI Malaysia.
Tan Min-Liang is a technology/telecommunications lawyer with Messers Tan & Tan and serves as an external legal counsel to a regional telecommunications company.
Gaurab Raj Upadhaya is the CEO of the Nepal Internet Exchange in Kathmandu, and is also a writer and radio journalist.

List of Figures

List of Tables

Preface

As this book goes to press, it is just about a decade since the roll out of second generation (2G) mobile services worldwide marked the rapid rise of a new medium. Numerous pressing questions are being thrown up today in the realm of wireless and mobile technologies.

How can news media and the entertainment industry harness mobile devices and technologies while also protecting their brands and encouraging innovation? Can publishers really make money by selling news alerts via SMS? How do WLANs and mobile access affect workspace design, workflow configuration and productivity in industry, education and government? What can regulators do to promote healthy competition between mobile and landline operators, and boost Wi-Fi and Wi-Max adoption? Can wireless media help bridge the digital divide? What merger and acquisition strategies work for investors, operators and startups in the wireless ecosystem? Can m-government amplify existing e-government practices? What location-based services and games work best?

How will the great 3G wars pan out in different markets of the world—for instance, will Wi-Fi be a serious contender? How will copyright laws work for mobile content? How can advertisers and marketers harness mobile marketing without compromising user privacy? How can wireless operators increase ARPU via content services like news delivered to cellphones? What new syllabi, technology partnerships and internship arrangements can help academia prepare students for the wireless newsrooms of the future? How will entertainment and content industries fare in the post-Napster world? What political impact will smart mobs and swarms have on society? Will payment by mobile phones displace cash and credit cards?

The Asia-Pacific region has its own twists and additions to such themes.

How effective will the new SMS anti-spam consortia be in Australia? What content-partnering relationships will help winners in the 2.5G and 3G wireless Internet race in South Korea and Japan, and will these models be easily exported to other markets in Europe and the Americas? How long will the Chinese government be able to monitor and control Internet and SMS traffic, or try to set new standards for Wi-Fi? Will the low-cost handheld device, the Simputer, help overcome India's yawning digital divide? What new ethical and legal issues are opening up along the mobile frontier?

How can knowledge management and business intelligence be better harnessed by Asia's wireless players? How are Asian governments dealing with the challenges of policy making and dispute settlement in the face of converging telecom, Internet, media and mobile sectors? How has mobile communication helped voters in Malaysia and activists in the Philippines? Can Grameen Telephone's successful micro-finance model of shared mobile device access in Bangladesh be replicated in other parts of the world? Will Asia be able to succeed in Wi-Fi diffusion as well as it has succeeded in consumer mobile? How are Asia's hardware, software and network services companies gearing up to face the heat of global market expansion and competition?

This book in your hands is the first comprehensive attempt to capture the complexity, opportunities and challenges of the impacts of wireless technologies (e.g., Wi-Fi, Wi-Max) and mobile networks (e.g., GSM, CDMA) across the Asia-Pacific region.

This book is actually part of a larger work in progress: the first book in the series was called *The Asia-Pacific Internet Handbook, Episode IV: Emerging Powerhouses*. It chronicled the rise of major Internet infrastructure, content and services markets in the region, especially Japan, South Korea, China, India, Australia and Singapore.

Drawing on the *Star Wars* movie series metaphor where the first released movie was called Episode IV as part of a larger saga, this Asia Internet series of books also began with Episode IV. The previous episodes (or 'prequels') have yet to be written; they would consist of the following phases.

Episode I (1960–1980) would mark the birth of early computing infrastructure in Asia, creation of premier research and educational institutes in the region (such as the IITs and IIMs of India), and early corporate structures such as the *keiretsu* of Japan and *chaebols* of South Korea. The New Economy in Asia is still shaped by these Old Economy forces.

Episode II (1980–1995) would mark the rise of the early internetworks, the academic Internet, and the Asia-Pacific Network Information Centre. Leading academics from this Episode continue to play influential roles in shaping the minds of the younger Net generation and of policymakers.

Episode III (1995–2000) would mark the rise of the commercial Internet, the beginning of widespread telecom and datacom deregulation, and the early wireless networks in Asia. Many of the leading players of this Episode are today's incumbent players.

The Internet constitutes a major market for Asia (as identified in Episode IV) but is also a major medium for news for domestic and diaspora Asian consumers (as identified in Episode V). The current book is called *Asia Unplugged: The Asia-Pacific Internet Handbook, Episode VI* as a reflection of the growing influence of the wireless Internet and mobile communications in a wide range of consumer bases and industry sectors across the Asia-Pacific. The book also draws out the contrasts between the mobile-savvy and mobile-emergent environments of the various Asia-Pacific countries, and highlights useful lessons to share with the rest of the world in areas ranging from forming content alliances to bridging the digital divide.

My journeys across the region have exposed me first hand to the spectacular dotcom boom and bust of the past five years, and the broader undercurrents of long-term Internet and mobile impacts on media, commerce, government, and consumer behaviour. My trips on consulting and speaking assignments to over 50 countries worldwide have convinced me of the unique texture of the Internet and mobile environment in Asia-Pacific, thus spawning the 'Asia-Pacific Internet Handbook' series to chronicle and analyse these developments.

With a background in IT, media and business, and work exposure in Asia, Europe, the US and Latin America, my interests over the years have coalesced into the fascinating and compelling intersection between technology adoption/creation and (1) national cultures, (2) organisational cultures, (3) sectoral cultures, and (4) professional/vocational cultures.

This has led me to launch two series of books, drawing on contributions from dozens of experts around the world. *The Asia-Pacific Internet Handbook* series explores the growth of the wired and wireless Internet and its impact on the technology sector, business environment, political climate, and cultural attitudes in Asia-Pacific. *The Knowledge Management Chronicles* consists of thematic compilations of essays about knowledge management (KM) practices in different organisational settings. Both themes—wireless and KM—come together in an interesting chapter in this current book, which focuses on wireless technology impacts on knowledge mobilisation to field workers, as well as the use of KM as a business discipline by wireless operators.

This book is divided into three parts: overall themes (like location-based services and m-government), country profiles (of wireless and mobile environments), and resources (reviews of relevant books on wireless and mobile media). The book provides a useful framework (based on the '8Cs' parameters: connectivity, content, community, commerce, capacity, culture, cooperation, capital) to assess wireless and mobile media impacts across countries. In terms of wireless environments, it classifies the countries of Asia-Pacific into seven types, based on a combination of these factors: restrictive, embryonic, emerging, negotiating, intermediate, mature and advanced.

The timing of the book also seems perfect. Asia's pioneers are widely regarded as the world's epicentre of the mobile revolution and the wireless information society. As early as 2002, Asia accounted for 36 per cent of the world's telecom market (up from 21 per cent in early 1991, and scheduled to hit 50 per cent in 2007), the world's largest regional user base of cellphones, 33 per cent of the world's Internet user base, 95 per cent of the world's 3G mobile users, 47 per cent of the world's ADSL broadband Internet users, and seven of the world's Top 10 most profitable telecom operators (according to ITU figures).

This book is required reading for all those involved in the entire spectrum of new media: journalists, publishers, broadcasters, portal players, designers, wireless operators, technology players, academics, educators, advertisers, public relations agencies, brand marketers, policymakers, legal experts, and students.

Putting together this informative book has drawn on the efforts and experience of over a dozen experts in Asia and Australia, and thanks must go to each and every one of them, their colleagues and families for their unique contributions and voluntary assistance. Special thanks go to Lunita Mendoza, editor of *Wireless World* magazine in Singapore, who joins me in this publication of the Asia-Pacific Internet Handbook series as co-editor. Her breadth of knowledge, personal contacts in the industry, business insights, depth of perspective, inexhaustible energy and graceful charm have been indispensable for this project.

The support from the Asian Media Information and Communication Centre (AMIC) is also gratefully acknowledged. And in true Internet style, the book also has an online companion on the publisher's Website at www.indiasage.com/asia-unplugged.asp. Questions? Fire away, and see you in subsequent episodes of the Asia-Pacific Internet Handbook series!

Madanmohan Rao
madan@techsparks.com
April 2004

Part One

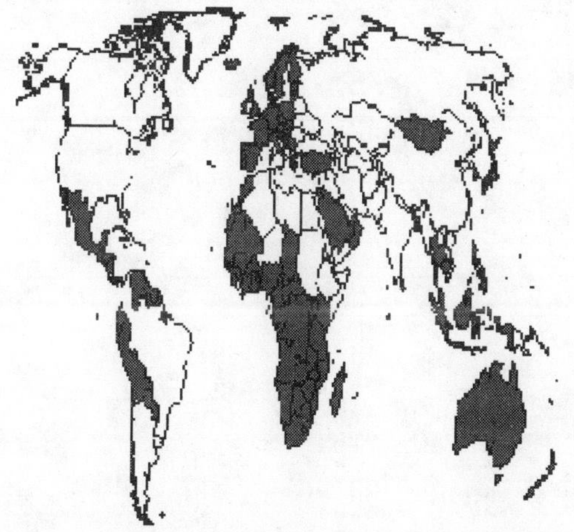

Overall Themes

1. The Wireless Tsunami

An Asia-Pacific Panorama

Madanmohan Rao

Editors' Note: This chapter provides an overview of the mobile ecosystem, highlights Asia's unique role at the head of the mobile race worldwide, identifies a wide spectrum of opportunities and challenges for different players, and ends with an insightful discussion and analysis. A wealth of collated data, research results and market forecasts is presented. Wireless environments in the Asia-Pacific are classified into seven categories, based on a combination of factors like ICT infrastructure, political culture, and international presence: restrictive, embryonic, emerging, negotiating, intermediate, mature and advanced. The key findings of all the contributors in the book are highlighted, and a set of 15 pithy lessons is identified.

Introduction

From dramatically changing business and political climates across Asia to grammatically changing spelling and sentence structures via short messaging service (SMS), wireless Internet and mobile communications are transforming Asian countries and cultures in unprecedented ways—with lessons, benchmarks and opportunities for the rest of the world. The Geneva-based International Telecommunications Union (ITU) predicts that by 2010, more than 50 per cent of all mobile-phone users in the world will be in the Asia-Pacific region, up from 35 per cent in 2000. In terms of manufacturing output, Asian nations like Japan, South Korea, China and Taiwan have become leading global players in hardware, while software and innovation hubs have emerged in India, Singapore and the Philippines.

The convergence of the three media ecosystems—traditional (print, broadcast), Internet/Intranet and wireless/mobile media environments—is sometimes referred to as the EverNet, SupraNet, MetaNet or PermaNet, and interesting variations are emerging in the diffusion and cultural acceptance of different forms of wireless media access around the world.

'The mobile screen has now become the "third screen" in consumer lives after the TV and PC screens. Mobile phones can become mobile companions when they are able to leverage the full power of the Internet, and not just voice,' according to Jay Highley, senior vice president at US carrier Sprint. (For many users, in fact, the 'third screen' may be the most important and relevant one!)

Along with the boom in mobile telephony around the world, a parallel wave of innovation in wireless corporate networking promises to usher in a new world of 'untethered knowledge workers' and 'flexible network organisations.' Wireless local area networks (WLANs) for 'plug and connect' applications are also rapidly emerging in offices, homes, hotels, warehouses, hospitals, stock exchanges, airport lounges, libraries, conference venues and cybercafes.

This chapter provides an overview of the wireless and mobile ecosystem, highlights Asia's unique role at the head of the mobile Internet race worldwide, identifies a wide spectrum of opportunities and challenges for different players, contextualises the rest of the chapters in this book, and ends with an insightful discussion and analysis.

The media analysis used in this book is based on a framework developed by the author over the years, called the '8Cs' of the Digital Economy: connectivity, content, community, commerce, capacity, culture, cooperation and capital (Rao, 2002). In other words, the power and success of the wireless Internet and mobile media depends on the size of the user base, nature of content services, interactive communities, sustainable business models, human resource capacity, progressive culture, cooperation between key sectors of society, and funding for entrepreneurs and startups in new media ventures.

Wireless environments in the Asia-Pacific can be classified into seven categories, based on a combination of these factors like ICT infrastructure, political culture, and international presence: embryonic, emerging, negotiating, intermediate, mature and advanced (explored in greater detail in Framework: The Wireless System).

Market Forecasts: Wireless Explosion in the Asia-Pacific

This section surveys some of the innumerable market forecasts being made with increasing regularity about the clout and innovation of the Asia-Pacific mobile market. The next sections deal with more qualitative analysis, but the numbers alone are quite a mouthful to begin with.

The results are in—and the numbers are impressive. Even in 2002, Asia accounted for 36 per cent of the world's telecom market (up from 21 per cent in early 1991, and scheduled to hit 50 per cent in 2007), the world's largest regional user base of cellphones, 33 per cent of the world's Internet user base (225 million Internet users, or an average of 6.3 users per 100 inhabitants), 95 per cent of the world's 3G mobile users, 47 per cent of the world's asynchronous digital subscriber link (ADSL) broadband Internet users, and seven of the world's top ten most profitable telecom operators (ITU, 2002).

Asia is also a pioneering player in cutting-edge 3G and 2.5G markets. Two of the world's top five mobile phone makers are from Asia (Samsung at No 3 and LG Electronics at No 5).

The Mobile Explosion: Youth Users, SMS

Two of the most unusual and interesting features of the mobile ecosystem are the explosive growth of SMS messages based on humble 160-character strings, and near-addictive behaviour in the youth user segment. Wireless communities—particularly communities for mobile youth (sometimes jokingly referred to as *Screenagers* or *Generation Txt* or *Generation Y-erless*!)—will determine the next 'killer application' that will drive growth in the wireless SMS market, according to Scott Long, general manager of a wireless solutions company.

The cellphone has become, variously, a status symbol, an office accessory, a communications convenience, an emergency alert tool, and even a high-fashion accessory for different consumers groups around the world—and an extension of life and personality for youth.

- In Japan, the penchant for the 'cutting edge, the cute and the compact' has given rise to a peculiar *keitai* culture. Many young people describe their cellphones as extensions of themselves, complete with identity icons, emoticons and personalised ringtones. Lost deep in conversations on mobile phones, users often blur their virtual private space and the real physical space. Keypad culture is so commonplace among youth that in some classes in Japan, students e-mail questions and comments to their professors via cellphones!

(*Box continued*)

(*Box continued*)

- In India, news reports address student use of SMS to cheat in exams or women's use of SMS to send risqué messages. According to a recent IDC India survey, women in India are the most 'vocal' and bold group when it comes to using SMS. 'When it comes to sending adult jokes, most Indian women find SMS a safe and comfortable option,' according to psychiatrist Samir Parikh. SMS is also seen as a covert way of propositioning.

- Nearly one in 10 polled Britons admits to dumping his or her partner by SMS rather than breaking the news face-to-face, according to a survey by market research firm NOP. Nearly half admitted to flirting by SMS.

- For about half of Dutch youths aged between 12 and 16, SMS is more popular than making phonecalls.

- Filipino operators Smart and Globe report that data services like SMS represent around 35 per cent of revenues compared with just 25 per cent for NTT DoCoMo, which lies in distant third place.

- Queensland University of Technology sociolinguistics lecturer Jo Carr warns that people using e-mail and SMS text messaging were unconcerned about grammar and punctuation. 'Grammar rules used to be an indication of social class and literacy in the past but today's society are doing things differently because language today serves the purpose of speed and social interaction,' she says.

- As with New York, Tokyo and London in earlier years, the Berlin youth fad in summer 2003 was SMS-related—the Flash Mob. Using SMS texting, scores of people would converge on, say, a Berlin department store and start bowing and kissing each other as if they were long-lost friends. Germans reportedly spend more time with their mobiles than in front of the TV or with their families. Over 70 per cent of commercial revenues on youth-oriented TV networks such as MTV Deutschland are SMS-related during certain time slots.

- In Malta, Italy, Ireland and the UK, the Pope's Thought of the Day is delivered directly by SMS to devout Christian users.

- The BBC recently reported that SMS has eclipsed voice calls as the most common use of mobile phones among young people in the UK.

- The Gartner Group reported in 2003 that SMS has exceeded e-mail as the most popular form of telecommunications among Europeans, particularly so in the UK where 49 per cent of adults use SMS compared to 39 per cent who use e-mail.

- In most countries with global system for mobile communication (GSM) networks, New Year revellers in 2004 smashed the record for the most SMS messages ever sent in a day. Around 111 million SMS messages were sent on all four main UK mobile networks between midnight December 31 and midnight January 1 of 2004. In Australia, Telstra customers sent 480 text messages each second between midnight and 1 a.m. on January 1, 2004. SMS traffic for some carriers in India shot up by upward of 400 per cent this past New Year. Hutch in Delhi recorded a high of 12.6 million SMS on December 31 and January 1 2004 as compared with 4 million the previous year. Globally, mobile subscribers sent more than 20 billion text messages in 2003.

- A survey conducted for Cingular Wireless reports that wireless phones edged out computers, software, and accessories as the top electronic gadget on the holiday wish list of 2003 among US youths aged 13 to 17.

The top market worldwide in terms of broadband Internet penetration is South Korea followed by Hong Kong, Canada, Taiwan, Belgium, Sweden, Ireland, the US, Denmark, Singapore, Austria and Japan. At the end of 2001, the market with the highest penetration of cellphones was Taiwan (a whopping 96 per cent), followed by Hong Kong (84 per cent), Singapore (69 per cent), New Zealand (62 per cent), South Korea (60 per cent), Australia (57 per cent), and Japan (57 per cent).

Asia also accounts for four of the top 10 markets in the world in terms of wireless Internet user base as a percentage of mobile users: Japan and South Korea are in the lead, followed by Finland, Canada, Singapore, the US, Germany, Italy, the UK, Taiwan and France.

Asia is now the largest and most innovative telecom market in the world—but it must demonstrate leadership in this role, said Yoshio Utsumi, secretary general of the International Telecommunications Union, at the ITU Telecom Asia summit in Hong Kong, titled 'From Recovery to Prosperity'.

Michael Minges, head of the ITU's data and statistics unit, says the Asia-Pacific region remains the world's largest telecommunications market, posting steady growth in telephone, mobile and Internet subscribers. 'This region continues to push the envelope for universal service,' Minges says, noting that the region had been able to add 'one new telephone user every second for the past decade'.

Growth in Asia-Pacific telephone subscribers was around 10 percentage points above the global average, with the gap increasing over 2003. There are more cellphone users than fixed line users in more than two dozen nations including Japan, China, Indonesia and Fiji.

From 1991–2001, China added over 300 million new telephone subscribers, half of the regional total, taking its combined teledensity from less than one in 1991 to 30 by mid-June 2002. 'Never before has a country added so many telephone subscribers so quickly and raised its teledensity so rapidly,' says Minges. China today also has the world's largest user base for mobile phones and the second largest base for Internet users.

In terms of SMS traffic volumes per average user, the Philippines (as of 2003) led the Asia-Pacific region with an astonishing average of 336 SMS messages sent per month per user followed by Singapore (75), Australia (42.9), South Korea (27.3), Malaysia (24.9), Taiwan (12), and Thailand (3).

An estimated 22 billion SMS messages were sent worldwide in 2003, compared with 16.5 million in 2002. In Europe, 18.3 million multimedia messaging service (MMS) messages were sent in 2003, with a 10-fold increase of MMS users during that timeframe and an average of over 4.5 MMS messages sent per user. Thirty nine per cent of all new handsets sold in Europe in 2003 were MMS-enabled, while 14 per cent were camera phones.

As anticipated, mobile gaming has been charging ahead with revenues of over $1.5 billion in 2003 (they were almost negligible in 2000). The compound growth is expected to be around 60 per cent per year, according to the ARC Group, with revenues of around $7.4 billion by 2007 (a large part of the total revenues of $25 billion in mobile entertainment expected by then).

Mobile gaming encompasses several categories of hardware, including cellphones (both online and voice-based), PDAs (which may also have phone capabilities) and mobile game platforms such as the GameBoy Advanced and the upcoming Sony PSP. The largest of these categories by far is cellphones, with an installed base of over 1.3 billion worldwide; almost all new phones can play games.

US gamers are far behind their Asian and European counterparts, with far fewer games being developed and less revenues being generated. For the global wireless market (not all mobile games are wireless since many are standalone or can be uploaded with removable media), the US represents about 5 per cent of the total, Europe 15 per cent and Asia over 75 per cent, according to Strategy Analytics.

In-Stat/MDR claims wireless gaming will grow to $2.8 billion worldwide by 2006. According to Internet Digital Software Association (IDSA), 31 per cent of game players claim to play games online. Japan and Korea lead the world in mobile gaming. US teens already spend more time on video gaming than watching television or using e-mail, according to game publisher Electronic Arts.

Gartner predicts that Asia-Pacific will have more than 1 billion phone connections during 2004, thanks to an addition of 125 million phone connections across fixed and mobile services during the year. A Chinese mobile phone maker will become one of the world's top-10 by 2005.

According to EMC, there are over 1.4 billion users of mobile phones worldwide. EMC data also suggests that the GSM subscriber base globally was close to 990 million at the end of 2003 (*and crossed 1 billion at the time of writing – Eds.*), accounting for over 71 per cent of the subscriber share with over 500 operators in 200 countries. The Asia-Pacific region added the largest number of GSM users as compared to other regions in 2003: 82.4 million. GSM/EDGE and W-CDMA are seen by many operators as a feasible migration path to 3G.

Launched in Japan in February 1999, DoCoMo's i-Mode wireless Internet service—a packet-based 'always on' data service operating at 9600 bps bandwidth—grew to a staggering 26 million users in just 30 months with over a billion messages being sent every day. More users in Japan today access Internet-related content via wireless devices than through the wired PC-based Internet.

Korea's mobile market is three to five years ahead of the Europeans, according to wireless consultancy firm Strand Consulting. Korean companies are aggressively creating intellectual property in the wireless domain: for instance, Korea's royalty income from MPEG (Motion Picture Experts Group) standard technologies is expected to reach $300 million in 2005. The carrier and content industries are increasing usage beyond 'killing time' applications; the domestic telematics market could grow to more than 700 billion won by 2006. 'Mobile Internet may be the biggest gamble in business history, but millions of people are using it now and the foundations for a shift to a true enabler of untethered life are already in place,' according to Nam Yong, president of LG Telecom. At a socio-political level, sites like OhMyNews have leveraged broadband and wireless connectivity and an army of citizen reporters to usher in a new form of participatory journalism which keeps communities alive and politicians in check.

In addition to consumer diffusion of mobile phones, the content and commerce services in the wireless ecosystem in Asia and the rest of the world have also been the focus of many market research studies and forecasts.

According to Jupiter Research, payments made with wireless phones, which will predominantly be for mobile content and applications, will total $812 million in revenue by 2006. Mobile entertainment revenues may actually cross $12 billion by 2005, according to McKinsey.

Strand Consulting estimates that over one billion ringtones may have been downloaded in 2002. Merrill Lynch predicts that the mobile phone content market will reach $25 billion globally by 2005, with a worldwide base of 1.6 billion wireless Internet users. Merrill Lynch estimates an average revenue of $1 per subscriber per month for content services. Worldwide, consumers spent $2 billion on ringtones, according to the Yankee Group. Of this, Americans spent $80 million in 2003, expected to shoot up to $1 billion in 2008.

Firms like Probe Research predict that the handheld device market will be conducive to 'spur-of-the-moment purchases during periods of boredom', including music and games. According to research firm Ovum, wireless micropayments—transactions of less than $10—could reach $200 billion worldwide by 2005, up from $3.5 billion in 2000. Mobile payments could reach $25 billion in 2006 worldwide, according to Frost & Sullivan. Tower Group Research predicts that 118 million western European and 145 million mobile users in Asia-Pacific will purchase low-cost mobile premium content by 2005. 'Wireless is the growth hormone for e-commerce,' says Gartner Group analyst Bob Egan.

According to research firm Ovum, location-based services (LBS) will deliver $19.5 billion of revenue by 2006. LBS applications include workforce tracking, fleet monitoring, equipment accounting in hospitals, emergency services, city navigation, traffic alerts, and basic information queries.

IDC predicts that Asian markets like China—the world's largest mobile phone market and second largest consumer PC market—will continue to drive semiconductor growth. China is a particularly 'fertile ground for disruptive innovation' as its low-cost suppliers move up the value chain. IDC predicts that the Asia-Pacific will become the dominant region in the world for PC and mobile phone demand and production. Pyramid Research says 3G mobile subscribers in Asia-Pacific will jump from 21 million in 2002 to 162 million in 2008.

By 2005, the worldwide WLAN (wireless local area network) equipment market is expected to approach $4.6 billion, according to research firm Cahners In-Stat; this includes network interface cards, inter-building bridges, and industrial embedded applications for indoor and outdoor usage. In-Stat/MDR predicts that wireless security will account for $4.8 billion by 2008 in three categories: infrastructure, devices and WLANs.

Gartner research projects that by 2005, 50 per cent of Fortune 1000 companies will have extensively deployed WLAN technology based on standards like IEEE 802.11. Laptops and handheld devices account for an increasingly growing proportion of computing units owned by US organisations.

The current leader in the WLAN marketplace in the Asia-Pacific region is Japan (with a whopping 60.5 per cent market share, according to Frost & Sullivan), followed by Korea, Hong Kong, China, Australia and Singapore (together accounting for 33 per cent of the market).

Research firm Aberdeen Group predicts that revenues for mobile middleware could reach $1.5 billion by 2006, up from an estimated $552 million in 2002. IDC forecasts a similar increase in mobile middleware from $227 million in 2001 to $1.7 billion in 2006. Mobile

CRM revenue will increase from a base of $300 million in the US in 2001 at a rate of 36 per cent each year till 2008, according to Frost & Sullivan.

Estimates indicate that by 2004-end, more people would have cellphones than TV sets. The number of mobile subscribers worldwide has exceeded the number of landline subscribers and the number of camera phones sold this year has outstripped the number of digital cameras sold.

A Spectrum of Opportunities and Challenges

Moving from quantitative to qualitative aspects of the wireless ecosystem, a wide spectrum of opportunities and challenges opens up for different segments of society: consumers, technology vendors, the media sector, enterprises, government agencies, regulators and non-governmental organisations (NGOs).

For a variety of reasons, this book does not cover cordless phones, pagers, infrared devices, Bluetooth, smart cards, wireless TV broadcast to buses, home networking, wireless accessories (e.g., earbuds), specialised devices (e.g., MP3 players) and activities like gambling and pornography on mobile devices.

Consumers

Despite the gloom-and-doom scenarios for the stock market valuations of some wireless players, consumers have never had it so good in terms of access options, pricing competition and communications convenience.

Consumers are rapidly taking to data services like SMS due to the low cost of messaging, ease of use in noisy environments, unobtrusive communication during meetings, communication options during urban traffic gridlocks and long train commutes, roaming agreements across nations, and the proliferation of premium services (e.g., downloadable ringtones, news alerts, sports scores, stock updates and targeted ads).

A recent survey conducted by Nokia reveals that individual SMS users fall into three categories: phonaholics (who cannot live without their mobile phones), enthusiasts (keen advocates of SMS), and pragmatists (who use SMS only as a basic tool).

'PCs are the dominant messaging device in the US, but mobile devices are more dominant in countries like Italy,' says John Whittington, senior business consultant for Nokia Networks in Asia-

Pacific. And while P2P messaging is strong among Asian consumers, business applications (e.g., via PDAs) are particularly strong in the US. Consumers in some markets now have special data services (e.g., safety information and emergency services) accessible via telematics devices installed in their automobiles.

Mobile subscribers turn to handheld devices and wireless communications for a number of needs and desires that are fulfilled via a range of applications: information needs (yellow pages, news), social needs (entertainment services), security (emergency alerts, monitoring homes), physiological needs (health indicator monitoring), personality development (m-learning, quizzes), business (productivity applications), transaction (banking), political participation (polling, voting) and even spiritual (e.g., horoscopes and prayer reminders via SMS). These needs further vary with demographic groups, as indicted in Table 1.1.

Each of these applications can be delivered wirelessly in differing ways and at various costs, depending on sophistication of the user device, sophistication of user interface, nature of subscriber plan, mix of voice and data services, roaming ability, time/length/frequency of service usage, and nature of user behaviour (sophisticated/normal/thrifty communication, transaction or information seeking).

These services can be billed through a combination of SIM cards, pre-paid cards, post-paid calling plans, kiosk payment or realtime payment gateways. For mobile payments to work successfully, key desired features include security, trust, cost (to vendor/bank/customer), control, learnability, convenience, anonymity, speed, mobility, and value-added features (e.g., totalling).

At the same time, concerns also arise in terms of acceptable levels of radiation from handheld devices, intrusion of privacy by aggressive telemarketers, the feeling of loss of personal space, excessive dependence on cellphones and SMS by children, and the lack of etiquette shown by mobile users who can annoy others with their ringing phones in theatres and restaurants!

Technology Vendors

Business travellers, managers at multiple sites, students on campuses, sales teams on large shopfloors, and even families using multiple online devices at home are the target market for the burgeoning industry of wireless networking. Patents for standards, protocols, chipset designs and wireless applications are the major sources of revenue on the intellectual property front. The challenge is to

Table 1.1: Mobile profiles of various demographic groups

Demographic group	Expenditure pattern	Usage pattern	Devices	Services	Socio-cultural needs
Teens, students	Low-Medium	Daytime, evenings	Cellphones	Pre-paid, SMS	Communication with friends; differentiation; entertainment; emotional support
Professionals	High	Daytime	PDAs, cellphones, laptops, data connectors	Post-paid, Internet access, voice, SMS	Managing financial affairs; company news; time management; lifestyle
Parents	Medium-High	Daytime, nights	Cellphones	Post-paid	Managing family affairs; managing household; security and monitoring

Source: Adapted from *Winning in Mobile eMarkets*, by Diebold Deutschland, 2000 TIMElabs Research Centre (www.timelabs.de)

generate and protect intellectual property rights (IPR) while also making it affordable for use by players at various points of the wireless value chain.

For instance, China—the world's largest cellular market with close to 300 million wireless subscribers—has a lot to gain by influencing existing wireless internet protocol (IP) standards and even creating its own ones, such as TD-SCDMA, which is less costly than CDMA2000 and W-CDMA systems.

Media Industry

Cross-fertilisation between wireless and other media like television and the Internet is providing new sources of data traffic, lucrative business models, user participation and even political mobilisation.

'Interactivity via SMS is becoming a huge draw for game shows and MTV,' says Francis Lee, vice president, product development at Singapore's wireless operator StarHub. The show 'Who Wants To Be A Millionaire' drew over a million SMS messages from viewers in just one day for a network operator in Europe. In some markets, MTV lets users vote via SMS for which video they want to watch next. At a P2P level, over 150 million SMS messages were sent in 2002 on Father's Day in the Philippines. Valentine's Day is another big draw for SMS users.

SMS and its successor MMS (with full audio and video) also share crucial synergies with the Internet. Many of the leading portals (e.g., Tencent in China, Yahoo in Singapore and India) have launched SMS services for instant messaging via gateways on their websites, and downloadable ringtones as well (e.g., Catcha in Malaysia). The *Indian Express* newspaper even tied up with controversial Arab news channel Al-Jazeera to provide users with SMS news alerts during the recent US-Iraq war.

'We offered SMS messaging services during the September 11 plane attacks in the US, so that our users in Singapore could communicate with people in the US. We received a lot of thanks for this,' says Ricki Mulia, senior producer of Yahoo Asia in Singapore.

'Among all our global properties, our Asia site has been the first to launch SMS features,' says Mulia. The SMS features were first launched as part of the 'Available Anywhere' campaign in Singapore, followed by Malaysia, Hong Kong and India.

At a brand marketing level, permission marketing via SMS can help a company reach its most valuable customers 24 hours a day. 'Mobile relationship marketing is good for in-store promotion and for driving traffic to the point of sale,' says Jerry Smith, interactive managing partner for OgilvyOne. McDonalds in Australia has effectively used permission-based SMS marketing to promote incremental sales during lean periods. *Cleo* magazine in Singapore also offers SMS ads for members who sign up for its Cleo Club service operated by N2N Consulting.

'SMS—and not WAP—is spearheading the mobile advertising industry,' says Andrew Ng, vice president of EdgeCom, a subsidiary of Ericsson. Such ads need to be effective for the advertiser, offer value for the operator, and be non-intrusive for the viewer. The opt-out feature for such permission marketing must be easily available, advised Ng.

Table 1.2: Asia-Pacific growth of MMS users as percentage of total subscribers

Country	Percentage of total subscribers in 2003	Percentage of total subscribers in 2004	Percentage of total subscribers in 2005
Australia	5.9	10.2	15.5
Hong Kong	4.8	8.3	16.9
India	0.1	0.2	0.3
South Korea	3.0	5.2	10.8
Malaysia	0.0	0.3	1.3
Philippines	1.2	1.9	3.0
China	2.2	4.5	6.8
Singapore	7.5	11.7	16.2
Taiwan	3.5	6.9	12.3
Thailand	0.4	1.2	2.3

Source: *IDC Asia-Pacific*, December 2003

An innovative SMS ad campaign was launched by Ogilvy One for a group of churches in Singapore who wanted to promote church-going activities among its members. They resorted to a permission marketing campaign nicknamed God (www.god.org.sg), where users signed up and received catchy SMS messages from a fictitious user called God—such as 'You don't need the phone for a direct line to me', 'Come to my place tomorrow' (sent on Saturday) and 'Thank me it's Friday' (sent on Friday evenings)!

However, technology, industry and regulator solutions for governance of problems like mobile spam will have to be worked out. For instance, NTT DoCoMo's i-Mode cellphone services will be able to automatically reject e-mails from any i-Mode handset that sends 200 or more e-mails per day. Industry associations for mobile marketing are also addressing the issue in countries like Australia.

Telecom Operators

Mobile operators are seeking to increase ARPU (average revenue per user) by promoting P2P traffic (which accounts for an estimated 90 per cent of SMS usage today) as well as premium SMS offerings (such as group chat, buddy lists, permission marketing, machine-to-machine communication). Payment plans can be per message, by data volume, by subscription, or some combination of the above. Pre-paid plans are taking off rapidly for budget-conscious parents who want to control children's usage of mobile phones, for travelling businessmen, and for premium services.

However, numerous questions still pose challenges to mobile operators. During hard economic times, will users cut back on SMS usage—or cut back on movies and other recreational spending? What is the optimal pricing ratio for SMS in comparison with voice tariffs? How can the SMS user base and SMS traffic be boosted while also increasing ARPU? How should the transition be managed between SMS and MMS, between GSM and 2.5G or 3G? What is the best revenue sharing model between content providers and mobile operators? Will Wi-Fi be a competitor to, complement of, or distraction from 3G?

Businesses

Wireless media has significantly transformed organisational communications and business workflow, ranging from calendaring and alerts to customer relationship management and logistics coordination (see reviews of relevant books in Part III).

At the multi-enterprise collaborative supply chain level, mobile applications can improve speed, cost and accuracy in procurement, supply chain execution (fulfilment) and measurement (asset visibility), and service management (field force automation).

While PCs and workstations have come under some criticism for 'tethering' knowledge workers to their desks, wireless technologies may be the perfect answer to 'mobilising' the workforce by letting them capture and harness key information and knowledge attributes wherever they are, whenever they want, and however they want (Rao, 2003b).

Strategies focused on knowledge mobilisation via handheld devices and wireless networks—ranging from Pocket PCs and cellphones to WLANs and Radio Frequency Identification (RFID) tags—can take knowledge management to an entirely new plane of performance, putting road warriors and field workers in the centre of the information and communications world via mobile portals and on-demand expert services.

Mobile content and applications should be workflow oriented, aligned with core business processes, developed in a standardised way for multiple devices, and designed with security and appropriate use in mind.

Government Agencies

Wireless transformations are taking place not just in the corporate sector but also in the government sector. These range from municipal functions and rural connectivity to citizen information alerts and emergency services. For instance, Singapore and Australia have developed wireless applications for parking solutions in garages. Burlingame, California, wirelessly regulates its parking meters to stop meter theft and monitor coin deposits. Colorado State Patrol has improved productivity on accident reports via wireless networks and portable computers. North Carolina police offers can tap into the FBI's National Crime Information Database wirelessly, within 12 seconds.

Governance: Regulation and Legislation

Beneath the hype and hope of mass consumer market adoption and enterprise uptake of wireless technologies, major tectonic shifts are taking place on another front: intellectual property rights (IPR), or the creation of technologies, content and applications that can be enforceably licensed for lucrative royalties at various points of the wireless ecosystem. The challenge for the content industry is to exploit new markets like downloadable ringtones while also safeguarding intellectual property rights and avoiding 'mobile Napsterisation'.

The rapid proliferation of multiple wireless technologies and the realities of scarce spectrum can lead to numerous challenges for regulators, ranging from licence allocation to the setting of tariffs (see case study of India by T.H. Chowdary in this book). Uninformed regulators and corrupt government agencies can be detrimental to public interest issues in the wireless ecosystem.

Academics

The wireless explosion opens up fascinating research questions for academics, media institutes and think-tanks (particularly in IT,

Table 1.3: The wireless alphabet soup

Acronym	Expansion, description
1G	First generation systems, analog, designed for voice.
2G	Second generation systems, digital, with voice and data (e.g., SMS).
2.5G	Evolving second generation systems, not yet fully multimedia.
3G	Third generation systems, multimedia, based on packet switching and internetworking. Speeds between 384 Kbps and 2 Mbps.
4G	Fourth generation systems, seamless integration of all wireless networks. Speeds of 54 Mbps, expected to be feasible after 2010.
AMPS	Advanced Mobile Phone System.
FDMA	Frequency Division Multiple Access. Dominant in 1G era (analog cellular).
TDMA	Time Division Multiple Access. Digital cellular technology that divides transmission into time slots. Dominant in 2G era.
CDMA	Code Division Multiple Access. Digital transmission technology based on spread spectrum and coding techniques. Dominant in 3G era.
CDMA One/IS-95	CDMA standard.
CDMA 2000	Transmission technology for evolution of CDMAOne/IS-95 to 3G.
CDMA 2000 1X	3G standard, double the capacity of CDMAOne, 1.25 Mhz capacity.
CDMA 2000 1xEV-DO	Evolution of CDMA2000, data optimised for peak rates of 2.4 Mbps.
MC-CDMA	Multi-carrier Code Division Multiple Access. Combination of three IS-95 carriers into one wideband carrier.
W-CDMA	Wideband CDMA (UMTS in Europe).
WCDMA-DS	Wideband Direct Spread CDMA. Data rate: 2 Mbps.
GSM	Global System for Mobile communications. European-developed digital mobile standard. 9.6 Kbps.
GSM *n*	GSM operating at different frequencies: 1800 MHz (PCN), 900 MHz, 1900 MHz (PCS 1900).
HSCSD	High Speed Circuit Switched Data. Intermediate upgrade technology for GSM; circuit-switched.
EDGE	Enhanced Data Rates for GSM Evolution. 384-473 Kbps.
GPRS	General Packet Radio Service. Packet-switched data enhancement for GSM-based 2G networks. Upto 115 Kbps.

(*Table continued*)

(*Table continued*)

Acronym	Expansion, description
IMT 2000	International Mobile Telecommunication; evolution of ITU's 3G standard.
IMTS	Improved Mobile Telephone Services (enhanced MTS, in the US).
UMTS	Universal Mobile Telecommunication System. European term for 3G standards. Based on WCDMA-DS. Different data rates for vehicular (144 Kbps), pedestrian (384 Kbps) and in-building (2 Mbps).
PCN, PCS	Digital mobile networks in Europe (PCN: 1800 MHz) and US (PCS: 1900 MHz).
PDC	Personal Digital Communication. Digital cellular standard used in Japan.
SIM	Subscriber Identity Module card.
Wi-Fi	IEEE 802.11b wireless Internet standard, transmitting at 11 Mbps in the 2.4 GHz spectrum.
Wi-Max	IEEE 802.16 wireless Internet standard, transmitting at 70 Mbps in excess of 45 kilometre radius.
Bluetooth	Short range radio technology for personal area networks (PAN). 781 Kbps, to a range of 10 metres.

Source: Author

journalism, communications, business, telecommunications, and policy) as well as challenges in designing curricula that are up-to-date, relevant in terms of industry needs, and loyal to issues of the public interest.

How can applications and educational content be designed for small-screen low-power handheld devices? How should news reports be re-purposed for PDAs and SMS? What business models can be used by publishers to make money from wireless content? What are the ethical issues of wireless marketing? How can wireless access help improve rural connectivity? Which country has the best regulatory model for mobile telephony? How can capacity and visibility be increased for local software industries developing mobile applications?

ICT4D NGOs

For non-profit agencies and NGOs in the field of ICT for development (ICT4D) working on improving access to underserved segments of society (particularly in rural or distaster-hit communities), wireless channels offer intriguing potential to bridge the digital divide, prevalent in much of the Asia-Pacific region (see Chapters 5 and 21 in this book for more details and case studies). This can range from wireless-in-local-loop (WLL) communication links for rural community centres in India (Rajora, 2002) to the grameen micro-finance models for shared cellphone services in Bangladesh. The role of telecoms in development has been well chronicled by numerous researchers (Gallagher, 1999).

In many developing countries of Asia, the penetration of cellphones has already exceeded the penetration of landlines (Cambodia was the first country in the world to cross this threshold in 1983). 'A whole new development paradigm will be unleashed in the next few years in Asia,' according to Yoshio Utsumi, secretary general of the ITU.

Telecom players and vendors need to focus on inclusion strategies and not just products, according to Noah Samara, CEO of Worldspace. The company beams radio and Internet content via satellite to parts of the world like Asia and Africa, including medical journal archives for healthcare workers. 'Content should become a major focus area for telecom players,' advises Samara.

A lot of data that is critically needed by the masses is in the public domain and a number of technologies are emerging that can help bridge the critical 'last-mile' problem such as Wi-Fi and satellite. However, regulatory obstacles are holding back services like Wi-Fi, Wi-Max and VoIP (Voice over Internet Protocol) in many developing countries of Asia, observes Heather Hudson, telecom professor at San Francisco State University.

Universal access goals are also becoming moving targets, evolving from basic landline connectivity and wireless access to Internet and then broadband. Developing nations should prioritise these services and targeted user organisations such as healthcare, libraries, NGOs, schools and governments. Sites like www.UniversalService.org have useful information resources and case studies in this regard.

'Technology is moving in the right direction. Human brokers—for instance, for operating telecentres and providing wireless access on a shared basis—are very important in this regard for developing nations,' says Hudson.

Framework: The Wireless Ecosystem

Steinbock (2003) divides the evolution of the wireless communications age into five eras: pre-cellular (prior to 1983: wireless telegraphy, AM/FM), analog cellular (1G: 1983–1992), digital cellular (2G: 1992–2001), multimedia cellular (3G: 2001–2010), and broadband cellular (4G: 2010 onwards). In terms of regulatory environment, there are three phases: monopoly, transition and competitive; each has its own defining market players, R&D pioneers and geo-strategic leaders, as summarised in Table 1.4. Steinbock also profiles leading wireless players like Nokia, Motorola, Ericsson and Qualcomm. A notable trend he identifies is the growing presence of PC-model players like Microsoft and Intel into the mobile handset, operating system and applications arena.

Kalakota and Robinson (2002) delineate the consumer and enterprise value chains in the wireless world, which includes data networks (GSM, GPRS, UMTS), telecom/datacom infrastructure (data centres for SMS and MMS), security, horizontal applications (e-mail, synchronisation), vertical applications (fleet management, banking, sales support, trading), content aggregators, client operating systems (WAP, PalmOS, Java), and end-user devices. Other players emerge in the world of the wireless Internet, e.g., for Wi-Fi chipsets, client radios, hotspot services, wireless ISPs.

A useful way to categorise the players in the wireless ecoystem would be along the lines of the '8Cs' framework (Rao, 2002) for the digital economy: connectivity (devices, access), content (e.g., SMS, 3G newsclips), community (group chat forums), commerce (m-payment), capacity (application development), cooperation (industry forums, regulation), culture ('instant' access, privacy challenges) and capital (venture capital, stock markets). These, in turn, can be further analysed by an additional dimension: the use of wireless technologies as an *instrument*, or the creation of wireless technologies as an *industry*.

As an *industry*, the wireless sector represents a major growing economic sector covering hardware, software, telecom/datacom and consulting services. The 'industry' dimension of the wireless sector, thus, typically involves ICT companies and standards-making institutes. As an *instrument*, affordable and usable wireless products and services can be used by a wide range of organisations: universities, manufacturers, retail outlets, and development agencies. These players do not develop wireless Internet products themselves (that is handled by the industry players), but use wireless and mobile tools for better supply chain management, workplace design, campus information services, or rural development projects.

As an industry (for techies) and instrument (for others as well), the wireless ecosystem can indeed transform the way societies work,

Table 1.4: The three eras of mobile communications

Phase	Technology	Markets and services	Players	Regulatory environment	R&D	Leading countries
Monopoly stage	Pre-cellular, 1G Wireless, FM, one-way and two-way cellular	Maritime, emergency services, war, industrial services	Monopoly PTT	Stifling, no dynamism. Focused on nation as a unit of analysis.	National PTT labs (e.g., AT&T)	US
Transition stage (1980s and 1990s)	1G and 2G	Corporate, mass consumer markets	Large private operators emerge	Privatisation, deregulation commenced. Regulators separated from national PTTs. Power shifts to competition policy authorities.	PTT labs, corporate labs	Northern Europe, Japan, US
Competitive stage (2000 onwards)	2G and beyond	Rise of data services, multimedia	Operators, IT + wireless players globalise	Industry competition is the norm	Globalised research and strategy	Scandinavia, Japan, China, Europe, US

Source: Adapted from *Steinbock* (2003)

entertain, study, govern and live—at the individual, organisational, sector, vocational and national levels. Not all nations can achieve success in the *industry* and *instrument* aspects of the wireless ecosystem. For instance, the US and Taiwan have companies which are global leaders in wireless Internet technologies and also have widespread diffusion and usage of Wi-Fi. India has low diffusion of Wi-Fi, but its software companies provide outsourcing support for many Wi-Fi product makers abroad. Singapore has strong diffusion and usage of Wi-Fi, but local companies are not global leaders in Wi-Fi product manufacture.

Coupled with these two aspects of wireless and mobile tools (usage and creation), the '8Cs' framework can be used to tease apart some of the key challenges in implementing the vision of wireless information societies such as increasing mobile diffusion and adoption, scaling up pilot projects, and creating wireless industries. This framework, summarised in Table 1.5, should assist industry players, NGOs and policymakers in assessing how they may best address the growing local potential of the wireless ecosystem.

Connectivity

The connectivity parameter of the wireless ecosystem includes technology components (e.g., network infrastructure, operating platforms, consumer devices, integrating software, applications) and service providers (e.g., wireless operators for consumer and corporate access services). Cellphones, PDAs, portable MP3 players, systems integrators, mobile operators, hotspots, Wi-Fi enabled laptops and base station technology providers fall in this category.

Content

Voice and data content providers are emerging as key success elements of the wireless formula. This includes content creation, syndication and aggregation. Content can be re-purposed from existing traditional media (e.g., downloadable cartoons from entertainment companies), new media (e.g., SMS news alerts from portals) players, or created by new startups (e.g., stock alerts from financial infomediaries). Challenges can arise in devising local-language cellphone keypads, tools and content for some of the diverse languages of the Asia-Pacific.

Table 1.5: The '8Cs' of the wireless information society

	Wireless as an instrument	Wireless as an industry
Connectivity	How affordable and wide-spread are wireless internet platforms and mobile devices (e.g., Wi-Fi cards, internet access, cellphones) for the common citizen?	Does the country have sizable demand and suppliers for wireless connectivity (i.e., clients and vendors for hardware, software, datacom solutions and services)?
Content	Is there useful content for citizens to use via wireless access channels? Is it relevant, and in the local language? Is it easy to use?	Is content being generated in local languages and localised interfaces? Is this being accessed/used abroad?
Community	Are there online/offline forums where citizens can discuss the wireless ecosystem and related issues?	Is the country a hub of discussion and forums for the regional/ worldwide wireless industry?
Commerce	Is there infrastructure (secure wireless gateways, payment agreements) for m-commerce for citizens, businesses and government?	Does the country have indigenous m-commerce technology and services for wireless channels? Are these being exported?
Capacity	Do citizens and organisations have the human resources capacity (tech, managerial, policy, legal) to effectively harness the wireless medium for daily use?	Does the country have the human resources capacity (tech, managerial, policy, legal) to create and export wireless products and services, and set standards?
Culture	Is there a forward-looking, open, progressive culture at the level of policymakers, businesses, educators, and citizens in opening up access to the wireless medium? Or is there nervousness and phobia about the disruptive aspects of Wi-Fi/ Wi-Max/SMS by incumbents?	Are there techies, entrepreneurs and managers pro-active and savvy enough to create local wireless product and services companies and take them global?
Cooperation	Is there adequate cooperation between citizens, businesses, academics, NGOs and policymakers to create a favourable climate for promoting the wireless medium?	Is there a favourable regulatory environment in the country for creating wireless companies, M&A activity, and links with the diaspora population?

(*Table Continued*)

(*Table Continued*)

	Wireless as an instrument	Wireless as an industry
Capital	Are there enough financial resources to invest in wireless infrastructure and players? What is the level of FDI?	Is there a domestic venture capital industry; are they investing abroad as well? How many international players are active in the local private equity market? Are there stock markets for public listing of wireless players?

Source: Author

Community

As with Web and e-mail-based community chat fora, wireless data services are also emerging for community-based communication, like, teen chat via SMS, wireless dating services, sports fan clubs.

Commerce

Despite early hype and missed opportunities, significant markets exist for m-payment and m-commerce services. Mobile ticketing, e-coupons, SMS-based payment at vending machines, handheld devices for accepting credit card payments, and e-signature devices for courier delivery are experiencing rapid growth.

Capacity

As with the PC and internet era, the wireless world calls for significant capacity building among technology solutions providers, managers, creative agencies, regulators and even consumers in harnessing the wireless ecosystem. This ranges from applications development and worklow re-design to creative communications and pro-active market regulation.

Culture

A company or a country gets a head start when it has the infrastructure and skillsets. However, true success in the wireless world can be achieved only by embracing a culture of change, lifelong learning and pro-active reform. An entrepreneurial culture and the presence of local champions are key enabling factors.

Cooperation

The success of the wireless ecoystem requires a significant amount of cooperation, partnering and alliances among the various players along the value chain, as well as public-private partnerships for research and development.

Technology and media players are forming consortia to establish standards and ethics in the wireless ecosystem. One such consortia includes Bluetooth for personal area networking. Players from the Ethernet world have gathered under the WECA alliance (Wireless Ethernet Compatibility Alliance), which offers Wi-Fi certification for vendor products. Other collaborative organisations that have sprung up include the GSM Association, UMTS Forum, Wireless LAN Association (www.wlana.org) and the Third Generation Partnership Project (3GPP) for standards development (see Table 1.6 for more details).

Table 1.6: Wireless and mobile organisations

Global industry associations	
Global Mobile Suppliers Association (GSA)	www.gsacom.com
3G Forum	www.3g.co.uk/3GForum/
UMTS Forum	www.umts-forum.org
GSM Association	www.gsmworld.com
CDMA Development Group	www.cdg.org
Wireless World Research Forum	www.wireless-world-research.org
Wireless Broadband Alliance	www.wirelessbroadbandalliance.com
Mobile Marketing Association	www.mmaglobal.com
WLAN Association	www.wlana.org
Wi-Fi Alliance	www.wi-fi.org
Wi-Max Forum	www.wimaxforum.org
Bluetooth (official site)	www.bluetooth.com
InfraRed Data Association	www.irda.org
Digital Enhanced Cordless Telecommunications (DECT) Forum	www.dect.ch
Global Wireless Education Consortium	www.gwec.org
Geospatial Information & Technology Association	www.gita.org
Mobile Data Association	www.mda-mobiledata.org
Wireless Location Industry Association	www.wliaonline.org
Smart Card Alliance	www.smartcardalliance.org
Global VSAT Forum	www.gvf.org
Regional Organisations	
Asia-Pacific Telecommunity	www.aptsec.org

(*Table Continued*)

(*Table Continued*)

Asian Media Information and Communication Centre	www.amic.org.sg
National industry associations	
Australian Mobile Telecommunication Association	www.amta.org.au
Cellular Operators Association of India	www.coai.com
Association of the Computer and Multimedia Industry, Malaysia	www.pikom.org.my
Communications and Information Network Association of Japan	www.ciaj.or.jp
Hong Kong Wireless Technology Industry Association	www.hkwtia.org
Korea Telecommunications Operators Association	www.ktoa.or.kr
National telecom regulators	
Office of the Telecommunications Authority, Hong Kong	www.ofta.gov.hk
Malaysian Communications and Multimedia Commission	www.mcmc.gov.my
Telecom Regulatory Authority of India	www.trai.gov.in
InfoComm Development Authority, Singapore	www.ida.gov.sg
Australian Communications Authority	www.aca.gov.au
Nepal Telecommunications Authority	www.nta.gov.np
Ministry of Information and Communication, South Korea	www.mic.go.kr
Ministry of Public Management, Home Affairs, Post and Telecommunications, Japan	www.soumu.go.jp
National Telecommunications Commission	www.ntc.gov.ph
Telecommunications Regulatory Commission of Sri Lanka	www.trc.gov.lk
Ministry of Information Industry, China	www.mii.gov.cn
Directorate General of Telecommunications, Taiwan	www.dgt.gov.tw
Market research companies	
IDC	www.idc.com
In-Stat/MDR	www.instat.com
EMC	www.emc-database.com
Pyramid Research	www.pyramidresearch.com
Strategy Analytics	www.strategyanalytics.com
Tarifica	www.tarifica.com
Frost & Sullivan	www.frost.com
Gartner Group	www.gartner.com

Source: Compiled by Editors

Capital

As in the case of the Internet economy in the last decade, the mobile economy is opening up a whole range of opportunities for startups and public listings along the entire value chain—content creation, content aggregation, user authentication, m-payment, horizontal applications (e-mail), vertical applications (m-banking), client operating systems (based on J2ME), terminals (colour screens), roaming services, mobile messaging data centres, m-advertising, MVNOs (mobile virtual network operators), Web-SMS gateways, and systems integrators.

For instance, Singapore-based ANTlabs (Advanced Networking Technologies Laboratories) has received funding from sources like the JAFCO Consortium. Other wireless R&D organisations in Singapore include Wireless Intellect Labs (with an MMS solution called Permodia). Siemens Mobile Accelerator (SMAC) venture funds have been invested in three companies in China and one in Singapore.

Learnings from Asia-Pacific: Findings of this Book

This section is structured into three parts: overall themes of the Asia-Pacific wireless ecosystem (like content and capital, as outlined in the previous section of this chapter), reviews of 10 country profiles in the region (based on the 8Cs framework), and reviews of 10 relevant books. All the chapter contributors are based in the Asia-Pacific region or are frequent visitors to the region, and were contacted in 2003 for this book project. The individual contributors for the country snapshots were given specific structures for each chapter which most have been able to adhere to, thus leading to a useful consistency in cross-country and cross-media comparisons. The individual writing styles vary, of course, and make for an interesting and lively read!

Overall Themes

Drawing on the 8Cs framework, individual chapters are devoted to issues like diverse connectivity options, mobile content and services, policy and regulation, and organisational impacts.

Lara Srivastava of the ITU sets the stage for the book, drawing on authoritative sources to show how Asia is at the cutting edge of the wireless information society. The age of the mobile Internet promises to be one of rapid innovation, which is evident from the success of 3G mobile services in Japan, South Korea and Hong Kong. Even the world of 2G is showcasing creative socio-economic applications in China (the world's largest cellphone market), India (with the most potential for growth), the Philippines (with SMS as a tool for political mobilisation) and Bhutan (where wireless technologies help bridge the digital divide). However, it should be noted with caution that a widespread shift to 3G mobile may take longer than expected due to high network costs and user satisfaction with earlier generation services. Regulators and industry players need to realise that there are a number of different options for providing mobile Internet services, billing plans, content alliances and migrating to next generation platforms. Other governance challenges will arise in terms of mobile spamming, user privacy, Internet addiction and mobile etiquette.

Chas Sweeting spells out the foundations, market potential and customer acceptance factors underlying the potential of location-based services (LBS), once regarded as the 'killer app' of the mobile world and still one of the more intriguing services in the mobile information society. These range from emergency and resource tracking to people monitoring and location-specific advertising services. To be truly successful, LBS must be seamless, accurate, timely, useful, affordable and not intrusive of users' pricacy. It also requires extensive cooperation between operators, technology providers and content players—which is why the uptake has been relatively slow. Written in a witty style, the chapter also highlights some LBS applications in Asian markets and contrasts them with those in the US and Europe. It may be enterprise applications of LBS that find greater acceptance and prove profitable as compared with consumer services.

While much attention has been focused on the high-profile 3G networks in parts of Asia, many markets and industry players favour a different, lower cost and sometimes complementary form of wireless communications: Wi-Fi. Leading markets here include South Korea, Japan, Taiwan, Australia and Singapore, though there is tremendous potential in other markets like Indonesia and India as well. **Subha Rama** covers the various alliances that are being formed across the region by ISPs, mobile operators, service aggregators and location-specific players like coffee chains. Challenges remain in finding profitable models, billing arrangements and offering seamless roaming. Interesting opportunities for widespread public access to the Internet are also emerging via community wireless networks.

Ashok Jhunjhunwala makes the forceful case that in many developing countries, wireless-in-local-loop (WiLL or WLL) technologies may be the most cost-effective solution for combined voice and Internet services in rural areas and less-dense urban clusters. For small offices, WiLL can actually be a more effective business communications solution than fixed line (which costs more) or mobile (with data services often limited to SMS). Whether based on standards like CorDECT or Wi-Max, fixed line wireless connectivity solutions need to be offered in conjunction with local entrepreneurs who can operate info kiosks and service points. Though not yet a major hotspot market, India can actually offer useful lessons and models for other developing countries via other forms of wireless networking like CorDECT WLL in homes, small offices and information centres, as pioneered by the Indian Institute of Technology in Madras and its incubated companies. Challenges can arise, though, in some regulatory environments.

Madanmohan Rao provides an overview of key impacts of wireless media on the news and entertainment sectors. From journalism tools and TV polling to music ringtones and mobile gaming, a vast new media ecosystem has been unleashed in the span of less than a decade. He covers the opportunities (instant news alerts, political mobilisation) and challenges (IPR violation, invasion of privacy) that arise in the mobile media ecosystem. Case studies are drawn from around the world, with special emphasis on the advanced markets of Japan and South Korea, the vast SMS potential in India and China, and alternative news and political movements in Malaysia and the Philippines.

Mobile commerce was once hyped as the 'killer app' of mobile devices, but as **Sunanda Sangwan** describes, there are significant challenges to be overcome with respect to size of the display screen, input characteristics, security and perceptions of trust. Still, the convenience and ubiquity of mobile devices makes for a compelling proposition, and m-commerce services are gaining acceptance in Hong Kong, Japan and South Korea. Content-to-person (C2P) sales of SMS and MMS data are proving to be lucrative and the trends to watch include the growth of P2P payments and the emergence of standards and consortia enabling mobile payment.

Madanmohan Rao further covers wireless impacts in business environments (sales force automation, supply chain management), government (use of SMS alerts for citizens, PDAs by government officials) and education (Wi-Fi infrastructure on campuses, PDA usage for corporate learning). While wireless solutions (such as Wi-Fi and RFID for inventory tracking), handheld devices (PDAs for sales staff) and mobile communications (SMS alerts by health authorities) open up vast new opportunities for productivity enhancement, more effective learning environments and innovative citizen services, challenges can also arise due to information overload and inadequate security. Careful attention must be paid to the metrics covering processes as well as employee impacts of wireless solutions and a useful classification of such metrics is provided in this regard.

There are also mutual synergies between knowledge management (KM), business intelligence (BI) and the wireless ecosystem. KM and BI are being used to improve efficiency and innovation by wireless players: operators, device manufacturers, network providers and software services companies. Wireless technologies are being used by a range of companies to enhance their existing KM strategies via 'mobilising' explicit knowledge and making it directly accessible to employees on the road. There are innovative practices on these fronts from Asia, such as NTT DoCoMo, LG Electronics, Samsung, Sun Microsystems Philippines, Hughes Software Services and Origin Exterminators (as well as examples from outside the region, such as Siemens and Ericsson Canada).

Table 1.7: Mobile penetration rates in Asia-Pacific

Country	Percentage of diffusion
Taiwan	112.8
Hong Kong	90.5
Australia	76.0
Singapore	74.7
South Korea	71.4
Malaysia	41.6
Thailand	35.1
Philippines	26.9
China	20.3
India	2.0

Source: *Far Eastern Economic Review*, March 18, 2004

Based on Wi-Fi connectivity, the Faculty of Arts at the University of Melbourne has implemented three flexible wireless classrooms for teaching and learning in the humanities. These classrooms are designed to support innovative teaching practices and enhance natural human interactions. **June Gassin** and **Som Naidu** present the results of a survey of the impact on teaching in these classrooms. Results indicate that the opportunities provided by these flexible wireless classrooms substantially influenced particular teaching practices. These included strategies for teaching, engagement of students with the subject matter, socialisation and interaction in class and providing feedback to learners. Attributes of the classrooms that had the greatest impact on teaching were the ability to configure space, move about in the classroom and use a variety of technologies.

Loudon Owen and **Paul Gardner** provide a sweeping overview of venture capital dynamics in Asia's wireless sector, with insightful comparisons to the US and Europe. Topics and case studies covered include industry lifecycles, management dynamics, regional/global opportunities, and alliance strategies—featuring giant established telcos and nimble startups along the entire wireless ecosystem. Though overall wireless penetration of Asia lags Europe, it is in areas of concentrated adoption that Asia is moving into the vanguard of the wireless industry, as in Korea and Japan. After the disastrous foray by 3G in Europe, Asian innovators and consumers may save the technology by proving it is commercially viable. Asian companies are rivalling historical Scandinavia for technological leadership in the wireless world. Areas ripe for investors include cutting-edge emerging technologies in the advanced markets as well as newer markets in developing countries where tried and true mobile technology can be diffused. Expansion in the North American and European markets presents an opportunity for Asian-based players with matured and proven solutions.

Michael Sainsbury provides a detailed case study of the SingTel Group, a diversified regional telecommunications services group with offerings in mobile telephony, fixed line and wireless Internet. It has accumulated minority interests in other mobile companies in the region—Globe in the Philippines, AIS in Thailand, Bharti in India, and Telkomsel in Indonesia—and acquired Optus, Australia's number two telecommunications carrier and number two mobile network. The case study covers content plans, tariffs, user adoption, growth rates, 3G trials, aggregated alliance savings, churn rates, roaming services, ARPU (average revenue per user), and regulatory issues like mobile number portability—issues of interest to media analysts, investors, regulators, technology players and other mobile operators.

The 21st century information society calls for a converged regulatory approach to traditional telecommunications, mobile telephony and the Internet. **Tan Min-Liang** provides a 'regional scorecard' of the diverse Asia-Pacific regulatory and legal scenario, ranging from Japan and Australia to India and Singapore. Concerns about security, management of competition, and convergence continue to dominate the discourse. Topics covered include IPR of content for mobile transmission, content filtering, electronic transactions, cyberstalking, wiretapping, database privacy, electronic evidence, Wi-Fi and VoIP.

Successive waves of innovation in digital technologies and communication models have posed increasingly complex challenges for policymakers, national regulators and international regulatory bodies around the world. As if the convergence between the Internet and telephony sectors in the 1990s was not complex enough to handle, new technologies like the wireless Internet and mobile telephony are proving even more tricky to handle in a unified framework. **Laina Raveendran Greene** highlights regulatory approaches being adopted in Asia-Pacific to handle the challenges of convergence, and identifies open questions surrounding public service models of Wi-Fi and mesh networks. Case material is drawn from countries like Singapore and Malaysia.

Governments around the world have often bungled the regulation of telecommunications services, radio spectrum and media content and India is no exception. To be fair, the rapid pace of technology change and the convergence between previously separate realms of technology have thrown up new opportunities (e.g., increasing bandwidth and switching efficiency) as well as challenges (e.g., maximising benefits to consumers while also promoting domestic/global industry players and investors). **T.H. Chowdary** highlights some of the regulatory challenges faced in India in terms of mobile operator licensing, interconnect agreements, tariff levels, managing the concerns of incumbents and new entrants, allocation of spectrum for Wi-Fi/Wi-Max, devising progressive regulatory schemes, and creating impartial regulatory bodies.

Vicente Rafael portrays a fascinating development in the mobile ecosystem: the use of SMS to coordinate popular movements and protests which eventually unseated the president of the Philippines in 2001. In the larger context, cellphones and low cost messaging services helped meet the pent-up communication demand in the Philippines. For 'Generation Txt' users in crowded cities like Manila, cellphones are almost treated like an additional limb. Like many other developing countries, the Philippines has access to some of the latest wireless devices while also being, paradoxically, mired in deteriorating terrestrial infrastructure. SMS helped spread rumours, jokes and information that steadily eroded whatever legitimacy President Estrada still had during impeachment proceedings, and cellphone users themselves became editors and broadcasters. This superbly detailed chapter captures the complexity as well as power of humble

SMS as a social tool and provides informative insights into the connections between new media and perceived political notions of empowerment.

Wi-Fi Nation: Man on a Wireless Mission

'Wireless activist' Onno Purbo of Indonesia and his students are on a mission to create a wirelessly-linked national network of Internet cafes to bypass government inaction and bootstrap access to the Internet for citizens of Indonesia.

Onno Purbo (onno@indo.net.id) is a former professor at the Institute of Technology in Bandung and self-confessed Indonesian ICT evangelist. He sees a self-financed, bottom-up wirelessly-powered Internet infrastructure as the key to achieving a knowledge-based society in developing countries.

'It is not the equipment, nor the legislation, nor the investment that counts; it is the ability to educate a critical mass of people to gain the information and knowledge that are vital to the establishment of such an infrastructure,' writes Onno in CERN courier (http://www.cerncourier.com/main/article/43/6/20).

'Community Wi-Fi' need not require the massive bureaucracy, highly skilled personnel and fancy equipment financed by bloated government ministries or multinational investors. Costs of Wi-Fi products are now so low that communities can set up, configure and use their own local Internet infrastructure for data and voice services. 'After seven years of trying to educate the Indonesian government about the basic idea of a community-based ICT infrastructure, in 1996 I succeeded in having it partially written into some sections in the Indonesian National Information Infrastructure policy known as Nusantara 21,' says Onno.

Now he has devoted himself fully to scaling up 'neighbourhood ISPs,' who currently support about 4 million individuals, more than 2,000 cyber cafés and more than 1,500 schools on the Internet, running on more than 2,500 Wi-Fi nodes.

Long-distance and local calls are routed through the Internet infrastructure without any telco interconnection, via a VoIP MaverickNet area code, +6288, that has been specifically assigned to this task! Tutorial files on implementing a bottom-up infrastructure can be downloaded free of charge from www.apjii.or.id/onno.

The brilliant, jovial and hyper-active Onno is the author of nearly 1,000 articles and close to 40 books on the setting up and use of ICTs. 'We need to educate people on what you can do with ICTs, how to set up a telecentre, how to set up a proxy server, in simple language,' Onno explains. He is also helping government officials draft an act for Wi-Fi diffusion. He is a research fellow with the ICTs for Development (ICT4D) programme area of Canada's International Development Research Centre (IDRC), meeting policymakers in developing countries to discuss new ICTs (http://web.idrc.ca/en/ev-45872-201-1-DO_TOPIC.html).

Wireless Environments: A Review of the Country Profiles

Based on the 8Cs framework, profiles of wireless environments are drawn up for 10 Asia-Pacific countries. Each country profile includes evolutionary steps, market statistics, content services, regulatory environment, local innovations and global competitiveness. Some countries are not featured in the book since the contacted writers either expressed an inability to compile the material, or unfortunately bailed out at the very last minute. Countries featured in this edition of the Asia-Pacific Internet Handbook include: Japan, South Korea, Singapore, Australia, India, China, Thailand, Malaysia, Nepal, and Bangladesh. This section briefly surveys the highlights of each country's wireless environments; the full-length profiles in Part II make for a very informative read.

Japan

Shinichi Saito provides a fascinating profile of the world's most advanced wireless information society—Japan. In March 2003, the number of cellphone subscribers in Japan was 75.7 million and the number of subscribers to cellphone Internet services reached 62.5 million, representing 82.6 per cent of all cellphone users. The cellphone user base exceeded the number of fixed phone subscribers in 2000. The m-commerce industry is expected to cross 3.2 trillion yen by 2006. Around 56.5 per cent of cellphone Internet users have used m-commerce. The e-Japan strategy is aimed at making Japan the most advanced IT nation within five years. A growing body of social science research is examining a wide variety of social impacts, ranging from gratification effects to digital shoplifting. A key challenge ahead is to reduce the digital divide in areas like access by disabled citizens.

South Korea

John Lee highlights South Korea's role as a world leader in broadband Internet–wireline and Wi-Fi. The country is also the largest producer of communications equipment in Asia. Broadband usage has 80 per cent household penetration. There will be an estimated 50,000 hotspots nationwide by the end of 2004. Mobile data services for cellphone users are rich in multimedia content. Ahead of the 2002

World Cup, the major wireless carries launched 3G wireless Internet services on the CDMA platform EV-DO, which enabled speeds of up to 2.4Mbps. M-commerce is alive and well in Korea, with over 20 million online banking accounts and wide usage of wireless as a transaction platform. Korean operators have also launched a platform called 'Wireless Internet Platform for Interoperability' (or WIPI). Other initiatives have been launched targeting RFID, wireless personal networks (WPAN), ultra wide band (UWB) and intelligent wireless sensor networks. The Korean government has also set up the 'Korea Venture Fund'.

China

Lunita Mendoza outlines the mesmerising potential of China's mobile market while also highlighting the IPR and political challenges that are opening up. China has the world's largest mobile subscriber base and second largest Internet population. Wireless subscribers outnumber wireline users. Markets are growing for products and services ranging from used cellphones to SMS gaming. A total of 220 billion SMS messages were sent in 2003. Chinese government officials are concerned about the use of SMS for spreading rumours and political messages. The WLAN market may cross $900 million by 2007. A number of home-grown wireless players are also venturing into foreign markets. Controversies are emerging as China flexes its muscles in terms of its own standards and requirements for wireless security, digital content and 3G protocols. Still, China is viewed as a powerhouse that is going to benefit Asia as a whole and will have repercussions for wireless players around the world.

India

India is now the world's fastest growing mobile market; CDMA and GSM subscriptions are finally taking off and the traditional regulatory mess regarding licensing seems to have been finally sorted out. With a rich content base, a wide range of premium SMS services have been launched. Gartner's telecom analysts expect that by the end of 2004, India will have more mobile phones than fixed lines. Market stabilisation will result in there being only four major cellular operators left in the field. Interesting initiatives have been launched to bridge the digital divide via wireless access, but much more scalability is called for. The market for Wi-Fi is also opening up slowly. Research funding for wireless technologies is pouring into Indian R&D

labs, numerous wireless startups have emerged, and outsourcing of enterprise wireless application development is playing to India's strengths as a software powerhouse, says **Madanmohan Rao**.

Singapore

Lunita Mendoza clearly profiles the pioneering and facilitative role of the Singapore government in developing the city state as a major infocomm player. As early as the end of 1998, almost half of Singapore's households had mobile phones. There is complete nationwide mobile coverage in Singapore which was one of the first countries in Asia to get an Internet connection. Its wireless operators are active regionally in cellular markets as well as Wi-Fi services. Singapore is also taking advantage of its ethnic and commercial ties to China and India in content and human resource areas. Singapore is consistently rated high in global benchmarking studies of e-readiness and productivity. Its policy machinery is pro-active and in step with trends in media convergence.

Australia

Culturally, Australians are early adopters of new media technologies. There are approximately 15 million cellphone users in Australia (more than 70 per cent penetration), far surpassing the number of landlines (10 million). There are six mobile phone networks, among whom Hutchison has launched '3', the country's first 3G mobile network. Australia's vast size also has political ramifications in terms of requirements for wireless services in remote rural areas, according to **Michael Sainsbury**. Content and marketing services companies as well as trade associations like the Mobile Industry Advertising Association are addressing mobile marketing strategies. An innovative urban hotzone, CitiLan, has been rolled out in Adelaide. M-commerce is still extremely nascent in Australia. The government has made more than $4 billion selling mobile spectrum since 1997. The mobile telecommunications industry now contributes more than $5 billion par year to the Australian economy.

Thailand

Srikanya Mongkonsiri provides an overview of the Thai mobile media market. There are four wireless operators in Thailand: AIS, DTAC, Orange and Hutch, accounting for approximately 21 million cellular users. SMS messaging and downloads are popular among young generations, especially teenage girls; MMS and GPRS services are growing as well. However, operators have seen ARPU decline and there is a perception that because the current prime minister is the owner of AIS, fair treatment to every operator is impossible. Wi-Fi access is provided by companies like Internet Thailand Public Company Limited (INET). The hotspot front has yet to gather momentum and are currently restricted to coffeeshop chains like Starbucks.

Malaysia

Though Malaysia is behind the high-adoption curves of Asian markets like South Korea or Singapore, it is way ahead in mobile penetration and infrastructure as compared with many other developing nations of the world. Rising affluence, lower handset and service charges, a large youth segment, market deregulation and low import tariffs are among the key drivers for growth in this market. Challenges remain in maintaining quality of service, and in managing the rocky transition to 3G. The 3G network operators are required to facilitate virtual network operators. M-commerce in Malaysia is still in its infancy. Hotspots have emerged in some urban pockets but have yet to see significant usage. The key drivers for private WLAN uptake in Malaysia will be its high population density, high mobility, and presence of intelligent office buildings. Innovative measures like the most creative local mobility application awards help spur software development on the wireless front. Global players like Alcatel and Ericsson have invested millions to develop Malaysia as a hub for 3G business in southeast Asia. **Lunita Mendoza** also provides informative insights and perspectives into Malaysia's wireless industry.

Nepal

Gaurab Raj Upadhaya makes the clear-cut case for the crucial role that wireless communications and mobile devices can play in a mountainous country like Nepal. Wireless technologies form the backbone of all major telecommunications services in the country. The private sector is becoming increasingly involved in small parts of the telecommunications environment, but the monopolistic environment continues to be a constraint. From WLL and Web-to-SMS gateways to Wi-Fi and VSATs, the success of telecommunications in Nepal hinges heavily on wireless.

Bangladesh

The story of Grameen Telephone in Bangladesh is an inspiring showcase of the fruits of mobile telephony services powered by local citizens and funded by microfinance programmes. The CityCell service provider is also a mobile pioneer in south Asia. Grameen 'phone ladies' provide villagers with a vital link to services such as calls to hospitals and to relatives both in the country and abroad, in a region with the lowest number of phones in South Asia. **Sayeedur Rahman** concludes that there is also a lot of potential for the next wave of communication services based on Wi-Fi, if regulatory obstacles can be overcome—a point echoed in the last book review in Part III.

Book Reviews

The third part of this book consists of reviews of useful resources for those interested in issues of wireless media, including some books focused exclusively on Asia. They will also be useful for students of mobile media in graduate business schools and communications programmes.

These books include *i-Mode Strategy* (by Takeshi Natsuno), *The Mobile Internet* (by Jeffrey Lee Funk), *DoCoMo: Japan's Wireless Tsunami* (by John Beck and Mitchell Wade), *The Freedom Economy* (by Peter Keen and Ron Mackintosh), *Wireless Rules* (by Frederick Newell and Katherine Newell Lemon), *Constant Touch: A Global History of the Mobile Phone* (by John Agar), *Mobile Usability* (by Christian Lindholm, Turkka Keinonen and Harri Kiljander), *Smart Mobs: The Next Social Revolution* (by Howard Rheingold), *Roam: Making Sense of the Wireless Internet* (by Bruno Giussani) and *The Wireless Internet Opportunity for Developing Nations* (by The Wireless Internet Institute).

Takeshi Natsuno, managing director of NTT DoCoMo's i-Mode, the world's largest wireless Internet service, traces how i-Mode was 'born with a sense of crisis'—wireless operator NTT DoCoMo wanted to tackle the flattening growth rate of voice-based wireless communication services by launching a new layer of mobile data services. Instead of just getting new subscribers, the game was to also get new usage and frequent re-use out of existing subscribers (the 'second S-curve'). Natsuno draws heavily on the successes and failures of the various business models of the wired Internet. Another significant source of foundational direction for Natsuno came from the interdisciplinary field of complex systems theory. For instance, the choice of partners was driven with a view for emergence and self-organisation. A balance also had to be struck between too much control of content (direct ownership) and un-mediated access to content (leading to inefficient services).

SMS and SARS: Using New Media for Health Communications

New media can effectively be used by government agencies as an additional communication channel to the citizenry, particularly for healthcare alerts during crises like the SARS outbreak in much of Asia. Unfortunately, they can also be misused via the spread of false rumours.

For instance, in April 2003, the health department of Hong Kong began publishing infection-site data on a daily basis. Hong Kong carrier Sunday Communications launched a messaging service in English and Chinese, which used this data to help subscribers pinpoint the locations of buildings with confirmed cases of SARS.

In another defensive move, the Hong Kong government had to send millions of mobile text messages to residents to debunk a hoax which caused many citizens to believe that the city was soon to be formally declared 'an infected place.'

'Chaos, fear and disorder spread faster than good news,' according to Hong Kong-based consultant Brett Elliot, in a report published by Wired magazine. 'Because SMS notes are terse anyway, disinformation seems to spread even faster because you don't get the whole story.'

Sunday Communications also announced it would donate 1,000 rechargeable SIM cards to the Hong Kong Hospital Authority to help hospitalised SARS patients stay in touch with family and friends.

Singaporean advertising executive Kim Lai said mobile messaging quickly became a key element of the SARS chatter that permeated daily life in this city-state.

Besides healthcare, other messages of importance to the public can be transmitted by government agencies via SMS broadcast services. For instance, announcements of water shortage and unscheduled power cuts are delivered via SMS alerts by municipal officials in Indian cities like Pune. In Bangalore, the police department sent a city-wide alert via SMS urging public calm during the events surrounding the kidnapping of a filmstar by a notorious bandit. Many government schools now disseminate their examination results by SMS as well as e-mail and Web.

Jeffrey Lee Funk, associate professor of business at Kobe University's Research Institute for Economics and Business Administration, provides a comprehensive treatment of the business dynamics underlying the success of NTT DoCoMo's i-Mode service, along with comparisons and advice for western markets. In a nutshell, the wireless Internet exploded in Japan due to a superb positive feedback loop between the initially chosen content, mobile device capabilities, phone prices, packet networks, business models, and user targeting. At first the users were young and the services provided were simple; then the user base expanded, content became more complex, and the devices much richer–which is not how the US and Europe are approaching this market.

According to **John Beck** and **Mitchell Wade** of Accenture, the real story behind DoCoMo's success is that it is all about feelings: passion, love, acknowledging and harnessing disparity, impatience, luck, fun and strength. It went from zero to over $30 billion in revenues, without acquisitions, in just a few years—in a floundering economy and in a product category that few users understood. DoCoMo succeeded because the right people in the company had the right mix of powerful, human emotions and succeeded in skilfully managing them. 'A company that understands the power of human passions, and manages those passions in its customers, employees and leaders, will create value faster than its competitors,' the authors contend.

Peter Keen, author of *The eProcess Edge* and **Ron Mackintosh** (founder of e-business integrator differentis) describe four kinds of mobile application categories: freedoms (which change the limits of the possible in the structures of daily life), conveniences (attractive and easy-to-use services), features (new functional elements), and untargeted/unfocused (no specific application). Location specificity, time sensitivity, and effective design of m-services are key to building m-commerce relationships in the mass market. At the level of logistics, m-commerce in organisations can help reduce inventories, cut purchase order costs, reduce the document flow, and speed up the supply chain. M-commerce strategies focused on knowledge mobilisation can take knowledge management to an entirely new plane of performance and effectiveness, putting road warriors and field workers in the centre of the information and communications world via mobile portals and on-demand expert services.

Frederick Newell (author of *Loyalty.com*) and **Katherine Newell** Lemon (co-author of *Driving Customer Equity*) explain that the wireless channel will impact everyone in the marketing chain, from the chief executive officer (CEO) and chief information officer (CIO) to sales forces and customer service reps. 'Wireless communication is the biggest change to hit telecommunications since Alexander Graham Bell called Thomas Watson 125 years ago. It is the biggest change to hit marketing since the development of the Internet,' according to the authors. But the message must add value or convenience or both, for recipients—there is also great opportunity for misuse and a backlash of over-regulation.

'Pocket watches provide the closest historical parallel to the remarkable rise of the mobile cellular phone in our own times,' observes popular science author **John Agar**. Like pocketwatches, cellphones also started off as expensive status symbols and cost as much as a car, but are now owned by over a billion people worldwide. 'City life is mobile. City life is fast. And in no region outside the Nordic countries have cellular phones become as culturally important as in the cities of the Pacific Rim,' says Agar. In many countries of Asia where the extended family is important and where youth are drawn to cities, mobile phones are important tools of cohesion. Interestingly enough, the power of texting was an accident—it was not considered to be important by telecom operators, but its power was discovered by users. Just as containerisation transformed the shipping industry and ports of the world in 1970s, the next generation of 3G mobile technology will transform existing cellular networks and devices at a heavy investment cost.

Christian Lindholm, **Turkka Keinonen** and **Harri Kiljander**, all with considerable design experience at Nokia, trace the complex design issues around features which allow users to handle multiple phone calls (e.g., US users prefer to switch between calls, Europeans would rather prioritise the first call), phonebook management, time management, rich media, messaging and office applications. 'The logic of PC interfaces does not scale downward. Small interfaces are essentially different from big ones,' say the authors. They segment users into six categories: experiencers, impressors, controllers, maintainers, balancers and sharers. Further differences arise across cultures, languages, age groups (e.g., youthful 'screenagers') and occupational/professional sectors. The book has its humorous moments as well, with discussion of the 'ugly duckling' technology (growth of e-mail/SMS), 'spouse test' (a spouse would never give praise for the sake of praise), 'mom test' (least common denominator of user sophistication), 'generation trap' (waiting for current generation feedback before moving on the next technology generation), 'bimbo trap' (a feature so easy to use that it is insulting to the user) and 'WAP is crap' phase.

The mobile phone, like the PC in the 1980s, is poised to make a jump from a useful toy adopted by a subculture to a disruptive technology that changes every aspect of society, according to new media guru **Howard Rheingold** (author of *Tools for Thought* and *The Virtual Community*). For instance, meeting times and places can be fluidly negotiated (or 'softening' of time and 'blurring' of place). The SMS subculture—the style, content and the very act of messaging itself—provides new raw material for identity formation, social bonding, individual personality and peer-to-peer networking. The interconnection of Wi-Fi networks to form neighbourhood area networks (NANs) or 'quilts' can bump into regulatory challenges. At the same time, Rheingold warns that the technology can be used to improve life and liberty and also degrade it, as both a weapon of social control and a means of resistance. 'Even the beneficial effects will have side effects,' he cautions.

The increasingly intimate relationship to the mobile phone and the connectedness it allows has led to descriptors like 'handy' in Germany and 'the small hand' in Scandinavia, observes European media analyst **Bruno Giussani**. But serious regulatory debates continue to revolve around the relevance of the auctioning model for telecom licences, revenue-sharing options, reconciling public-private interests, impact on long-term national economy/infrastructure, scheduling of licence fee payments, the constraints raised by the scarcity of available spectrum, and alternatives like a 'beauty contest' based purely on merits of bidding companies' proposals. As for the early disappointments with WAP, it sparked off numerous puns and digs at the acronym such as 'Waplash', 'Wapathy,' 'Wait and Pay' 'Wrong Approach to Portability', 'Where Are the Phones?', and even 'WAP is crap'. But new opportunities are opening up for voice portals and SMS-based portals.

Wireless Internet technologies have a key role to play in helping developing countries leapfrog other platforms due to the considerable speed and relatively low cost of deployment, according to UN Secretary General Kofi Annan. 'The Internet's biggest advantage over other communication technologies is that it can provide an array of solutions and products,' according to the **Wireless Internet Institute**. For instance, the ISP WorldLink provided Internet access at the base camp of Mount Everest. ISP Africa Online provides wireless access. DakNet in the south Indian state of Karnataka uses a government bus as a mobile access point for store-and-forward data services. Internet Project Kosovo (IPKO) launched broadband wireless connectivity in Kosovo in 1999 in the weeks after the war, to UN agencies, NGOs, cybercafes and businesses. UNDP uses WLANs at telekiosks in the airport and post offices in Kabul. However, future issues need to be resolved in terms of interaction between Wi-Fi/Wi-Max and 3G, pro-active peer agreements and creating affordable yet workable interconnections between various telecom service providers and ISPs.

Discussion and Analysis

The Asia-Pacific region is bubbling with activity on all manner of fronts in the wireless ecosystem. Scans of the trade press reveal a wide range of hectic 'deal flow' for companies in areas like WLANs (Infonet), predictive text (Zi Corporation), wireless networks (Nortel), chipsets (Infineon), MMS solutions (LogicaCMG, ACL Wireless), SMS alerts (NexLabs), GSM roaming services (MobileWay), hotspot roaming (GRIC, iPass), operator alliances (OTEGlobe), interface design (Codacom, Intercube), home networks (NetGear), transmission towers (AsiaSpace), wireless bridges (Orthogon Systems), smart cards (GemPlus, NamITech), network administration (ANTlabs), network-based storage solutions for consumers (PurpleACE), optical transport (Tellabs), content applications (Landmat), enterprise access solutions (Onset Technology), billing systems integration (SchlumbergerSema), Wi-Fi switches (Trapeze), ruggedised handhelds (Psion Teklogix), hotel hotspots (XPossible), VoIP at hotspots (Inovas Telecommunications), mobile network operating system (Wysdom), broadband wireless (Navini), RFID (Innoxius Technologies), mobile translation services (MetaConcepts), mobile workflow (SMP Solutions) and mobile video telephony (Dilithium Networks).

Clearly, Asia-Pacific displays a broad diversity in terms of diffusion of wireless markets and industries, over and above the existing matrix of social, political and economic diversity. On the one hand, markets like Japan and South Korea are at the cutting edge in terms of advanced domestic wireless industries and export of these technologies globally; on the other hand, countries like Afghanistan and East Timor are only beginning to roll out wireless infrastructure. Based on the '8Cs' framework, the wireless environments in Asia-Pacific can be classified into seven types, depending on a combination of factors like ICT infrastructure, political culture, and international presence: restrictive, embryonic, emerging, negotiating, intermediate, mature and advanced (see Table 1.8).

Countries at the embryonic stage include Afghanistan and East Timor, where Wi-Fi and mobile infrastructure is just being rolled out. Nepal and Bangladesh have passed this stage but still have significant digital divides in rural areas. Countries at the negotiating stage include China, which have large domestic Internet and wireless infrastructure but conduct surveillance of digital content and have strict policies of control. Countries at the intermediate stage in Asia include India and the Philippines, which have less of an issue with

Table 1.8: Classification of wireless environments in Asia-Pacific

Type of wireless environment	Characteristics	Examples
Restrictive	1. Government exercises tight control over media and telecom environment 2. Diffusion, awareness of IT/Internet/mobile communications among population is very low	North Korea, Myanmar
Embryonic	1. Wireless infrastructure is just being rolled out 2. Donor agencies are active in funding and providing human resources	Afghanistan, East Timor
Emerging	1. Wireless infrastructure exists in urban areas 2. Widespread digital divide exists, m-commerce is not yet widely prevalent	Nepal
Negotiating	1. Strong Internet/wireless infrastructure exists 2. Local capacities and markets exist for online news, e-commerce, m-commerce 3. Government is 'negotiating' benefits and challenges of new media; authorities exercise strong control over online content, search engines, SMS; political and cultural censorship of Internet is practised	China
Intermediate	1. Sizeable markets for Internet, SMS, wireless services exist 2. Strong local capacities in wireless software design and outsourcing 3. Digital divide is still an issue, donor agencies are active	India, The Philippines
Mature	1. Large-scale penetration of Internet, wireless 2. Mature business models for online and mobile content	Australia, New Zealand
Advanced	1. Large-scale penetration of broadband and wireless Internet, as well as 2.5G, 3G 2. Some wireless technology companies are major players in global markets; wireless content models are being exported	Japan, South Korea

Source: Author

censorship, but still have a notable digital divide; they are strong players in industries like outsourcing for wireless software, design and business services. Countries at the advanced stage include Japan and South Korea, with enviable levels of new media penetration and bandwidth, and flourishing content environments for Internet and wireless, and global players in wireless products and business models.

South Korea and Japan are estimated to be a year or two ahead of the rest of the world in terms of sophistication of wireless Internet environments. Asia-Pacific will be the world's test-bed for 3G services for the rest of the decade, says UMTS Forum chairman Bern Eylert. The region will account for $118 billion of global 3G revenues of $320 billion in 2010; the top three 3G markets in the world will be the US, China and Japan.

Mobile voice alone will account for huge markets in countries like India and Vietnam. By 2005, half of the world's population will be using cellphones, according to Amarendra Narayan, executive director of the Asia-Pacific Telecommunity, in Bangkok.

Carriers should also keep an eye on newer devices created by the IT industry, such as PDAs and PCs with GSM and 3G chipsets fully integrated. 'Completely new devices will emerge in five years. Handsets will be as powerful as today's laptops,' predicts Bosco Fernandes, vice president at Siemens Information and Communication Mobile Networks.

Challenges remain for many operators in Asia-Pacific in terms of figuring out how to migrate to 2.5G or 3G for competitive reasons without losing out on fully recovering costs of GSM networks that have already been rolled out. The entry of new wireless players in some Asian markets has not been a smooth process either (WLL in India and Indonesia), and other markets like China are slowing down. Confusion is also growing as companies and countries announce new Wi-Fi and 3G standards. Uncertainty in liberalisation rules has hampered international carriers from investing in markets like China. There is also confusion in some markets as to whether Wi-Fi will compete with 3G, when they may actually be complementary services.

Significant obstacles in this Brave New Wireless World for Asia include ensuring interoperability and migration between a bewildering array of standards, products and bandwidths, ensuring level playing fields between a host of operators, developing local-language interfaces, tools and content, increasing capacity for wireless strategies in its educational institutions, and overcoming the wireless divide.

Despite these challenges, however, the concluding note for the growth of the wireless medium in Asia-Pacific continues to be one of hope and optimism. With proper planning and creativity, the region should be able to retain its position at the head of the global wireless economy.

Conclusion

This book marks an important and valuable step in studies of the wireless ecosystem, and in regional communications discourse of Asia-Pacific. It provides a useful framework (based on the '8Cs' parameters: connectivity, content, community, commerce, capacity, culture, cooperation, capital) to assess new media impacts, and classifies online media environments in Asia-Pacific into seven types: restrictive, embryonic, emerging, negotiating, intermediate, mature and advanced. The Internet and wireless technologies are transforming countries and cultures in unprecedented ways—especially in Asia-Pacific.

In conclusion, there are about 15 key lessons that this book provides, based on the analysis of the writers across the region.

1. A tight, close-knit relationship between device manufacturers and wireless carriers is necessary to be able to fine tune an evolving wireless ecosystem to meet the needs and aspirations of the users. This is where markets like Japan, South Korea and Scandinavia lead.
2. A progressive business model for growing high-value mobile content and applications involves generous revenue sharing on the part of the operator, with content and application providers. Japan and South Korea again excel in this and, as a result, have a very healthy and competitive mobile content ecosystem.
3. Location-based services will indeed be a 'killer-app' in the mobile ecosystem, but it may actually be the enterprise market that will be the main driver in the early stages.
4. Knowledge management and business intelligence will be key initiatives and disciplines that wireless players will have to nurture if they expect to continually maintain an innovative edge in the wireless race.
5. Research methods for determining metrics and monitoring performance impacts of wireless environments (RoI) will become more complex but will eventually improve.
6. The age of 'convergent marketing' is finally dawning as amazing synergies emerge between traditional media, the Internet and mobile communications systems—but governance approaches will also arise to deal with contentious issues like SMS spam.

7. Some of the key disruptive technologies to watch out for include Wi-Fi and Wi-Max along with VoIP as severe challenges will arise for incumbent voice carriers and ISPs. Until then, hotspots in small locations will continue to proliferate.
8. Despite much excitement about 3G and Wi-Fi, other wireless technologies like WLL (e.g., DECT) will continue to play an important role, particularly in developing countries.
9. Despite much excitement about the prospects of 3G in many new markets, a substantial portion of mobile subscribers are still quite satisfied with basic voice and the humble SMS though competitive pressures will make it hard for wireless operators to scale up ARPU.
10. While Asia leads in consumer mobile markets and innovation, much of this is in the domain of cellphones. The US leads in PDA adoption, radio frequency identification (RFID), Wi-Fi and enterprise wireless solutions which expand on PC-based data and e-mail applications (though South Korea is a notable exception and leads in mobile as well as Wi-Fi, and smartphones blur the distinction between cellphones and PDAs). However, the coming decade could upset some of these trends.
11. The growing sophistication and 'smartness' of wireless networks and devices as well as the needs of enterprise wireless integration will spawn huge markets for wireless hardware, software services and design—playing to the strengths of Asian players like India as an outsourcing hub and China as a manufacturing giant, and continuing generations of entrepreneurs.
12. Wireless connectivity and affordable handhelds offer a tantalising opportunity to bridge the digital divide and go beyond the 'PC mindset' of the early generations of ICT4D practitioners. Although costs have dropped, they will need to drop even more along with increasing availability of supporting infrastructure like electricity.
13. Regulators will face an increasingly complex task in juggling the needs and priorities of incumbent and emerging wireless operators as disruptive technologies and blurring market boundaries become the order of the day. These complexities include creating level playing fields for operators, spectrum allocation for Wi-Fi/Wi-Max, legalising VoIP, and ensuring compliance with safety and security standards.
14. Empowered users of the Internet and mobile devices will face

unprecedented opportunities for self-expression, community formation, content creativity and even political mobilisation. In the 21st century, every business is a publisher, every Internet or mobile user is a reporter and photographer, every citizen is an editor, and every community is a political party. At the same time, government and private sector opportunities will increase for surveillance and tightening of IPR; consumers too will have to become ever more vigilant about hoaxes, scams and frauds.

15. Citizens (especially youth) and employees will try ever harder to harness new media and also maintain the elusive work-life balance, opening up a wide array of research opportunities for academics and analysts on issues ranging from consumer attachment and identity evolution to productivity studies and media effects. The book in your hands is, thus, a notable step in this regard.

References

Easton, Jaclyn (2002). *Going Wireless: Transform Your Business With Mobile Technology*, Harper Business, New York.

Funk, Jeffrey Lee (2001). *The Mobile Internet: How Japan Dialed Up And The West Disconnected*, ISI Publications, Bermuda.

Gallagher, Lynne & George Turnbull (1999). *Telecommunications in Action*, The Regency Foundation, London.

ITU (2002). *Asia-Pacific Telecommunications Indicators*, International Telecommunications Union, Geneva.

Kalakota, Ravi & Marcia Robinson (2002). *M-Business: The Race to Mobility*, McGraw-Hill, New York.

Keen, Peter & Ron Mackintosh (2001). *The Freedom Economy: Gaining the m-Commerce Edge in the Era of the Wireless Internet*, McGraw-Hill, New York.

Lindgren, Mats, Jorgen Jedbratt & Erika Svensson (2002). *Beyond Mobile: People, Communications and Marketing in a Mobilised World*, Palgrave, New York.

Newell, Frederick & Katherine Newell Lemon (2001). *Wireless Rules: New Marketing Strategies for Customer Relationship Management Anytime, Anywhere*, McGraw-Hill, New York.

Rao, Madanmohan (2002). *The Asia-Pacific Internet Handbook, Episode* IV: *Emerging Powerhouses*, Tata McGraw-Hill, New Delhi.

Rao, Madanmohan (2003a). *The Asia-Pacific Internet Handbook, Episode* V: *News Media and New Media*, Times Academic Press, Singapore.

Rao, Madanmohan (2003b). *Leading with Knowledge: KM Practices in the Global InfoTech Sector*, Tata McGraw-Hill, New Delhi.

Rajora, Rajesh (2002). *Bridging the Digital Divide: Gyandoot, the Model for Community Networks*, Tata McGraw-Hill, New Delhi.

Settles, Craig (2002). *Wireless, Inc.: Using Mobile Devices and Wireless Applications to Connect with Customers, Reduce Costs and Maximise Profits*, AMACOM, New York.

Steinbock, Dan (2003). *Wireless Horizon: Strategy and Competition in the Worldwide Mobile Marketplace*, AMACOM, New York.

2. Mobilising the Internet
Asian-style

Lara Srivastava

Editors' Note: This comprehensive chapter sets the stage for the book, drawing on authoritative sources to show how Asia is at the cutting edge of the wireless information society. The age of the mobile Internet promises to be one of rapid innovation, as evinced from the 3G mobile services of Japan, South Korea and Hong Kong. Even the world of 2G is showcasing creative socio-economic applications in China (the world's largest cellphone market), India (with the most potential for growth), the Philippines (with SMS as a tool for political mobilisation) and Bhutan (where wireless technologies help bridge the digital divide). However, it should be noted with caution that a widespread shift to 3G mobile may take longer than expected due to high network costs and user satisfaction with earlier generation services. Regulators and industry players alike need to realise that there are a number of different options for providing mobile Internet services, billing plans, content alliances and migrating to next generation platforms. Other governance challenges will arise in terms of mobile spamming, user privacy, Internet addiction, and mobile etiquette.

Introduction

In recent years, the world has seen an explosive growth in telecommunications, particularly mobile communications. It has not been uniform and the poorer areas of the world are not necessarily the least developed in this regard. At the end of 2002, there were more mobile phones in the world than fixed-line telephones—over 1.15 billion (see Figure 2.1 on p. 71). Mobile technologies, together with the Internet, were the primary demand drivers for telecommunications over the last 10 years. Their growth patterns have been largely similar since the start of the 1990s, separated by a gap of about two years (see Figure 2.2 on p. 72). It would seem that the eventual coming together of these two high-growth industries should give rise to somewhat of a phenomenon—a brave new age of the mobile Internet.

This age of the mobile Internet is set to be one of rapid innovation, giving rise to applications and services that would not otherwise be possible. For instance, location-based technologies, can help police and parents protect children from abduction or other forms of crimes. Such technologies, combined with customised advertising, can enable local businesses to attract users who are passing by (within a certain radius). Multimedia messaging services and streaming mobile video will open up more exciting person-to-person services and customised entertainment.

Before the mobile Internet hits the mass market, however, significant long-term investment is required. Investors are still wary due to the excessive prices paid for 3G licences, particularly in Europe, and the relative failure of early mobile Internet services in the GSM world, such as WAP (Wireless Application Protocol). As earlier waves of technological innovation and convergence have shown, the real fruits of the mobile Internet may not be reaped for another decade or more. Not surprisingly, wariness and foresight are detectable in the business plans of new mobile Internet businesses. Early evidence of the potential of mass market penetration is already visible, however, through data services such as short messaging service (SMS).

This chapter provides an overview of the mobile Internet while paying particular attention to the Asian region, which currently leads the world in mobile penetration and the use of mobile Internet technologies. At the end of 2002, the Asian region boasted some 448.2 million mobile subscribers (up from 333.4 million a year earlier) accounting for 39 per cent of the world's total (see Figure 2.3 on p. 73). Moreover, two of the top three mobile economies in the world (in terms of the number of mobile phones per capita) are from the region: Taiwan, China and Hong Kong (China). Japan and Korea boast the world's largest number of mobile Internet users and for a single economy, while China has the world's largest number of subscribers. The market with the greatest potential for growth, in terms of the number of people without mobile phones, is also in Asia—India.

Advances in Mobile and Wireless Networking

Although the shift to 3G is a radical one, in some cases involving the construction of new networks, the transition from 2G to 2.5G

Fixed lines and mobile subscribers (millions) and countries in which mobile has overtaken fixed

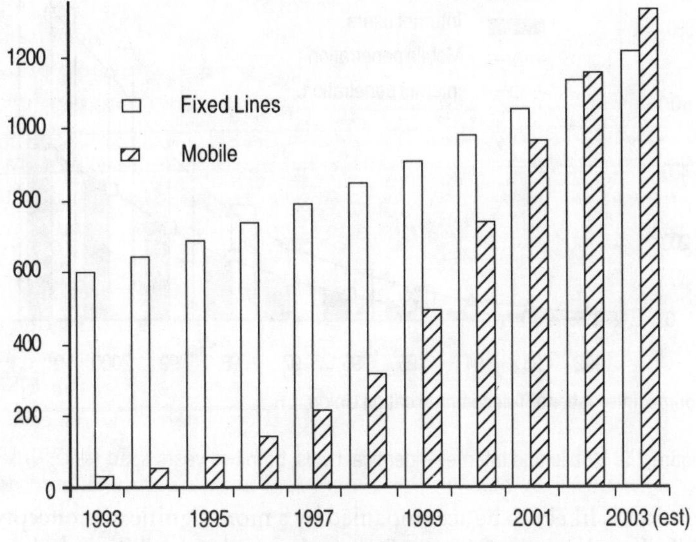

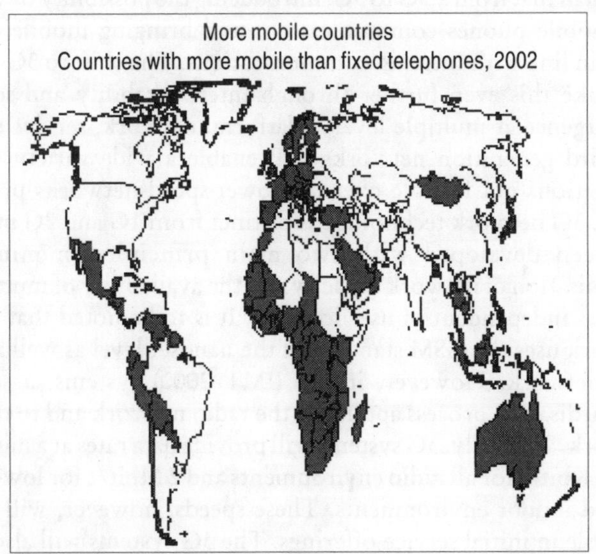

Source: International Telecommunication Union

Figure 2.1: Mobile overtakes fixed

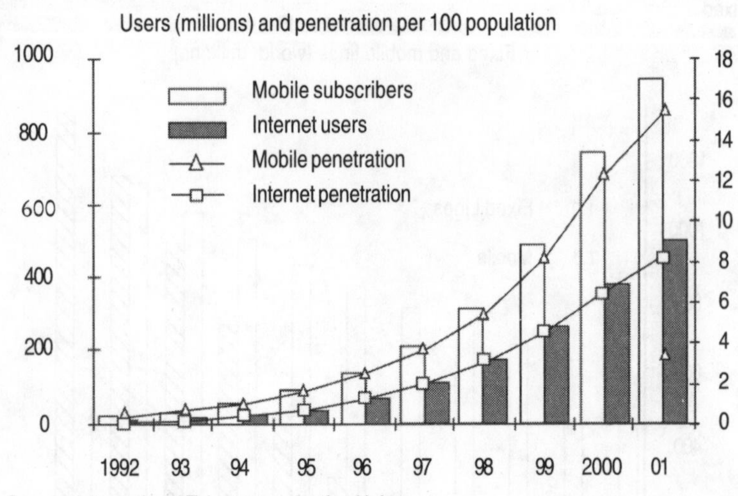

Source: International Telecommunication Union

Figure 2.2: Mobile and Internet—identical twins, born two years apart

services is likely to be accompanied by a more significant conceptual shift than that from 2.5G to 3G, introducing the possibility of 'always-on' mobile phones communications and bringing mobile phones more in line with Internet communications. The shift to 3G mobile will take this even further through interoperability and seamless convergence at multiple levels (platform, network, service etc.)

Third generation networks will enable a wide variety of data applications not feasible over the lower-speed networks prevalent today. 3G network technology is distinct from 1G and 2G in that it has been developed with two main principles in mind: the maximisation of network capacity and the availability of multimedia services independent of user location. It is to be noted that a GSM network uses the GSM standard at the handset level as well as in its core network. However, in 3G (IMT–2000) systems, a separate standardisation process applies to the radio network and to the core network. Typically, 3G systems will provide data rates at a minimum of 144 Kbit/s for all radio environments and 2 Mbit/s for low-mobility and indoor environments. These speeds, however, will not be available in initial service offerings. The 3G systems will also allow for symmetrical and asymmetrical data transmission, which means that they can be used either for real-time communication or for data downloads.

Mobile subscribers per region (1997–2002)

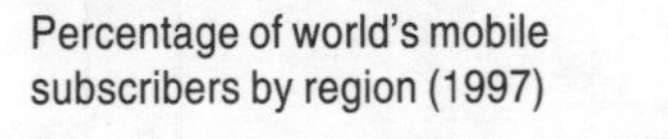

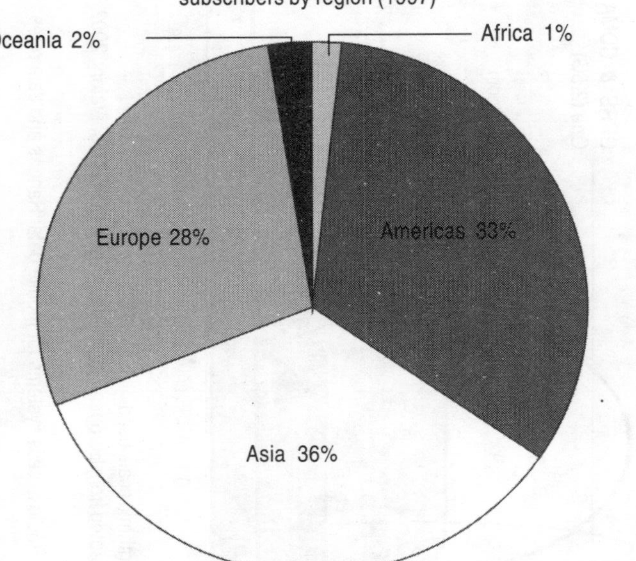

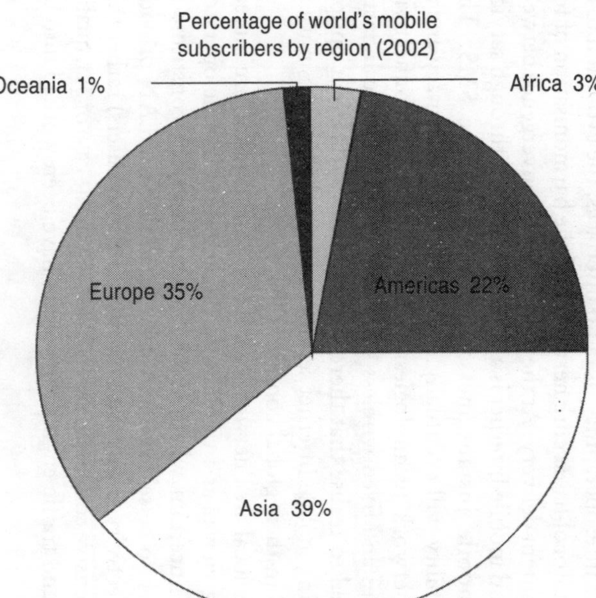

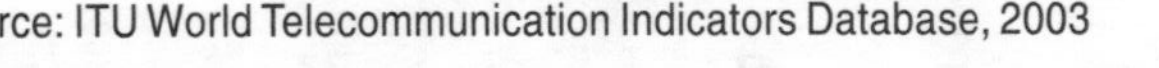

Source: ITU World Telecommunication Indicators Database, 2003

Figure 2.3: Upwardly mobile in Asia

The viability of future 3G services will rely on continued efforts towards the interoperability of radio interfaces, the evolution to an Internet Protocol-based core network and the harmonisation of formats for content delivery. At the service level, convergence between the fixed and mobile Internet is already happening, through services such as mobile instant messaging and fixed-line SMS. This interoperability will eventually encompass complementary and alternative network technologies, such as short-range connectivity technologies and fixed wireless solutions. Regulators and industry players need to realise that there are a number of different options for providing mobile Internet services and that 3G services must be considered in their global context (Figure 2.4).

Wireless local area networks (WLAN), for instance, are increasingly being viewed as a viable solution, and most mobile operators are as yet uncertain whether they pose a threat or constitute an opportunity to be seized. It must be noted that a WLAN provides mere connectivity at this stage (mostly for laptop users) rather than a comprehensive connected service like 3G. In the case of 3G, quality of service and traffic engineering issues have been worked into the

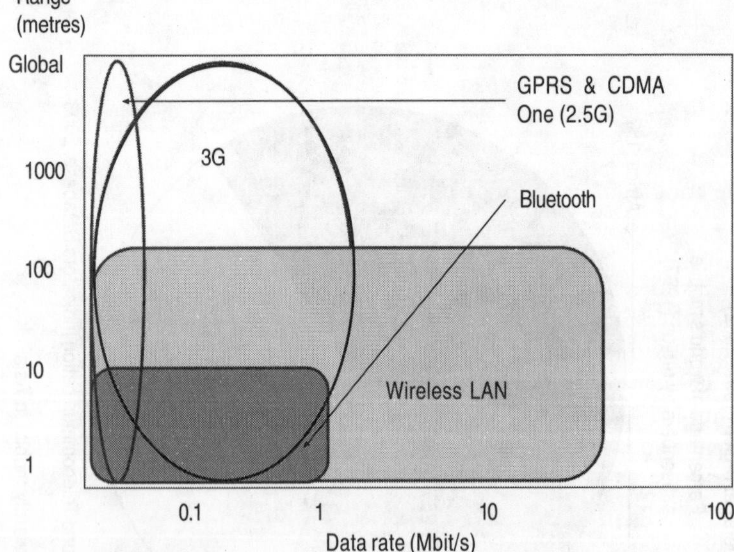

Note: Logarithmic scale used
Source: Adapted from European Information Technology Observatory, 2002

Figure 2.4: Radio access systems for mobile data: Ranges and data rates

design. Moreover, 3G is truly a 'mobile service' whereas WLAN is a fixed-mobile service, offering wireless connectivity within limited geographical areas or 'hotspots'.

In Japan, one of the first Asian 3G economies, high-speed mobile services have been struggling. Given the success of 2G mobile Internet services in countries like Korea and Japan, users do not feel the need to upgrade to 3G until prices drop significantly. On the other hand, WLAN hotspots have been more successful in terms of take-up. One of the main reasons is cost. In many coffee shops and airports, users can even use the network free of charge. The popular 802.11b WLAN standard offers speeds of up to 11 Mbit/s, much higher than the speed of 3G systems. It is not surprising, therefore, that some operators have chosen to embrace WLAN rather than compete with it. Many of the 3G pioneers are moving towards business models that bundle WLAN services with 3G, facilitated through interoperability and roaming agreements. In Japan, manufacturers began integrating WLAN cards into their 3G mobile handsets in 2002. WLAN technologies also have their application in the developing world where WLAN is being used for the most part by individual users or companies. In the Asian region, India's Satyam Infoway introduced WLAN in some apartment blocks and Thailand's Bangkok airport began offering a wireless hotspot in early 2003. In Bhutan, WLAN technology is being used to extend access to basic communications (see Box 2.1 appearing on p. 77).

A Host of New Applications and Services

The combination of mobile and Internet technologies has already transformed the way people interact and has had a profound impact on operators' business plans. There are a number of different applications currently available on mobile handsets, some of them in wide use, others are still developing their full potential. To illustrate: Some 30 billion SMS messages were sent worldwide during December 2002 alone. Operators hardly expected this simple text messaging technology to become such a widespread service and a significant revenue booster. The phenomenal growth of SMS was predominantly user-driven, rather than the result of any targeted marketing efforts. Once the potential of SMS became clear, companies began exploiting its broadcast mode and offering a wide array of billable information services. These services include local and international news, stock updates, weather forecasts, banking and travel information. Asian countries are no exception to the SMS phenomenon—in Hong Kong, China, mobile users received 71 million messages in December 2002 alone. In Japan, a 2002 survey found that 75 per cent of all mobile users and 92 per cent of youth users (aged 12–18) send an average of over five mobile messages a day.

The SMS phenomenon is not restricted to the industrialised world. In China, for instance, during one week in February 2003, mobile phone users sent 7 billion text messages, amounting to $12.6 million in revenue. The Philippines also provides an interesting example. In 2002, an average of 75 million text messages were sent per day, amounting to 7.5 SMS per subscriber. In fact, SMS played an important part in recent Filipino politics. When President Joseph Estrada refused to step down following a corruption scandal, Filipinos used SMS to organise mass demonstrations that eventually led to his downfall. The Philippines' Civil Service Commission is now encouraging Filipinos to send tip-offs about corrupt government officials and employees via SMS. Unique applications of SMS have been popping up all across Asia. In India, since March 2003, subscribers can order candy from vending machines via SMS. BPL Mobile has tied up with Cadbury's to set up 100 machines dispensing chocolates to its cellular subscribers. The charge for the candy is billed to the user via his or her mobile phone bill and then passed on to Cadbury. China has opened up its annual parliamentary session to suggestions via text messaging in an effort to amplify the voice of the people, or at least the voice of the country's 200 million mobile phone owners. Since 2002, the Dubai police have been sending SMS to drivers urging them to observe lane discipline on highways as part of an awareness campaign organised by the traffic and moral guidance departments.

Messaging services have also brought information technology closer to groups that have had limited access to it, such as children and the deaf community. High-speed data services combined with additional functionality, such as location technologies and improved security, will further enhance the user experience. However, users have also been receiving unwanted SMS and the prevention of mobile spamming is now high on the agenda of many regulators.

Although mobile data services are available on second generation mobile platforms (such as GSM), it is through the advent of 2.5 and 3G that users will begin to reap the benefits of the mobile Internet, through high-speed communications and multimedia applications. There are a number of factors that will enable the rapid and successful development of the mobile Internet over such networks. First and foremost, the availability and affordability of adequate Internet-enabled handsets will be a prerequisite for users. Second, unrestricted and non-proprietary mobile Internet content needs to be fostered.

The Mobile Internet as a Means to Extend Access—Pilot Project in Bhutan

Wireless connectivity for developing countries via WLANs

Wi-Fi, or Wireless Fidelity—local area networks using radio technologies based on the IEEE 802.11b or 802.11a standards—is generally thought of as a technology for mobile Internet users in developed economies. Wi-Fi is commonly perceived as wireless broadband for jet-setting business people, students and trendy dot-comers and hotspots are located in busy, urban places such as airports, college campuses and coffee shops. But Bhutan may prove this elitist image of Wi-Fi wrong. It is almost as though the predominant religion there, Buddhism, extends even into communications: where 'Gross Domestic Happiness' is as important as Gross Domestic Product: a source of delight for some, who appreciate the aspects of Wi-Fi that run counter to commercial and economic systems (Wi-Fi is often free, and typically operates in unlicensed frequencies).

Bhutan is probably one of the most difficult countries in the world to 'wire.' It is predominantly rural, with mostly mountainous terrain. Some 79 per cent of the population lives in rural areas—and many are without telephone service. The nation's telecom operator, Bhutan Telecom (BT), has been looking at alternative technologies for providing telephone service. Wi-Fi meets a number of BT's goals. It supports Internet Protocol that can both meet voice telephony needs (using Voice over Internet Protocol, [VoIP]), as well as data communications requirements. As BT notes, delay in implementing telephone service in rural areas *may have been a godsend as recent technological developments may offer a much more cost-effective alternative for the provision of rural access.* Another motive for moving to new technology is that manufacturers are discontinuing the Digital Radio Multiple Access Subscriber System (DRMASS) that BT has been using for rural telecommunications and equipment will no longer be available.

A project in Bhutan has been exploring the feasibility of Wi-Fi for this mountain kingdom. There is ample spectrum available in the kingdom. Results of a first pilot project were very promising. There was scarce radio interference and power requirements using solar power and batteries were sufficient due to the lower energy requirements of the Wi-Fi equipment. The pilot project has now been extended to 14 villages as part of $ 300,000 project. Depending on the evaluation, the technology may be extended to more villages.

*Note: See Dorj, Thinley (2001): IP-Based Rural Access Pilot Project. http://www.bhutan-notes.com/clif/bt_rural_access_pilot.html.

Source: ITU Asia-Pacific Telecommunication Indicators, 2002.

Third, simple and transparent billing models are required, taking into account the difference between voice and data services and the growing importance of content. In this regard, valuable insights can be gleaned from the evolution of the fixed-line Internet, notably relating to its early development. Initially, users were charged per minute for browsing the Internet. In most countries, operators then moved to a subscription model with call charges for time spent online. In countries where local calls are unlimited, like the US and Canada, flat-rate plans were introduced in December 1997. Countries in Europe and Asia soon followed suit. It can be said that the mobile Internet is following a similar trajectory. In the early days of WAP over GSM, users were billed for every minute they spent online. Services such as i-mode combine monthly subscription with volume or packet-based billing. Always-on GPRS billing models have evolved with the introduction of volume-based charges. With high-bandwidth applications and increased spectrum efficiency, will flat rates become the norm for mobile data, as they have done for fixed data? Flat-rate schemes for data services are already being considered by a number of mobile operators. Hutchison Whampoa is offering flat-rate schemes for voice and multimedia on their 3G networks in Italy and the UK. Still, users will have to wait some time before this type of charging models become widely available. On the whole, what we have learned from the success and failures of 2G technologies is that person-to-person messaging, simple interfaces and timely content delivery will be the key to future service development and revenue generation. A mere simulation of the fixed-line Internet experience will not compel users to access mobile data services.

First-movers and Early Adopters

In the area of network deployment, the major Asian economies are the clear first movers in licensing high-speed always-on 3G mobile services, with Japan and Korea being the first to actually deploy 3G services (Figure 2.5). NTT DoCoMo introduced its W-CDMA 3G services as early as October 2001 under the brand name FOMA (Freedom of Mobile Multimedia Access). At the utset, the company was aiming for a 150,000 user base by the end of 2001. It was not able to reach this target until a year later. This was due to a number of factors including short battery life, limited coverage area, a lack of new and compelling applications, and the incompatibility with the operator's 2G system, PDC. On the other hand, its competitor KDDI had more success with its 3G service, based on CDMA 2000 1x, a standard which allowed for a smoother migration path from its 2G and 2.5G services (CDMAOne). By the end of 2002, the company boasted almost 4.7 million 3G users, despite a late launch. In fact, CDMA 2000 1x is gaining momentum worldwide as it has been

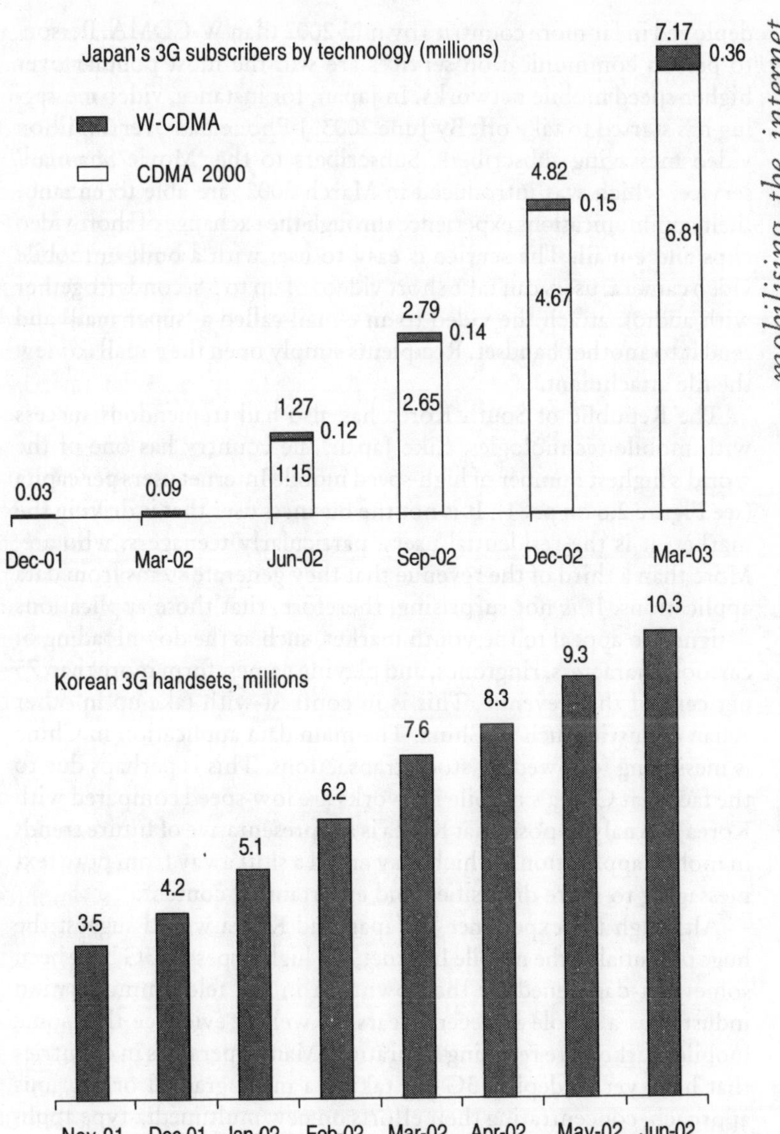

Note: In Korea, there are no subscription charges for 3G mobile Internet services (CDMA 2000 1x and 1x-EV-DO). The figures show the number of handsets theoretically capable of supporting these services.
Source: MPHPT (Japan), ITU.

Figure 2.5: Early 3G pioneers

deployed in far more countries by mid-2002 than W-CDMA. Person-to person communication services are still the most popular over higher-speed mobile networks. In Japan, for instance, video messaging has started to take off. By June 2003, J-Phone had over 2 million video messaging subscribers. Subscribers to the 'Movie Sha-mail' service, which was introduced in March 2002, are able to enhance their communications experience through the exchange of short video clips and e-mail. The service is easy to use: with a built-in mobile video camera, users can take short videos of up to 5 seconds (together with audio), attach the video to an e-mail called a 'super mail' and send it to another handset. Recipients simply open the e-mail to view the file attachment.

The Republic of South Korea has also had tremendous success with mobile technologies. Like Japan, the country has one of the world's highest number of high-speed mobile Internet users per capita (see Figure 2.6 on p. 81). It is not the business user that is driving the market: it is the residential users, particularly teenagers, who are. More than a third of the revenue that they generate stems from data applications. It is not surprising, therefore, that those applications designed to appeal to the youth market, such as the downloading of cartoon characters, ringtones, and playing games, form more than 75 per cent of that revenue. This is in contrast with take-up in other Asian countries such as China. The main data application in China is messaging followed by stock transactions. This is perhaps due to the fact that China's mobile networks are low-speed compared with Korea's. Analysts posit that Korea is a representative of future trends in mobile applications, which may entail a shift away from pure text messaging to more diversified and entertaining content.

Although the experience of Japan and Korea would suggest the huge potential of the mobile Internet, the high hopes for 3G have been somewhat dampened by the downturn in the telecommunication industry as a whole in recent years, as well as evidence that some mobile markets are reaching saturation. Many operators in countries that have yet to deploy 3G are taking a more gradual or cautious approach, concentrating their efforts on new multimedia-type applications over existing 2G platforms. Singapore, for example, has not made the move to 3G as rapidly as might be expected for a country with such dynamic ICT policy-making. Although all three mobile operators, SingTel (the former incumbent), MobileOne (M1), and StarHub, obtained 3G mobile licences when these were auctioned in April 2001, none rolled out 3G networks as of late 2003. In this respect, Singapore lags behind Japan, Korea and Hong Kong. Some of

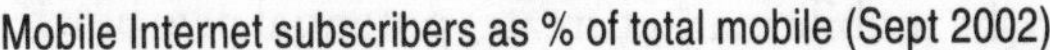

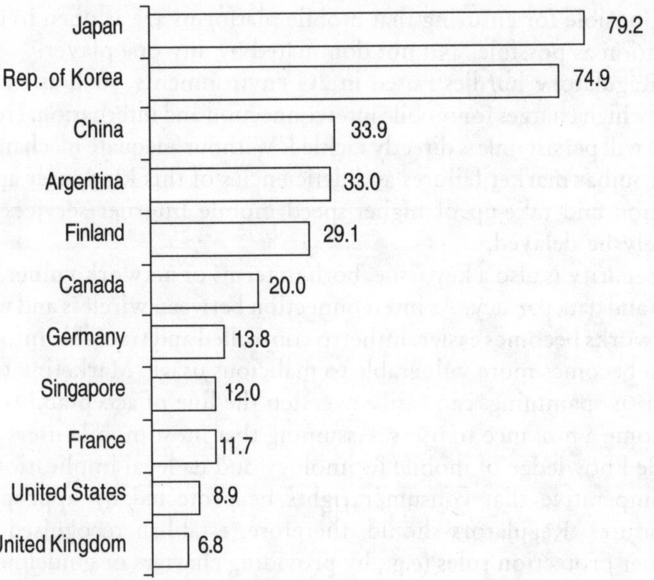

Source: Ministry of Public Management, Home Affairs, Post and Telecommunications (MPHPT), Japan, based on materials from 3G mobile

Figure 2.6: Mobile Internet around the world (Mobile Internet subscribers as a percentage of total mobile subscribers in selected countries)

the less developed Asian economies, such as China, the Philippines, and Thailand, are also less ripe for 3G roll-out, and are in no rush to award 3G licences owing to their particular market contexts. But in these countries, low or medium-speed mobile Internet technologies are being taken up at a fast rate.

Policy-making and Regulation for the Mobile Internet

The policy of fair competition has worked best in developing the mobile and Internet markets. It seems likely that this will be no different with the age of the mobile Internet. Fair and non-discriminatory licensing processes must be guaranteed for services such as 3G and, in some cases, WLAN. But the role of regulation remains just as crucial after licensing. The mobile market needs to be monitored and dominant operators should not be allowed to abuse their market position over less established operators. The introduction of Internet access into the mobile market creates potential new bottlenecks such as portals, billing and walled gardens of content. Regulators are responsible for ensuring that mobile platforms are as open to competition as possible, and not dominated by any one player.

Regulatory hurdles raised in 2G environments, such as exceedingly high charges for mobile interconnection and international roaming, will persist unless directly tackled. Without adequate mechanisms to combat market failures and deficiencies of this kind, user appreciation and take-up of higher-speed mobile Internet services will surely be delayed.

Security is also a key issue, both in terms of network vulnerabilities and data privacy. As interconnection between wireless and wired networks becomes easier, hitherto controlled and traceable information becomes more vulnerable to malicious usage. Marketing tools, such as spamming, can easily overstep the line of acceptability and become a nuisance to users. Assuming that most mobile users have little knowledge of mobile technology and its legal implications, it is imperative that consumer rights be protected by appropriate measures. Regulators should, therefore, establish recognised consumer protection rules (e.g., by providing charters or guidelines on protection of consumers' legal rights). Educating users about the threats and opportunities of new technologies should be high on the government's agenda.

The Call for Content

Licensing and deploying 3G services and driving mobile growth alone is not enough to guarantee a mature and successful mobile Internet market. Internet content also needs to be developed. Service providers should be discouraged from imposing commercial restrictions on content providers or establishing 'walled gardens' of content. The fixed-line Internet established a tradition of largely free and non-proprietary information access. This is now changing, but at a relatively slow pace. Over the Internet, messaging (e-mail) is still available free of charge. Mobile communications, by contrast, have always come at a premium. Users seem quite willing to pay per message for SMS, per packet for i-mode content, and premium rates for voice calls while roaming. Moreover, a direct relationship is established between the individual user and the mobile operator, facilitating billing for a variety of add-on services. This was not typically the case with fixed Internet access. On the whole, this bodes well for the future of paid digital content services on mobile devices. Combined with high worldwide mobile penetration and short-range technologies, it may mean greater success for mobile business-to-consumer commerce than has been seen over the fixed-line Internet. However, in order to be valuable, content needs to be accessible and relevant and efforts need to be made to lift language barriers. In Thailand, for instance, the limited knowledge of English among users was an obstacle to the acceptance of messaging technologies. This is in sharp contrast to Korea, where significant efforts were made to develop Internet content in the local language: in fact Korean was the sixth most common language on the Internet during 2002. Moreover, the availability of handsets that support Hangeul characters have contributed significantly to the success of mobile data in that country.

It is clear that at this time, the majority of revenues from the mobile Internet do not stem from Internet browsing. As mentioned earlier, messaging services are generating maximum excitement. And it is increasingly being used to deliver targeted content to users, rather than letting users simply browse Internet content. In Hong Kong, Hutchison Orange's 'UK premiership football service' is a good example. Users would pay about 0.65 US cents per match to receive timely SMS-alerts of starting line-ups, goals scored and red/yellow cards issued during the match. Given the success of SMS, operators are keen to promote multimedia messaging services (MMS), which will allow users to send audio and video messages. CSL (Hong Kong) was the first operator in Asia to launch MMS services. In December 2002, Hong Kong's six operators signed agreements for the interconnection of MMS traffic, and this has largely opened up the market. In Japan, sending and receiving still photos is proving to be very successful over high-speed 3G networks. Korea's KTF offers a Magic 'multi-message' service which allows messaging from phone-to-phone, phone-to-PC (e-mail) and PC (web)-to-phone. It is interesting to note that in Korea, users tend to send character animations with accompanying text more frequently than real photos of themselves. In terms of non-messaging content, ringtones and video games seem to be the most popular: in Japan, over 80 per cent of the respondents of an MPHPT 2002 survey use their mobile to play video games on a regular basis (Figure 2.7).

Mobile Ubiquity: Anyone, Anyplace, Anytime... Anyhow?

The major uses of the mobile Internet over 2.5G networks are likely to be messaging, but the extra bandwidth of 3G networks will allow

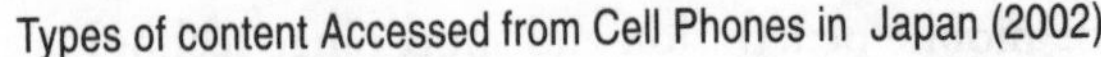

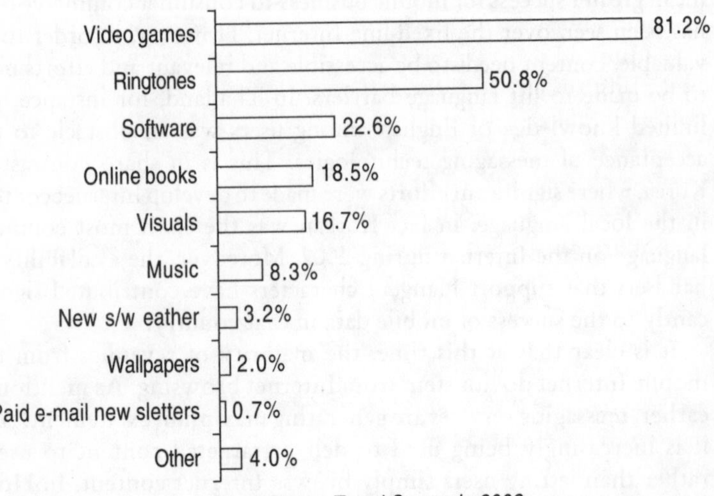

Source: MPHPT *Communications Usage Trend Survey in 2002*

Figure 2.7: Content for the mobile Internet (Popular content on Japanese cellular mobile phones)

for download, video streaming and multiplayer games. Early experiences with 3G mobile Internet services, in Korea and Japan for instance, indicate that it is the teenagers who are driving the market. In Korea, although teenagers have lower disposable incomes than older age groups, they are spending around three times more per user on mobile data services. In Japan, video messaging has proved immensely popular among young people. What this suggests is that the younger the user, the more likely they are to be comfortable with the intrusive nature of mobile communications. Youngsters also have more time for playing games and sending frivolous or flirtatious messages. The key question is whether they will continue to use the mobile Internet when they are older, and have more disposable income? If they do, then the mobile Internet gamble will certainly seem like money well spent for the operators, particularly if the mobile Internet becomes increasingly present in every aspect of our daily lives. In the 1980s and 1990s, the microchip spread from the computer into hundreds of other devices, from computers to washing machines to cars. The average car, these days, has as much computer power as some of the early Apollo rockets, and many new models now come equipped with wireless navigation systems. Most families in the industrialised world already own dozens of microchips embedded in different devices. The next stage in this process of ubiquitous communications is for those microchips to gain the ability to communicate and report on their status and location. The technology to make this happen is already available—for instance, nanotechnology, software radio, cellular communications, cheap processing power, location-tracking systems—but the networks and the billing systems are not yet in place. The advent of the mobile Internet may indeed make this a reality.

But are we prepared as a society, and as individuals, to live in a world of technological ubiquity, a world in which an intelligent fridge sends out for milk, or the mobile phone tells users that their loved ones are on their way back home? Consider the use of tiny Radio Frequency ID tags embedded into clothing to help retail businesses track inventory. Will these remain active once the item has been purchased? What kind of information will be collected and for what purpose? The new generation of always on, anytime, anyplace technologies may allow for levels of convenience but also of surveillance unknown and unimagined by earlier generations. Furthermore, at the dawn of this age of the mobile Internet, it is important to consider what effect these technologies are having on the way we grow, interact, socialise and learn. In Korea, young Internet 'addicts' are being treated in specialised centres, to control their use of the Internet: some of these youths are spending over 12 hours a day at their PC, stopping only to eat. In other countries, such as Japan, the generational gap between children and parents is widening through a growing technological sub-culture among teenagers. The adult world is no exception to this social transformation. In a world where mobile phones follow everyone everywhere, a new mobile etiquette is emerging, in which it is perfectly acceptable to be late for an appointment, or receive loud-ringing calls during intimate dinners, business meetings, or visits to the toilet. In this respect, Asian countries such as Japan have taken active steps to limit the use of mobile phones on trains, buses, and other public places. In other regions, however, awareness of the nuisance factors associated with mobiles is being discussed only peripherally. Whenever a new development (such as a widely-used technical innovation) arises amidst society, proper account has to be taken of it, and new adjustments have to be made to accommodate it. The mobile phone, too, requires some such set of rules to permit its rational usage. Alas, as in many matters, some degree of freedom has to be relinquished, so that we may *all* exercise a measure of freedom.

Endnote

This paper is drawn in part from *ITU Internet Reports 2002: Internet for a Mobile Generation* (www.itu.int/mobileinternet). Lara Srivastava < lara.srivastava@itu.int >, Policy Analyst at the International Telecommunication Union (ITU), was the lead author of this report. ITU's *Asia-Pacific Indicators 2002* also covers mobile communications in the region (www.itu.int/ITU-D/ict/publications/asia/2002/index.html). For more information on these and other ITU publications, see the website of ITU's Strategy and Policy Unit at www.itu.int/spu/ or ITU's home page at www.itu.int.

3. Location-based Services

The Killer App of the Mobile World?

Chas Sweeting

Editors' Note: This detailed chapter spells out the foundations, market potential and customer acceptance factors underlying the potential of location based services (LBS), once regarded as the 'killer app' of the mobile world and still one of the more intriguing services in the mobile information society. These range from emergency and resource tracking to people monitoring and location-specific advertising services. To be truly successful, LBS must be seamless, accurate, timely, useful, affordable and not intrusive of users' privacy; it also requires extensive cooperation between operators, technology providers and content players–which is why the uptake has been relatively slow. The chapter also highlights some LBS applications in Asian markets, and compares them with those of the US and Europe. It may be enterprise applications of LBS that find greater acceptance and prove profitable, as compared to consumer services.

Introduction

In 2000, location based services (LBS)—applications which make use of a cellphone's physical location—promised to be the saviour of wireless data services, with research companies valuing the LBS market at tens of billions[1]. Analysts and industry pundits proclaimed LBS the next 'killer app'—a label which has proven more often than not the kiss of death for any spanking new technology, setting the stage for unrealistic expectations and subsequent disillusionment.

Sure enough, a scant two years later, the same industry experts[2] revised this by several orders of magnitude to tens of millions and, today, there are few industry insiders who would claim that LBS will account for the majority of mobile data revenue streams.

Given this less than euphoric market outlook, however, LBS deserves to be on the radar screen. First, because LBS represents one of the few real opportunities for mobile network operators to differentiate themselves. Unlike SMS, GPRS, MMS and WAP which sooner or later become generic commodities, LBS will inherently differ from one operator to the next. Second, because you should not believe everything that analysts tell you!

At the support-group for disillusioned wireless technologies, LBS sits alongside WAP, 3G and, to a lesser extent, GPRS. In fact, the WAP and 3G auction fiascos attracted so much attention and criticism that LBS actually managed to avoid the spotlight and has since made slow but steady progress. Paraphrasing the late Mark McCormack ('Beware the big ideas, and the best ideas usually start as small ideas'),—LBS is a small idea growing steadily in what is still a nascent industry. As a disruptive technology, this is perhaps the first time that peoples' physical locations can be used as data points for applications on such a wide scale, which has the potential to affect our daily lives and is reason enough to continue reading this chapter.

Market Drivers

Mobile operators around the world face the same major challenges:

1. Commoditisation of services—since there is little quantifiable difference between operators offering comparable service plans with similar coverage.
2. Steadily decreasing voice ARPU (average return/revenue per user) as operators are forced to compete on price in a commodity market.
3. Increased churn (loss of existing customers) and the associated high cost of customer acquisition. (Interoperability and mobile number portability remove the final barriers to switching service providers at will.)
4. Balancing capital expenditures against uncertain return on investment, especially with competing investment priorities (e.g., GPRS, MMS) and since the recent downturn of the telecommunications industry (e.g., 3G licence auctions).

Faced with this, the mobile operator is driven by a desire to:

1. Increase ARPU from wireless data services, which tap a new revenue stream as opposed to cannibalising voice revenue.

2. Reduce churn, since it costs less to retain existing customers than acquire new ones.
3. Differentiate themselves in an otherwise commodity market driven primarily by price.

According to Angela Finney, senior product marketing manager of Openwave: 'We're finding that the location platform is the operator's one big opportunity to differentiate.' Enhancing the brand experience and perceived value are in turn key to increasing customer loyalty and retention.

While most of the world is driven by capitalism, mobile operators in the US are implementing LBS primarily to comply with government legislation. The Federal Communications Commission's Enhanced 911 (E911) rules require that network operators must be able to locate 67 per cent of 911 (emergency service) call within 100 metres and 95 per cent of calls within 300 metres[3], beginning October 2001 and to be completed (with total coverage) by December 2005.

The Technology—Foundations

All LBS relies on some form of triangulation (i.e., the determination of an unknown point by the formation of a triangle using two known locations along with geometric projections). These technologies can be split into handset-based systems whereby the position calculation is performed by the handset, and network based methods where equipment in the network performs the positioning functionality. Both methods can be made to look the same to the end user, but the latter (network-based) has the advantage of lack of dependence on handsets. Several mechanisms exist for locating mobile handsets, each with its own advantages and disadvantages in terms of accuracy, consistency, speed and cost.

Cell-ID

Cellular networks are divided into 'cells' (hence the name), each with their own base station which transmits and receives radio signals to and from the mobile handsets (i.e., cellphones) within that cell's coverage area. Cells can vary in size from 500 metres in cities to several kilometres in rural areas; calls are handed-off from one cell to the next as the user physically moves between cells.

The network operator knows which base station a particular mobile handset is registered to at any given moment, as long as the phone is switched on; that is simply how cellular networks function. Since operators also know where their cells and base stations are physically located in the real world, the cell-ID can be used as an approximate measure of the handset's location. Accuracy, anywhere from 500 metres to several kilometres, is dependent on cell density (which in turn is a measure of the distance between base stations).

Sectored cell-ID

Many base stations use 'sectorised' antennas which take a traditional cell area and subdivide it into 'sectors,' thereby splitting transmissions into three or more arcs (depending on the antenna); a mobile device can be identified as being within a particular sector, which is smaller than a cell, thereby allowing greater accuracy.

Enhanced cell-ID

Enhanced cell-ID uses measurements of radio wave signal-strength at one or more antenna as an indication of distance from each base station. However, signal strength can fluctuate and be affected by diffusion, reflection and refraction on account of anything from changes in weather to dense concrete structures in urban areas. Enhanced cell-ID can be costly in network investment, yet only adds an increase in precision of 50 per cent over sectored cell-ID (e.g., 250 metres within high cell density, urban areas).

Cell-ID + TA (Timing advance)

Timing advance is a GSM specific parameter which provides an approximate measure of the time taken for signals to travel from the base station to the mobile handset.

Since radio waves travel at the speed of light, it is easy to calculate distance if you know the time taken for radio waves to travel between two points; in this case, between the handset and the base station. Since this information is already built into the network, it is relatively inexpensive to implement and can provide accuracy in the 250 metre range in small cells as found in urban areas.

U-TDOA (Uplink Time Difference of Arrival)

A signal sent from a mobile handset will arrive at several base stations at different times; the network operator can use the difference in arrival times at these base stations (which are themselves synchronised by an atomic clock) to calculate the distance of the handset from each base station, which is sufficient for hyperbolic triangulation.

Since it is a network-based technology, U-TDOA supports legacy handsets. It is also accurate enough (less than 50 metres) to comply with E911 phase II requirements. The one major drawback is that it requires monitoring equipment to be installed at virtually all base stations and is potentially the most expensive for operators to implement.

James Rhyu, former CFO for Sunday Telecommunications in Hong Kong, lived and breathed LBS for a year....

Depending on the service, for LBS to work, you will often need a pretty high degree of collaboration. For example, if you want to provide a restaurant finder (which by the way I am not sure is a good idea), then in order to make it compelling you should not just be able to identify where a restaurant is (although that might be a good start) but it would be more compelling if people had some incentive to use it like a coupon attached to the location. That requires signing up restaurants—which is a painstaking activity—and likely not worth the effort unless it is combined with something else like a loyalty program which has other revenue streams. And then you need to integrate this with some form of premium billing.

In Hong Kong, there are privacy issues and resistance from men who do not want to be tracked (and I guess some women as well). Inventory tracking is also nice for insurance companies, cars, shipping, etc. And there is demand, but the price points (used to be) too low. Just like the Internet, giving away something for free is an easy proposition but finding something people are willing to pay for is not so easy.

Advertising models do not work. We tried them numerous times and they are very difficult to pull off profitably, consistently and successfully.

On delivery vs. execution: If a mobile operator wants to do both, they need to understand the costs in execution. For a mobile operator, I would say almost categorically, you only want to be the transmission vehicle. Take a toll for using your road, and maybe get involved in some co-marketing but that is all. So then you need to work with a partner on execution but finding a partner to execute properly is very difficult. And vice versa, if you have a service, finding an operator and dealing with that operator can be difficult. Mobile operators can be notoriously cumbersome to deal with unless it is a plug and play application model like China Mobile or DoCoMo.

In all the time I have been looking at LBS, my belief is that the enterprise and commercial applications seem to have a better business case than the consumer ones. In applications such as fleet management, the logistics are easier because you have a specific targeted solution. You can work out the business case and decide whether the investment is justified. Or you can develop a generic application and apply it to a situation. If you can sign up customers (a big IF), then the returns are normally greater.

(*Box Continued*)

(*Box Continued*)

On the commercial/corporate side, fleet management, inventory tracking and logistic support seem to have generated interest. The real question is what exactly can you provide that will offer a decent return to all? With the advent of GPS systems in cars and handhelds, much of this opportunity is slipping away.

Cargo tracking for insurance management is another good example. There is apparently lots of lost cargo which the insurance companies receive claims for. But outfitting each container does not always work because the thieves unload the containers. So you know where the box is but not where the goods have gone.

Profitability per customer is obviously low with individual customers vs. commercial clients, and scaling is difficult with individuals. In Italy, there is a tourist attraction service that works reasonably well—you can request information about tourist sites close to your present location—but I am not sure how much revenue is generated.

In terms of challenges, the first factor (that comes to my mind, not necessarily the first thing to consider) is the technology: a search engine that maps and integrates to databases of information. This is quite a task. The individual pieces are relatively simple but combining them is the challenge; and the upfront costs and investment would be high for uncertain return. Each incremental step of increased service requires work with uncertain demand/return. For example, 'Which is the closest Thai restaurant?' That requires some engineering to do. Again investment with uncertain return. If you expand it once more and say: 'Are there any coupons for restaurants close to me?' then you need some coupon mechanism and the marginal return is questionable.

The people who will benefit are the restaurants in this example but they are unlikely to pay for it.

Customer awareness and adaptation is another challenge facing LBS. This includes education on what the service can do, so that customers are not frustrated when they discover the first services are not as good as online Google. This requires marketing efforts with some associated costs of course. But it is one type of service that once people do begin to use, it will quickly become second nature. Just like the early Internet days with search engines. People adopt quickly because they have always used and liked directory services.

In conclusion, to mobile operators, I would say be open to new applications, do not spend money and time on development, and try to make money through volumes not by getting a larger cut of each transaction.

To developers, I would say first find a real demand and do not try to create a demand that may not exist. Then make sure there is a business case with a real opportunity.

E-OTD (Enhanced Observed Time Difference)

E-OTD works in the same way as U-TDOA, by measuring the time difference of the handset from several base stations. However, unlike U-TDOA, E-OTD performs the timing measurement not at the base stations but at the mobile handset.

Though more accurate (less than 125 metres) than Cell-ID mechanisms, E-OTD requires changes in both network infrastructure (though less than U-TDOA) and extra functionality in mobile phones, which is a problem for network providers who are looking for a more ubiquitous and pervasive solution.

Conventional GPS

Global Positioning System (GPS) is a satellite-based navigation and positioning system, originally designed by the US government for military purposes; the first satellites were deployed in 1978 and GPS now comprises 24 satellites. A relatively inexpensive GPS receiver can locate its position anywhere in the world, based on the difference in time taken for signals from each of the three or four satellites to reach the receiver. The location and trajectory of the actual satellites is also encoded in the radio signals; armed with the distance from three or four known locations (i.e., the positions of the satellites), triangulation can be used to derive the location of the GPS receiver. GPS is a stable, tried-and-tested technology which can provide accurate positioning within 20–100 metres, increasing to 1–6 metres using differential GPS.

However, GPS signals (originating from outer space) are weak and cannot penetrate buildings or even thick clothing; line-of-sight is therefore one of the main issues, particularly in urban areas. Integrating GPS with a cellphone also adds cost and requires additional battery power. Although accurate, GPS has a slow time to first fix and can take as long as 30–60 seconds.

Assisted GPS

Assisted GPS (A-GPS) couples GPS-enabled handsets with additional information about the GPS satellites, calculated in advance and sent to the handset over the cellular network; this additional information, called assistance data, allows handsets to make timing measurements much faster and reduces the time to first fix from 30–60 seconds (for conventional GPS) to as little as 1–10 seconds. As well as faster positioning, A-GPS also provides improved sensitivity (less than 50 metres), cost, and battery life.

On the downside, A-GPS requires network investment and significant modification to handsets (with associated costs), so market penetration and ubiquity remain an issue for mobile operators. Although A-GPS improves indoor coverage, results can still vary and are inconsistent.

So, which technologies are mobile network operators opting for?

Almost all mobile operators contacted in Asia (excluding Japan and Korea) who are rolling out location based services have opted for Cell-ID based positioning systems. The reasons are simple:

1. Risk management: With few applications and uncertain demand/acceptance by consumers, operators wish to experiment without exposing themselves to excessive capital investment.
2. Ubiquity and handset independence: Operators have been burned before. In 2000/2001, the handset and network equipment manufacturers hyped first WAP then GPRS to consumers, encouraging mobile operators to invest heavily in network infrastructure; unfortunately, consumers were disillusioned with the miserable WAP experience, whilst GPRS handsets simply did not arrive in time. The big losers? The operators. Cell-ID works with all legacy handsets and requires no modification to the handset.

Whilst Cell-ID based positioning may not be the most accurate technology available, it does allow mobile operators to start experimenting with services now. Operators are deploying LBS in phases, starting with simple, ubiquitous applications which will introduce customers to the concept without causing privacy concerns. As consumers become accustomed to the idea and demand more advanced functionality, mobile operators will be able to justify further investment and possibly offer hybrid solutions (e.g., a mix of A-GPS and Cell-ID). Cell-ID is not sufficiently accurate for the E911 requirements in the US, where GSM carriers tend to select U-TDOA as standard, and CDMA carriers opting for A-GPS.

Setting up a Location Based Service

To understand the components of an LBS system, let us consider a mobile subscriber (the user) who wishes to find the nearest Korean restaurants, using an LBS directory service: the directory service is provided by a third-party content provider (CP), which is working with the mobile network operator (mno).

1. The user sends a request (say, using SMS) to a special short-code. The MNO routes all traffic arriving at this short-code to the content provider.

2. The CP's application server sends a request to the MNO's location server, asking for the location of the user's handset, as identified by the caller ID received. The location server should provide an API (e.g., using XML and/or HTTP Post) which protects the CP from the underlying location positioning systems.
3. The location server requests the location of the user's cellphone from the network. Specifically, the location server talks to a piece of equipment called the GMLC[4] (Gateway Mobile Location Centre) which returns the co-ordinates (longitude and latitude) using the Location Interoperability Form—Mobile Location Protocol (LIF-MLP) protocol or, in worst cases, a proprietary protocol.
4. The location server could optionally also query a GIS using the OpenLS protocol, to turn the coordinates into a physical address (aka reverse geo-coding).
5. The location server returns the co-ordinates (or physical address) to the content provider's application server which then queries its own content database and returns relevant matches to the original subscriber. Since it could take as long as 30–60 seconds between the first and last steps, an intermediate confirmation receipt could also be sent to the user to let them know that their request has been received and that they should receive the results soon.

The key component is the location server (aka location middleware or location platform), which could be provided by a network equipment manufacturer such as Nokia or Ericsson, or a middleware provider such as Openwave; some operators are even devising their own proprietary systems in-house.

From a software and systems architecture standpoint, building LBS middleware is not a trivial task. According to Angela Finney of Openwave: 'From a middleware perspective, you really want the ability to integrate with any network, any positioning intelligence and any application; and from the oldest, legacy handsets to the latest devices, which means you also need to detect handset profiles and deliver the respective services.' For example, a points-of-interest application may use Cell-ID while an emergency service may use A-GPS.

Then there is the question of integration: 'Whilst there are GSM standards, the reality is that operators have not always upgraded their equipment so we often have very complicated customisation work, which has actually become one of the big parts of Openwave's business, since we have the experienced integration teams', according to Finney.

In addition, billing (such as premium or reverse SMS billing), triggering, privacy and personalisation services could either be provided centrally at the location server or on an application-by-application basis by each content provider.

The Value Chain

As with most wireless data services, the stakeholders comprise the usual suspects: mobile network operators, network equipment manufacturers, middleware providers, third party content and application service providers, handset manufacturers and the end users (business and consumer).

Financial and personal resources are required to develop location middleware, along with development tools and APIs. Several companies including Openwave which has acquired SignalSoft; Ericsson and Nokia over the past few years are already in the race for this. The remaining opportunity is for third party content providers to develop the actual services, which is after all where the value lies from a consumer's perspective.

The entire location platform—from handset to positioning equipment and location middleware—is just the infrastructure upon which value-added services can be developed. Location by itself is not a product and is of limited value. Combining location with relevant, qualitative information is where the value lies, meaning that LBS offers an opportunity to local and regional content providers; this is in contrast to the Internet world.

As long as it adds sufficient value and is priced reasonably, corporate and consumer customers will pay; that applies to any business. A seamless billing mechanism such as premium SMS is crucial, as is compensation to the third party content and application developers (a common-sense[5] approach which has proved successful in Japan, China and some parts of Europe). Possible billing models include per transaction (disliked by corporate customers since it cannot be budgeted), by minutes of usage (which makes less sense), by volume of data traffic (on packet-switched networks, such as GPRS), on a monthly subscription basis (like i-Mode) and/or enterprise contracts and and licences; the choice of model depends largely on the nature of the application.

Sample Applications and Services

Most location based services fall into one of the following categories:

Category	Description and examples
Emergency services	Presumably you do not know or are unable to communicate your location. "I'm having a heart attack–send an ambulance' *'My car's broken down and I don't know where I am'*
Location based billing	Cheaper rates when calling people within a certain distance on the same network. (We already have the reverse of course: inflated rates for international roaming.)
Traffic information	Telematics—integrating mobile phone networks with in- vehicle monitoring systems. *'You are driving on the A41—there are road works 10 km ahead'*
Finder services	Identifying available options within a certain category and region. Where proximity is not the only consideration (e.g., choosing a restaurant), a catalogue of content is also required to help with the decision process. *'Where's the nearest petrol station/ATM/Korean restaurant?'* *Note: Sex services offered via wireless could potentially benefit from this!*
Directions	You know where you want to go, but need instructions to get there. You may or may not know where you are. *'Bugger, I'm lost. Where is the nearest subway station?'.* *'How do I get to Times Square from here?'*
Tracking services	Locating something that is lost or checking up on something for peace of mind. For example, automatic vehicle location; tracking pets, assets and errant teenagers or spouses; fleet management. (This makes a huge assumption–that the cellular device is attached to the tracked target.) *'Where's grandfather wandered off to this time?'* *'My car's been stolen! Where can I find it?'* *'Has anybody seen my Pitbull?'* *'Let's find out where all our staff are at 5pm.'*
Resource management	Improved efficiency and/or increased competitive advantage through improved routing, such as field-force automation. *'Do we have a dispatch rider close enough to the pick-up point?* *'Send the nearest available tow truck.'*

(*Box Continued*)

(*Box Continued*)

Category	Description and examples
Peer-to-peer community	Peer-to-peer, social-enabling. e.g., Date finder, friend finder, location-based chat (or flirting) and location-based games. These differ from straight finder services mentioned above since peer-to-peer applications benefit from the network-effect; that is, the value of the community (network) increases exponentially as the number of participants (nodes) increases.
Advertising/marketing	Push advertising and pull marketing. (discussed later)

Caution must be exercised, however, rather than applying LBS to every business problem when simpler or alternative solutions may exist; always consider what value LBS adds to the equation.

For example, RFID (radio frequency identification) chips, which are far cheaper[6] and do not have the power requirements of cellular devices, are simpler to implement and more suitable for inventory management; indeed Walmart requires its top 100 suppliers to use RFID tags on cases and palettes of consumer goods shipped to its distribution centres and stores by January 2005. However, RFID has a limited range (50–85 metres for active RFID) and requires readers to be installed, which leaves an opportunity for cellular LBS to offer wider, more ubiquitous and pervasive coverage, presumably for higher-value products which can justify the cost.

In telematics, GPS and GIS (geographic information systems; i.e., mapping) car navigation systems have been available for years. 'I recently used one and the key was that the map was big and constantly changing; and there is a voice guiding you at all times,' says James Rhyu, former CFO of Sunday Telecommunications (HK). 'The phone is just too small to do this with a map'. Again, this leaves a possible niche for cellular LBS in the low-to-mid markets, since the cost of in-car GPS systems is high.

Even for the simplest applications, there are alternatives to LBS: why use a friend finder service, when you could just call your friends and ask where they are?

Considerations for Building Applications

Once the decision to launch an LBS has been taken, there are a number of factors to be considered whilst designing and developing the application. These are presented in the non-exhaustive lists below:

Technical Considerations

Whilst technical considerations (such as accuracy, latency, consistency and coverage) are a function of the location positioning technology employed by the mobile operator, and hence out of the developers' hands, you should still bear the following in mind (if only from the point of view of educating end users of the limitations and implications):

Accuracy The question is not 'what's the greatest absolute accuracy we can attain?' but rather 'what level of accuracy is required for a given application?' When building or designing applications, always start with the requirements as opposed to the technology.

Latency Latency in an LBS context is the time required to calculate the position of the cellular device. If you are on a subway and using a directory finder service to help decide where to stop, a minute's delay can be the difference between one station and the next. Also allow for delays outside of the LSB—such as, message queuing in the SMS centre. Receiving a coffee discount coupon via SMS just as you are leaving Starbucks with your latte will probably be met with less enthusiasm than if the coupon had arrived a few minutes earlier.

Time-to-live Once you have determined the location of a cellular device, the time-to-live refers to the duration of time for which that location information is valid. Again, start with the requirements and ask 'how often do I need to update the positioning? Is it sufficient to obtain the position once or do we need to track the handset constantly?' For example, step-by-step driving directions could require updating the location every 10 seconds.

Coverage Is multi-operator and/or multi-national coverage required? Enterprise applications for multi-national companies may require workforce ubiquity. For media companies, the desire to make offerings available to all viewers or readers often flies in the face of the mobile network operators' obsession with proprietary services.

Consistency Call them demanding, but customers probably do not want to be told 'please stand by a window so that we can locate you', especially for mission-critical or emergency services. Meanwhile, more frivolous applications may have less stringent requirements.

Triggering If you are building the 'guy walks past a Starbucks and is sent a discount coupon' service, will you have to track the location of every single cellular device on the network at regular intervals to check for matches? This would be a monumental waste of computing resources. Event driven applications which respond to a user's request are straightforward by comparison.

Handset-based location systems, such as GPS, allow for applications residing on the handset to periodically check the device location and verify if the location is in a zone which triggers a proximity service. Distributed triggering such as this would be more scalable but is dependent on the handset.

Soft 'human' Considerations

Geography (urban vs rural areas) Some 500 metres may be the epitome of precision to the nomad tracking his prize yak on the plains of Mongolia (population density four people per square km), but does not quite cut it in central Hong Kong where it's the difference between shopping at Times Square and trawling the bars in Wanchai.

From a cellular network perspective, urban areas have smaller, more densely packed cells to accommodate more users and calls, whereas base stations could be several kilometers apart in rural areas. Conversely, dense concrete structures, reflection and refraction, and radio wave fluctuations can adversely affect the consistency and/or accuracy of location positioning in urban areas.

Whom or what are you really competing with? Mobile applications are not necessarily competing with a desktop application on your PC. For example, a directions service is potentially competing with the passerby on the street, or the guy pumping fuel at the service station—both of which provide fast, reliable information that does not require trying to read a map or entering your destination on an input device built for pixies.

Culture Repeated surveys in the US have shown that a large proportion of people own a cellphone primarily for personal security; people also spend more time in their cars, making telematics a higher priority. In comparison Asians have very different reasons for owning a cellphone (fashion, friends, communication needs, lack of alternatives and freedom amongst them) and these may vary on a country to country basis.

Usability At the end of the day, it basically comes down to the end user experience, from the handheld device to the application itself. For example, consider a direction/mapping service. Entering a destination address such as '1600 Pennsylvania Avenue' could be frustrating, especially if you are unsure of the spelling or in a foreign country which employs a different alphabet; if it takes an additional 45 seconds to get a location fix, would it not be quicker to ask a passerby? Are the returned results—either step by step directions, or a map—clear and understandable?

Personalisation Personalisation provides an opportunity to streamline user interfaces, providing only the information that is required at the time that it is required.

Adding Marketing and Advertising to the Mix

At the height of LBS hyperbole, there was an unhealthy obsession with the 'coffee coupon': Imagine you are walking down the street and come within 50 metres of a Starbucks—wham, suddenly a discount coupon for a Starbucks coffee pops up on your cellphone. This is push advertising, advocated for the most part by analysts and media pundits who have never sold a dollar of advertising in their lives.

Let us put aside the technical aspects for a moment: the issue is simply that whilst consumers may realise they can be tracked, they are not too happy about operators selling this information to advertisers and random hot deals appearing continually on their phones.

Deluded admen will point to the 20–30 per cent response rates of the first wireless advertising campaigns as evidence that the public will accept and even desire this ('deluded' does not even come close, in the latter). Mobile operators should be more concerned with the 70–80 per cent who did not respond and whose lifetime customer value far outweighs any short-term advertising revenues. From the consumers' perspective, a relationship exists between them and their service providers, not the advertisers; and it is the operators who are likely to experience the brunt of any backlash.

As with e-mail marketing and other media, the saturation of advertising only diminishes the value of the medium, and the effectiveness of the message. The intensely personal and intrusive nature of the mobile phone which makes it the advertising equivalent of the laser-guided missile also has the potential to backfire with results as predictable as a Scud's flight path.

Spam—unsolicited text messages are spam—is brand suicide for a mobile operator. In fact, just as with Internet e-mail, spam control will actually be of more value as a service differentiator.

This leaves us with pull-marketing: allow consumers to create profiles, configure privacy settings, and specify preferences to control what they receive and when. From a direct marketing perspective, there is probably insufficient value to warrant doing this for individual advertisers or marketers; instead, look to incorporating a whole range of them, in say a loyalty programme. Within this context, time-sensitive offers on perishable goods—from concert tickets

> **What of the Social Implications?**
>
> Steve Martineau of Maxis Malaysia ponders the finer points of LBS ... and married life.
>
> It could well create social problems with a wife saying 'Honey, wouldn't it be great if we could use this?' and the husband replying 'I'm not so sure about that...'.
>
> In Korea, there's the story told by SK Telecom of a woman who bought a second cellphone to put in the trunk of her husband's car.
>
> What if the service gives incorrect information and tells the wife that her husband is on the opposite side of the town than what he claimed to be? We're going to have to educate customers of the limitations.
>
> *The question we operators have to ask is 'will it strike a chord or will it alienate users?'*

to food stuffs—sent to potential consumers within a certain radius combines permission, time-sensitivity and location information in a win-win situation that uses LBS to add real value.

So, What is Out There Right Now?

Hong Kong Sunday Telecommunications has been one of the most aggressive GSM operators in its approach to LBS, investing earlier and positioning LBS as a key differentiator in what is otherwise one of the most competitive and cut-throat telecommunications markets in the world. Using the maximum six neighbouring cells plus the serving cell, related parameters such as transmission power, timing advance, received signal strength, antenna height and known cell locations (latitude/longitude) are all fed into an LBS server to calculate handset location to an accuracy of 150 metres.

During the SARS crisis in 2003, Sunday launched an alert service which allowed subscribers to instantly know which buildings within their calling area (up to 1 kilometre) had confirmed a typical pneumonia cases. Admittedly, this service was probably of more value as a PR exercise, garnering substantial media attention, than as a revenue-generating cash-cow.

Sunday has also launched a family watch service allowing users to monitor the location of a family member's mobile phone. Use your own mobile phone to register the target handset to be located, then confirm authorisation using the target handset.

Japan It is no surprise that Japanese operators were already rolling out location based services in 2000 and 2001: NTT Docomo's i-area uses PDC Cell-ID to provide location-specific weather and rain alerts, local guides and maps, and traffic information. DoCoMo also offers the DLP (DoCoMo location platform) based on GPS but is more expensive and far less popular, with limited network support. Meanwhile, competitor J-Phone uses PDC Cell-ID to provide location-based news, weather, yellow pages, coupons and maps.

KDDI's EZ@NAVI system provides door-to-door directions and maps, localised weather reports, city guide information and friend finders, all using Cell-ID. KDDI also added a GPS service in 2002, though a common complaint with the GPS systems seems to be that they are slow (over a minute to position).

Recently, location positioning has been integrated with the built-in cameras in certain phones to allow location-stamping of photos; for example, to prove that a photo was taken from atop Mount Fuji at a certain date and time.

CoCoSeCom is a company offering a tracking device, using GPS-1 chips from Qualcomm. Originally positioned as a means to monitor children and elderly people with mental problems, the device is also small enough to be used for tracking items in case of theft and now GPS-enabled mobile phones are marketed with CoCoSeCom support/compatibility. In the first two weeks after the launch, the service received 70,000 orders and shipped 10,000 units (at about $70 per person or $100 per vehicle). With regular monthly charges of $5 per person and $9 per vehicle, and additional charges for location requests ($1 via the Internet and $3 via a telephone), this reveals a veritable opportunity in other markets.

Singapore In 2002, SingTel launched a find-the-nearest-ATM service whereby a subscriber dials *612 to receive a list of ATMs in the user's immediate vicinity; a local-weather service was also launched. SingTel is using Cell-ID positioning with an accuracy of 500 metres (which is the average size of its cells).

SingTel's *SEND-A-TAXI service is of more interest since it demonstrates how the accuracy limitations of Cell-ID can be circumvented. The taxi-booking service requires a user to dial *654, whereupon s/he will receive a list of taxi-stands or landmarks in the immediate vicinity. The user selects a pick-up point from the list and confirms the booking by pressing '1'. Upon confirmation, a taxi is deployed to the specified pick-up point, whilst the taxi number and ETA (estimated time of arrival) are sent to the user. SingTel charges 20 cents per *Send-A-Taxi request, and another 30 cents for a confirmed booking, whilst the taxi booking fee is $3.20.

USA Perhaps unsurprisingly, the US seems to show a predilection for PDA-based devices and/or the integration of existing GPS applications to the cellular network.

Don Rich, CEO of Acorn Solution in the US, was technical manager for an LBS traveller service developed for a Palm OS PDA in Boston in 2000. The service included auto-location with a map of the surrounding area; finder services for businesses, restaurants, night clubs and subway stations; filtering of results by distance and type of product, venue, entertainment, cuisine; maps and turn-based directions to all locations. Users could also call a cab to their current location. GPS was used for the positioning and messaging was over a pager network (remember those?), so that data was synchronised over-the-air.

Don offers some sage insights: 'None of these services is that unique by itself, but the combination of several traveller needs into one package was very powerful. Pretty good for way back in late 2000 anyway. Most consumers view location as an enhancement to, or prerequisite of, the service or piece of information that they actually care about. That's how the consumer differs from the LBS-fanatic—we just love to see those latitudes/longitudes and the 'dot on the map'.

'LBS application providers need to remember that location is not the goal for the customer: it's the suite of services built on top of that location information. I think that's why the 'packaging' or 'bundling' aspect is so important to LBS: if the bundle of services appeals to a certain set of needs (say those of urban tourists) and it's priced right, then maybe you've got something.'

'Our team was small (four developers, one system administrator and myself) and we produced it in under six months. That said, it took significantly more "human-factors people" to set up and manage the focus groups, trial user selection, press releases, user training, device acquisition, post-trial interviews, etc., than it took developers to create the solution. A good LBS field trial is a significant project.'

Europe Several European operators have taken tentative steps into LBS, usually based around friend-finder and directory services; in the UK, O_2 customers can track the location of standard GSM phones on their account, somewhat useful in a country where belongings go missing with wanton abandon. LBS-enabled SMS and chat are also promising since it has been shown that SMS traffic between subscribers increases with physical proximity.

Black cabs (taxis) in London have also launched a nearest-taxi service though it circumvents the Cell-ID accuracy limitations by putting the customer in direct voice contact with the nearest cab driver in order to complete the transaction. A number of third party companies have developed personal tracking and enterprise fleet management applications; for example, FleetOnline has managed to launch a multi-operator, multinational LBS fleet management application in the US, Europe and Asia.

Challenges, Learnings and Recommendations

Various scapegoats have been suggested for the slow rate of LBS adoption: an early Logica report suggested that GPRS would be necessary for LBS to succeed; whilst others repeatedly blamed form factor (small screens, input mechanisms, PDA vs cellphone), which reflects a preoccupation with maps and navigation-systems.

As demonstrated by some of the simple applications described earlier, it is possible to create LBS using nothing more than text messaging. A 'friend finder' application may not place your friends on a map but could lead to a voice call which generates additional revenue. A 'nearest taxi' service can be implemented with nothing more than SMS. Vehicle tracking does not even necessarily require a screen on the mobile device.

Therefore, given that some form of location/positioning technology is now available, we cannot escape the fact that nobody is using them because there are no compelling applications. As one industry insider puts it, 'We are all dressed up, but have nowhere to go'.

The major challenges still facing LBS are:

1. Privacy Consumers realise that they can be tracked—they just do not like mobile operators using and sharing this information without their permission. From a legal aspect, who actually owns 'location' and who is going to get access to this information?
2. Usability of compelling applications If the best technologies are transparent to end users, we have a long way to go.
3. Manage expectations Two words: 'WAP' and '3G'. Fortunately, the industry appears to have learnt lessons from these disasters: Do not rush to market and release before it is ready, do not over-promise, educate users, and allow consumers time to adapt.
4. Interoperability and roaming Steve Martineau put it succinctly: '*If we are providing a friend-finder service, do we want to be saying "Are my friends nearby? Well, only if they're with Maxis." Probably not in the long term. Openness of LBS across operators can only benefit everybody, as it has with SMS.*'

5. Third-party developers Tip for any network operators: You do not have to invent a new business model, just copy one that works and transplant it to your local market. Clue: NTT DoCoMo and China Mobile—incentivise third-party developers by sharing revenue (as much as 90 per cent for the developers) and everybody benefits. It does not have to be a zero-sum game. Make this process as seamless as possible, from signing up as a third party developer to the billing and revenue-split.
6. Pricing An IBM employee who would rather not be named subscribed to the WAP and MMS services of his local cellular company in Hong Kong. He used them for 15 minutes and was unpleasantly surprised to receive a bill for over HK$1000 ($120) at the end of the month. He promptly cancelled the service. Why is it assumed that the innovators and early adopters will pay extortionate rates, and should be milked accordingly? All too often, this simply results in an unpopular product.
7. Costs The operators' wish-list includes low-cost, high accuracy, easy installation and low maintenance; As the great Steven Wright said 'You can't have everything (after all, where would you put it?)'. For operators in the US attempting to comply with E911 Phase II legislation, the costs are substantial and include incremental handset costs (for A-GPS), network hardware and software, licensing fees, support and maintenance, as well as associated personnel and integration costs.

Conclusion

To sum up, location based services are one of the main avenues for mobile operators to differentiate themselves, which in turn creates opportunities for local content and application providers. The technical obstacles are surmountable, as are the business challenges mentioned above which means that the ball is now in the consumers' court: How much do they really want LBS? From mobile network operators deciding that LBS is a sufficiently high investment priority, to third party content developers regarding it a promising business opportunity, and end consumers willing to pay for them—that will be the balance to strike. Certainly do not rush into LBS with unrealistic expectations—but definitely do not discount LBS either....

Endnotes

[1] *$40b by 2006*—Allied Business Intelligence; *$20b by 2005*—Strategis group.

[2] *$5m in 2002, growing to $167 in 2006*—InStat.

[3] This is for network-based location systems. Handset-based location systems require even more accuracy: the operator must locate 67 per cent of calls within 50m and 95 per cent within 150m.

[4] Note that the GMLC has to page the network to establish that the subscriber is available, which could take some time depending on the load of the GMLC. If Cell-ID is all that is requested then the GMLC can promptly respond to the location server.

[5] The remarkable thing about common sense is that it is not that common: Some mobile network operators really do not provide a viable business environment for third party content and application developers. No wonder they have no compelling data services!

[6] Approximately $0.10–0.15 depending on volume.

Acknowledgements

Angela Finney, senior product marketing manager, Openwave.
Mah Bing Zeat, former chief marketing officer of Sunday Telecommunications, HK.
James Rhyu, former CFO of Sunday Telecommunications, HK.
Craig Ehrlich, chairman of the GSM Association (GSMA) CEO Board.
Bill Best, CTO of the GSM Association.
Alvin Lau, wireless consultant.
Steve Martineau, GM, Mobile Data Business Unit, Maxis, Malaysia.
Don Rich, CEO of Acorn Solutions.
Karen James, Autodesk.
Jaap Groot, FleetOnline.
The members of the LBS and Keitai-l mailing lists.
Thanks also to SingTel, Ericsson Malaysia, NTT DoCoMo.

4. Wi-Fi
Will the Asian Tiger Roar?

Subha Rama

Editors' Note: While much attention has understandably been focused on the high-profile 3G networks in parts of Asia, many markets and industry players are favouring a different, lower cost and sometimes complementary form of wireless communications: Wi-Fi hotspots. Leading markets here include South Korea, Japan, Taiwan, Australia and Singapore, though there is tremendous potential in other markets like Indonesia and India as well. This chapter covers the various alliances that are being formed across the region by ISPs, mobile operators, service aggregators and location-specific players like coffeeshop chains and hotels. Challenges remain in finding profitable models, billing arrangements and offering seamless roaming. Interesting opportunities for widespread public access to the Internet are also emerging via community wireless networks.

The Asian public hotspots market saw a burst of activity in 2003 when the Wi-Fi showed signs of taking off. The market was no longer restricted to a handful of pioneers but had started attracting interest from incumbent carriers and mobile operators who jumped on to the band wagon in an effort to make amends for their tepid initial response to the new wave technology.

Now why would the carriers be interested in a technology that they thought was in no way comparable to 3G, was good only for data services and little else, was plagued with problems of interference, spectrum issues and security, and suitable for very small localised footprints?

To understand the stirring interest of carriers, one needs to scrutinise the wireless landscape over the past couple of years. For starters, 2G services with voice and SMS continue to be hugely popular in much of Asia. Many carriers are still finding ways to stump up the huge resources needed for 3G network rollouts outside a few markets like Japan. Wireless hotspots based on 802.11, aka Wi-Fi, have thus a viable option for fulfilling some of 3G's promises in many Asian markets.

The Wireless Past... and the Future

The year 2003 could be termed as a transitional year for Wi-Fi with the wireless hotspots movement showing signs of spreading its wings to hitherto unexplored territories. Although Korea and Japan still dominate the wireless landscape in the region, there were other countries that were showing keen interest. Australia and Singapore were both witnessing a shakeout. While SingTel's new-found enthusiasm was in part influenced by peer pressure, Telstra's move to acquire SkyNetGlobal's hotspots was indeed a resounding acknowledgement of Wi-Fi's potential.

In Indonesia, which has just 7.5 million fixed line connections, nine million mobile phone subscribers, and meagre access to fibre or ADSL for a population of 210 million, Wi-Fi has been identified as a technology that can offer a cost-effective means of communication.

However, there are exceptions. In India, Wi-Fi is still mired in regulations, with the government allowing indoor WLAN usage only early in 2003. In the Philippines, Wi-Fi usage is riddled with confusion. But apart from select countries where bandwidth restrictions have bogged down adoption, there has been a proliferation of wireless hotspots across a number of geographies in Asia-Pacific.

Figure 4.1 shows the relatively low degree of maturity of the public hotspots market in Asia-Pacific. It must be admitted that the public hotspots market in the region had a longer gestation period as compared with North America and Europe. However, one can expect rapid take-up of Wi-Fi services by 2005 as standards firm up and carriers begin to display a keen interest in offering these services.

Adoption of Wi-Fi

Though Wireless Local Area Networks (WLANs) find their origins in the enterprise, there has been a startling increase in the use of WLAN applications in the public environment as well, especially in densely populated areas with limited physical scope, such as convention centres, hotels and airports. A number of telecommunications carriers in Asia have been actively deploying WLAN hotspots. Among

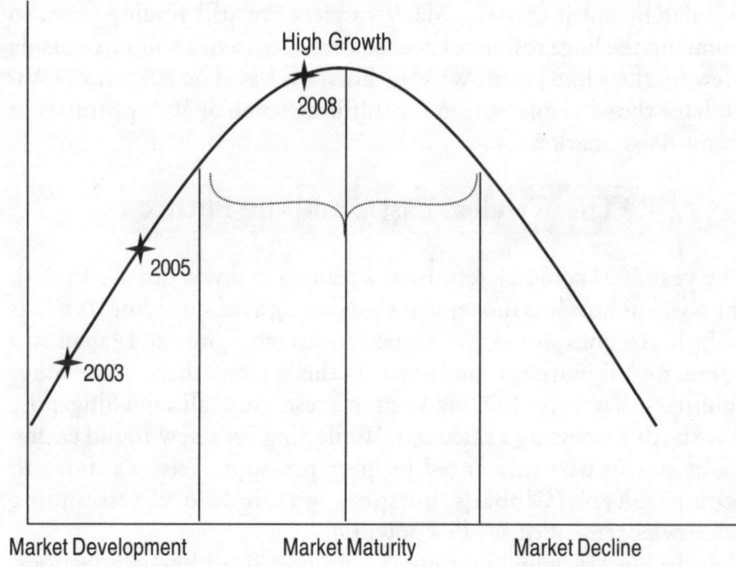

Source: Frost & Sullivan, 2003

Figure 4.1: The Wi-Fi hotspot market lifecycle in Asia-Pacific

these are early adopters like South Korea, Hong Kong and Japan where the incumbent carriers have taken the initiative to deploy the services country-wide. Carriers that have entered the WLAN services segment have demonstrated that it can be a viable business proposition that can bring in incremental returns for small additional investments.

However, in a number of Asian countries, WLAN services are still not a priority for carriers. In these countries, the movement was largely driven by pure-play wireless ISPs (WISPs) who did not enjoy the advantages that large carriers had—an existing backbone and consequently the ability to price the service really low to achieve critical mass. Without these, coupled with limited access to funds, these WISPs were struggling to make Wi-Fi a profitable business proposition.

They had to rely heavily on WLAN equipment vendors to source equipment at a subsidised cost or for free (depending on the kind of profit-sharing agreements they had with these vendors) and buy bandwidth from fixed-line service providers. They were unable to price the services high for fear of a low take-up among subscribers.

Most of these WISPs had to confine their functions to just billing and customer care.

In countries like Australia and Singapore, incumbent carriers played a wait-and-watch game while the WISPs tried to establish a successful business model out of Wi-Fi services. However, carriers are beginning to realise that WLAN services can add value to their existing fixed-line services. Figure 4.2 shows the revenue opportunity from Wi-Fi services in Asia-Pacific.

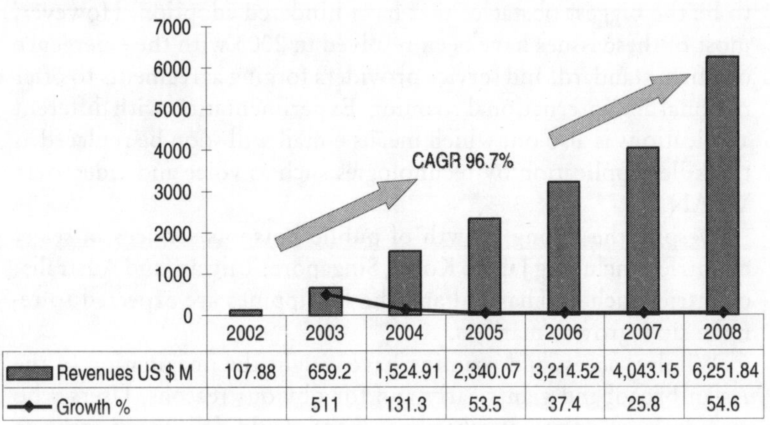

Source: Frost & Sullivan, 2003

Figure 4.2: Wi-Fi services revenue forecasts in Asia-Pacific

Table 4.1: WLAN subscriber base in Asia-Pacific, 2002–2008

Year	Users	Growth Rate (in %)
2002	417,499	–
2003	3,671,082	779.3
2004	11,567,878	215.1
2005	21,220,079	83.4
2006	36,065,374	70.0
2007	54,853,877	52.1
2008	93,947,130	71.3
Compound Annual Growth Rate (2002–2008): 146.6%		

Source: Frost & Sullivan, 2003

Countries like Japan, Korea and Taiwan, which have high ADSL or broadband penetration, have reported active Wi-Fi usage. The broadband experience has made it easy for carriers to extend it to wireless broadband. Most of the operators in these countries—Korea Telecom, Chunghwa or NTT—offer Wi-Fi as a value-added service to their existing ADSL customers. They have also priced the service low, sometimes as low as $10 a month for unlimited wireless Internet access. The other promising markets are Australia and Singapore where the carriers have been a little late in realising the potential of Wi-Fi.

Lack of coverage, uncertainty about the WLAN standards, and the absence of roaming agreements between service providers have proven to be the biggest obstacles that have hindered adoption. However, most of these issues have been resolved in 2003 with the emergence of a firm standard, and service providers forging agreements to offer national and international roaming. Experimentation with different applications is also on, which means e-mail will soon be replaced as the killer application by technologies such as voice and video over WLAN.

Despite the strong growth of public hotspots services in select countries, including Japan, Korea, Singapore, Taiwan, and Australia, countries such as Thailand and the Philippines are expected to remain slow growth markets.

Wireless access and roaming have caught the imagination of the Asian broadband giants early and for obvious reasons. Users who already have a taste of what broadband could deliver are eager to extend its scope to include wireless access. However, apart from large carriers, none of the smaller WISPs are even close to profitability. In Taiwan, though the hotspots movement was off to a great start, the initial enthusiasm seems to have definitely waned. A number of projects, which are still in the trial phase, do not seem to have a definite business model. This includes the one set up by Cisco and Infoserve for the VIP rooms of China Airline at the Chung Cheng Airport. ISPs like Infoserve, which was offering wireless access as a complimentary service to its ADSL subscribers, have been scaling back operations in the face of dwindling revenues.

Of all the countries in the region, Korea seems to have the most ambitious plans for promoting access through 802.11. Unlike their counterparts in other countries, Korean carriers began to pursue WLAN services actively immediately after the government slackened regulatory controls over the 2.4 GHz band. Korea Telecom (KT), which has branded its WLAN service as Netspot, and Hanaro Telecom with its HanaFos have announced plans to install 25,000 hotspots throughout the country. Dacom, Korea's second largest ISP, has also jumped into the fray with its AirLan service. There are also other aspirants like Thrunet and Onse Telecom that are eyeing the market with keen interest.

KT, like Chunghwa Hinet in Taiwan, plans to leverage its existing 500,000 strong DSL customer base to buttress its WLAN initiative. KT's existing customers need to pay just $10 a month to enjoy unlimited wireless access at the carrier's existing 5,000 hotspots.

The launch of Centrino, the embedded WLAN chip from Intel, is expected to influence the growth of Wi-Fi capable devices which will lead to an increasing number of people accessing Wi-Fi services at public hotspots. There are a number of other factors that drive the deployment of public Wi-Fi networks. For mobile operators, wireless data services are complimentary to voice services. With carriers looking to integrate General Packet Radio Service (GPRS) with WLAN services, the market for public Wi-Fi seems poised for some exponential growth.

The Wi-Fi Business Proposition

It is true that many of the initial efforts at offering Wi-Fi services did not prove to be profitable. In fact, there are some wireless experts who subscribe to the view that some of the hotspot providers do not have a business model at all. While this may be a case of stretching it a bit too far, it is a fact that in countries like Taiwan hotspot operators who were giving away the access 'free' are busy scaling down their operations and taking a long, hard look at their existing business models.

Much of the early initiatives for public wireless access came from small pure-play wireless Internet service providers (WISPs) while the large carriers opted to wait and watch. Just as in the US, where Boingo and MobileStar made big-time news and little money, hotspots are hardly proving to be a cash cow for the WISPs in Asia.

There are a number of different elements that make up the WLAN value chain. The value chain starts with the end-user, typically a mobile professional, but potentially anyone with a device able to connect through a short range wireless technology to a public access network. The user will typically be billed by either a service provider (which may or may not be a network operator) or the property owner. Either of them may also want to set up a portal. The end-user clearly needs a device, but they can obtain this from a variety of sources. Content and applications developers may be used for extra services other than just Internet and VPN access. Finally, infrastructure is required in order to be able to offer services.

These different elements all have different value to add to the complete picture. It is not always clear how the different elements will interact, and how the revenues will flow. Value chains also tend to change over time, with the different parts enjoying periods when they add significant value and are well positioned, and others where they may be sidelined or in weaker positions.

Service providers and operators are at the core of networks, largely because they will have a billing relationship with the end-user. In many cases they own the network, although the property owners will also have a stake in this. Service providers may be mobile network operators (SingTel, Telstra, Telia and Sonera), wireless service providers (BluEngine, SkyNetGlobal, Yellowspots), corporate Internet service providers (Worldcom, BT Ignite and Cable & Wireless), as well as a range of other companies who already have an installed base of wireless LAN users. There are also companies that specialise in providing broadband access to the hospitality industry such as Intertouch.

The service provider typically purchases the network infrastructure equipment and installs the network, although a systems integrator is likely to do the actual installation. They are also responsible for maintaining and managing the network, billing and providing support services. One of the major selling points will be the number of sites available to the user. In order to maximise this, roaming agreements must be put in place. The service provider may also want to offer specific content and applications, aggregated in a special portal. This portal enables the user to choose between the different services available, including Internet, VPN access, information, entertainment and local promotions and offerings.

For this, the service provider must strike deals with applications and content developers. Revenue sharing deals must be created on a case by case basis between service providers and property owners. If the service provider is a mobile network operator, the property owner may also negotiate cellular call revenue sharing into the equation. The service provider may also provide devices or device add-ons and strike deals with device manufacturers.

Billing Types

Service providers can choose to bill subscribers from a number of different options. The end-user will not always pay the service provider; in some instances, it is the property owner who bills the end-user directly at the point of use. Either way, the one receiving the money will have to share some of this revenue with the others. How this is divided varies and depends on the nature of the relationship between the two.

There are a few different ways in which a service provider charges the end-user: per minute, per day, per packet, monthly flat fee, per service type, pay as you go, pre-pay or a combination of these.

Per minute billing, per day billing and flat fees seem to be the most popular at present. Typical cost per minute stands in Asia is at $0.11 in Asia-Pacific, $0.5 in Europe and $0.2 in the US. Flat rate monthly fees average about $150 to $1,000 worldwide, depending on whether there was an installation or set-up charge, the number of users enabled, and so on. Many service providers offer monthly flat fees with a limit on time or data usage, with supplementary costs when going over the agreed limit.

A number of hotels offer the service on a daily basis. Typically prices are in the region of $15 to $25, which provides the user with 24-hour access. The hotel also offers the wireless PC card for this time, and the user must leave a credit card or deposit until it is returned. Although this price seems very high, hotel telephone and Internet charges tend to be very high anyway, and from this perspective the price is appropriate.

Early feedback shows that during a day a user will access the network for around an hour, and so the rate is not a bad price compared with $0.5 per minute charged in European countries. We believe that this method of payment will be important as the number of enabled devices increases rapidly. Although there will be many heavy users wanting a monthly subscription, there will be others who only want to access these networks a few times a year. For them it is better to pay per day or per usage, or alternatively have a pre-paid amount which they can use as and when they like.

Per packet billing is popular in the GPRS world and is likely to be implemented in the wireless public access world as well, particularly if such services are integrated. StarHub in Singapore has already started implementing per packet billing.

Cellular pricing schemes are also being copied by more advanced service providers. Per service billing is unlikely to appear for some time. This type of billing is likely to dominate the consumer space as applications such as music, games and so on become more popular.

Non-billed Revenues

Although we expect the vast majority of wireless access revenues to come from billing related to airtime usage in some way, there are a number of other ways in which revenues may be accrued. Essentially these boil down to advertising and commerce. The two are closely interrelated, in some ways more so than for other portal business types. The reason is that an online transaction and an offline one (i.e., at the point of sale) are closely related. The simple reason is the local nature of the public space. The portal can provide a summary of discounts, offers, promotions, deals and so on to be found in the vicinity.

It's All About Partnership

Though there are a number of players in the WLAN services market, all the market participants have realised that collaboration between different service providers is a key success factor in the market. Almost all the players have entered into multiple tieups. For example, under the much-publicised agreement with Telstra, SkyNetGlobal's existing customers will continue to have access to all its hotspots in Australia for a period of two years. Several ISPs are entering into roaming agreements with WISPs to extend wireless roaming to their customers. OzEmail, the third largest ISP in Australia, is working with the CKW Wireless consortium to deploy iBurst, a wireless network that can be accessed from a moving vehicle, unlike other 802.11-based networks.

Azure Wireless has a tieup with iPrimus Broadband Wireless, while Yellowspots works with Air Portal. Telstra foresees a number of agreements with WISPs in the next few months.

Both Azure and SkyNetGlobal have tieups with Enterasys (formerly Cabletron) for equipment and cards. XONE has a tieup with Cisco and HP. The XONE business model is interesting as it involves exclusive, long-term hotspot venue negotiations with major retail/office complex developer/owner/managers such as the Mirvac Group, the Macquarie Office Trust and AMP Henderson Global Investors, rather than individual cafe or store owners. XONE also received co-marketing support during the launch period from the Mirvac Group, HP, Cisco, Intel, Microsoft and others.

Tying up with property owners has proved to be by far the most critical issue in making the hotspot business model a success. Large café and restaurant chains such as Burger King and StarBucks in Singapore have backed the big brand name of SingTel while early movers such as BluEngine have roped in certain strategic locations like the Olio Dome and Raffles Marina. SingTel has also targeted community centres throughout Singapore to establish its wireless surf zones. For carriers, partner acquisition has been merely a question of leveraging their existing relationships with facilities-based and service-based operators. In fact, in places like the Changi airport, these operators have become satisfied with providing access for a percentage of the revenues earned while the property owners manage the service.

The Wild, Wild Wireless East

The reasons for the growing interest in public access hotspots are obvious. Apart from offering tremendous cost benefits to the user and the service provider when compared with 3G (you just have to look at the cost of 3G licences and the incremental expenses for the rollout), Wi-Fi's promise of integration with 2.5G technologies like General Packet Radio Service (GPRS) can open up enormous market opportunities. Integration with GPRS is bound to be complementary. While it can resolve issues associated with the range and geographical coverage of WLANs, the integration will also overcome problems of lower data rates for GPRS.

In Australia, community wireless networks have been growing at will, causing a significant amount of headache to ISPs offering public wireless access as well as fixed line service providers. Community network operators, some of whom have set up large networks spanning hundreds of nodes, operate without a carrier licence. Apart from problems of interference, the regulators have found it difficult to ensure that none of these networks are exploited commercially. However, these networks demonstrate the growing enthusiasm for broadband wireless access in the country.

WISPs are obviously facing a Catch-22 situation. On the one hand, they know that usage will increase dramatically only if they are able to set up hundreds of hotspots in cities so that users can roam and still stay connected. On the flip side, unless the take-up improves and hotspots prove really profitable, fresh investments could become close to impossible. The market clearly beckons the big carriers, who already have the network infrastructure and a ready subscriber base that is waiting to be converted to wireless. In Taiwan, South Korea and Hong Kong, carriers have already started doing this. Telstra, SingTel and StarHub are testing the waters. After all, who would not be comfortable if their primary carrier could also offer wireless roaming? Billing and consolidation worries disappear and so do doubts about differential pricing.

The Road Ahead

Mobile network operators are desperate to increase the average revenue per user (ARPU) as cellular penetration is peaking and voice prices are falling. Therefore, they need to drive up data revenues by offering higher speed services to deliver a whole range of different applications and content. To do this, they are rolling out GPRS services. A symbiotic relationship between GPRS and WLANs can target two distinctive needs better than the 'universal' 3G. WLAN is a low-mobility service, well suited for surfing the Internet with a laptop and downloading large files and e-mail. Its lack of geographical coverage can be compensated by the ubiquity of GPRS networks, which is in fact an upgrade of the GSM (Global Standard for Mobile Communications) network that covers over 150 countries. The lower data rates of GPRS (about 40 Kbps) can be offset by the high-speed WLANs. Over the next three to five years, WLANs are expected to cannibalise some 3G data revenues. Already, studies show that WLAN accounts for more than 10 per cent of the worldwide data revenues.

The main opportunities for Wi-Fi will arise from:

Attracting High Mobile Data Usage Subscribers

Wireless public access services currently offer high rate wireless data services. This is an attractive proposition to corporate executives who spend a lot of time on the road. They have time to be productive at airports, hotels, trains and other places while away from the office. They do not need to carry cables and can connect simply with a notebook (or other devices) and a wireless adapter. The technology is available today, is reliable and is seeing falling prices.

Driving Revenues at Relatively Low Capital Investment

This is essentially a new revenue stream. Early feedback shows that pricing is relatively low as the service is of a high bandwidth and is new. Prices of $0.11 per minute or $150 per month are attractive revenue streams. As the number of sites increases and awareness increases, so subscriber numbers will increase, further driving revenues. In 2006, it is estimated that the total revenue opportunity in Asia-Pacific alone will be $3.4 billion. This may be substantially higher if the installed base of devices is higher than estimated, ARPU and usage is higher, and new revenue streams such as per service billing, advertising and m-commerce become available. At the same time, the actual cost of installing these networks is relatively low. Typically, a single site installation will cost around $30,000 to $50,000. The average revenue per site per year is expected to be around $80,000.

Offering a Seamless, Integrated Service

Handover technology development and the increasing number of roaming agreements will drive the possibility of offering a seamless service. This will enable high data rates in public hotspots, as well as network access elsewhere. There are already a number of companies looking at handover technology and this is expected to be commercially available within two years. In addition, hotspot services can be integrated with corporate ISP and residential Internet access. In essence, therefore, subscribers can access the network wherever they are.

Gaining Experience in Offering High Speed Wireless Data Services

By driving into this market early, mobile network operators in particular can have early learning experience in offering high speed wireless data services. This will be essential in understanding user patterns, billing and pricing, marketing, application development and network management. This, in turn, will give them an advantage in driving 3G services. Operators can hope to gain a subscriber base of several hundreds of thousands of users. It will be relatively easy for operators to upgrade these wireless hotspot users to a fully integrated high-speed service.

5. Wireless in the Local Loop

Towards Connecting India

Ashok Jhunjhunwala

Editors' Note: This chapter makes the forceful case that in many developing countries, fixed wireless in the local loop (WiLL or WLL) technologies may be the most cost-effective solution for combined voice and Internet services in rural areas and less-dense urban clusters. For small offices, WLL can actually be a more effective business communications solution than fixed line (which costs more) or mobile (with data services often limited to SMS). Whether based on standards like CorDECT or Wi-Max, fixed line wireless connectivity solutions need to be offered in conjunction with local entrepreneurs who can operate infokiosks and service points. Though not yet a major hotspot market, India can actually offer useful lessons and models for developing countries via other forms of wireless networking like CorDECT WLL in homes, small offices and information centers, as pioneered by the Indian Institute of Technology in Madras and its incubated companies. Challenges can arise, though, in some regulatory environments.

Introduction

The last 10 years have had a major impact on the telecom scenario in India. It started with reforms that brought in competition to end monopoly operation of telecom services. Today, four or more operators are firmly established in most telecom circles of the country. The tariffs have fallen substantially and have become affordable to much larger sections of the Indian population. The number of telephones has grown from under 10 million lines to almost 60 million lines: but the numbers reveal only part of the story, as the real acceleration took place only in the last couple of years. Despite competition in, there was no major impact felt till around 2000. It was still the incumbent government-owned Bharat Sanchar Nigam Limited (BSNL) and Mahanagar Telephone Nigam Limited (MTNL) who accounted for most of the lines till then, with the competition busy in cherry-picking a few high paying subscribers. The tariff was still cost plus and unaffordable to most.

If one looked at the capital expenditure (Capex) required to install a telephone line till then, it hovered around Rs 30,000 per line ($1 = Rs 44). The prevailing finance cost, depreciation and operation and maintenance cost were such that unless an operator earned 40–50 per cent of Capex as yearly revenue, there would not even be a break-even point Jhunjhunwala, 1998. Who in India could afford to pay Rs 1,000 plus towards telephony? India has a large population and, therefore, a potentially large market—but what is the affordability? Figures 5.1 and 5.2 show the estimated household incomes in India in 2003. Even if one assumes that 4 per cent of the income could be spent on communications, hardly 10 per cent of the over 60 million urban households and 2.5 per cent of the over 130 million rural households can spend Rs 1,000 per month on telecom. India is a large market but only at the right price point. The average revenue per user (ARPU) has to decline to under Rs 300 per month for 100 million Indian homes to become consumers. Now, if the ARPU is to decline to this level and the operator has to still make money, the Capex needs to go down to under Rs 10,000 per line.

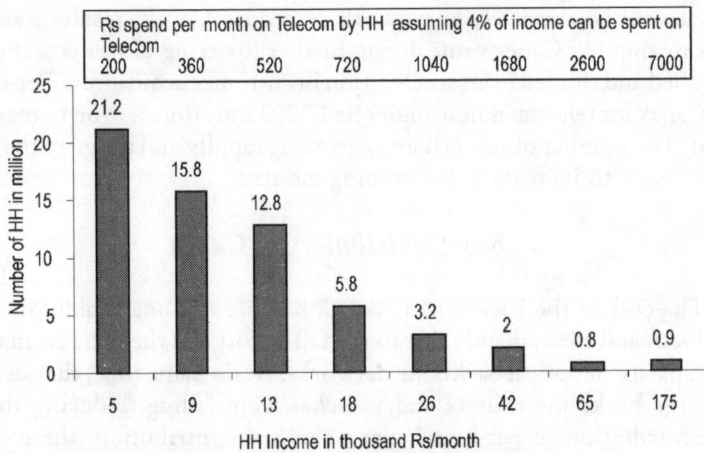

Figure 5.1: Number of urban households in India in 2003. Different income levels and amount that can be spent by these households on Telecom assuming 4 per cent of income is spent on telecom

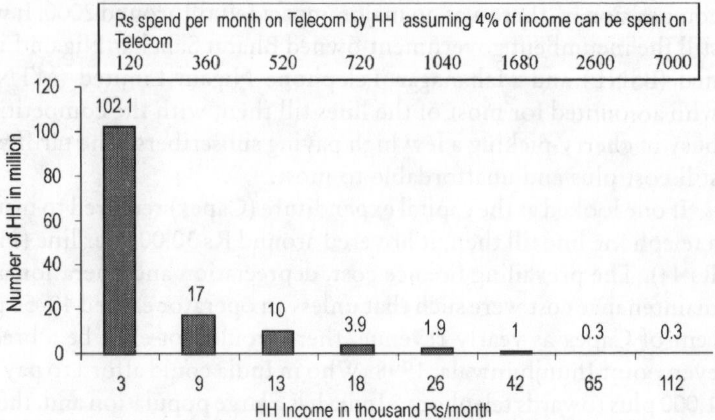

Figure 5.2: Number of rural households in India in 2003. Different income levels and amount that can be spent by these households on Telecom, assuming 4 per cent of household income can be spent on telecom

Targeting Capex

As long as tariffs were cost plus, there was no great urge to push down Capex. Telecom growth remained low. With competition coming in and driving down tariffs, the realisation has dawned that Capex needs to be pushed down. As efforts were made to look at each element of the telecom network to reduce Capex, results started showing. As Capex came down, further lowering of tariffs was justified and this led to an acceleration in customer acquisition. Per-line Capex for telecom is now under Rs 17,000 and efforts are on to reduce it. The number of subscribers is growing rapidly and the growth rate is likely to increase in the coming months.

Key Contributors to Capex

The cost of the backbone network had been falling rapidly in the 1980s and 1990s mainly due to the fall in cost and the increase in the capacity of optical backbone networks. At the same time, the cost of large backbone routers and switches were falling, reducing their contribution to per-line Capex. The key contribution, therefore, came from access network (or what is often called the local loop), the part of the network that connected subscribers to the large switches and routers in the backbone. Traditionally, it was based on copper, and even though innovations like RLUs and RSUs (Jhunjhunwala, 1995) had brought down the cost of the local loop, the access network remained a major bottleneck both in terms of cost as well as fast expansion of the network.

Wireless in Local Loop (WLL) emerged as a concept in the early 1990s precisely to overcome this problem. If the access network used wireless communication to reach the subscriber, maybe the cost would be lower than that involved in laying copper lines. Moreover, the expansion would be much faster. The key was to design a wireless communication link that would provide all the functions carried out on the local loop and would still cost less than copperlines.

Mobile Communications

Mobile communications, of course, was already using wireless communication in the local loop. However, the primary purpose of mobile communication is to provide mobility or connection to subscribers on the move. It did not start with cost-saving as the primary reason. Moreover, it did not begin with equating itself with wireline; the voice quality that it would offer could be slightly inferior, and a lower calling rate (Erlang traffic) could be assumed. Many of the other services like fax and data communications were considered less important. The key was to reach a subscriber anytime, anywhere. It came in as a premium service.

But as mobile technology evolved and mobile standards like GSM and IS-95 (CDMA)[1] become mature, they did provide reasonably good quality voice, data communication capability like SMS and MMS, and the costs kept tumbling. (Figure 5.3). The mobile technologies in fact became the key to bring down Capex in India and may make-up over 50 per cent of the new market in the near future.

Yet, mobile technologies are different from WLL. As mentioned earlier, WLL is to be a replacement of wired local loop to homes and small offices and, in that sense, must provide all or more services provided by a wired line. While toll quality, voice and ability to handle high Erlang traffic is a must, what is more important today is services like Internet. Dial-up Internet is available on wired lines in bit rates of 33.6 Kbps or even 56 Kbps (dedicated non-shared connection). Even the 2.5G mobile technologies, maturing in the near future, would at best provide 100 Kbps *shared* connection to all the users in a sector.[2] Thus, as soon as 10 or more users in a sector log on, the bit rate available will fall under 10 Kbps. It is a far cry from dial-up connectivity. Further, the copper wire to homes

Figure 5.3: CorDECT wireless in local loop

can now support high bit rate data communication on DSL. Mobile communications in most parts of the world cannot yet dream of this.

The key objective of WiLL is to provide wireless (instead of copper or fibre) connectivity to homes and small/medium offices. It is, therefore, a fixed connection. It needs to provide voice connectivity as well as Internet connectivity at least at dial-up speeds (33.6 Kbps or 56 Kbps unshared) and hopefully better. In the Indian context, it must bring down the Capex of fixed connections.

While several systems have attempted to provide such connectivity in the past, the one that dominates the Indian market today is CorDECT WiLL (Figure 5.3) developed by the Indian Institute of Technology at Madras (IITM) and Midas Communications, a company incubated at IITM based on DECT standard (DECT, 1991). CorDECT provides two connections to a subscriber, a telephone connection and a 35 Kbps always-on Internet connection (and a premium rate connection at 70 Kbps). It is designed to provide high density connections in urban homes with base stations located typically every couple of kilometres; in rural areas, it serves subscribers in a 25–30 km radius of a base station (using a relay base station). The Capex for CorDECT installation works out to be under Rs 8,000 per line today in urban areas and under Rs 12,000 per line in rural areas.

Role of WiLL in India

What role does a WiLL fixed connection have to play in India where mobile phones are capturing the imagination today? One of the most precious resources in wireless communications is spectrum. At any point of time, given a certain spectrum, technology can allow fixed wireless to have almost three times as much capacity (total data handling capability) as compared with mobile. Medium-to-high speed Internet connections at homes and offices have become important today in addition to voice. The option for such connectivity comes either from fibre, DSL on copper, data on coaxial cable, or fixed line WiLL.

In countries like India, data on co-axial cable has not worked primarily due to the existing cable being one-way and the somewhat poor maintenance and installation practices designed to serve one-way low revenue cable TV market. Even when the cable is replaced to provide two way communication, it becomes difficult to maintain it because of existing practices. Fibre in the local loop with the last hundreds of metres of copper is emerging as one of the most cost-effective mechanisms to connect homes and offices in dense parts of of urban India. But as soon as one moves out of the town centre, fibre in the loop is not cost-effective. Larger copper loops have already priced themselves out. DSL on copper is, therefore, an option only for the incumbent who has buried copper.

How does then one provide connectivity to homes in suburban areas of large cities, small towns and in rural areas? The only cost-effective option is WiLL.

If one carries out a detailed cost analysis, one finds that in most cities and towns, there will be some part of the area where fibre access network would be the best cost option. The area would be 4-km radius in cities like Chennai, 1-km radius in cities like Madurai and probably 400-metre radius in towns like Salem. But beyond these areas, WiLL would be the best option to provide connectivity to homes and offices.

Where would WiLL Go?

The WiLL services available today are only the first generation technology. In the near future, WiLL would provide 100 Kbps or more unshared (and about 2 Mbps shared) Internet connection to each subscriber, coming somewhat close to DSL services. It is also important that cost comes down continuously—the Capex target would be Rs 5,000 per line in urban areas and Rs 10,000 per line in rural areas. Further, a variety of technologies are likely to be available in future. Technologies like 802.16 (IEEE, 2001) are promising and may make a major impact in fibre.

WiLL and Rural Areas

It is, however, the rural areas of countries like India where WiLL is going to make a major impact in the future. It is today recognised that Internet is not just a communication means but Power. Providing Internet to all villages is therefore not a matter of choice but a prime necessity. WiLL is by far the lowest-cost option to provide this Internet connection, especially in countries like India. In typical rural Indian plains, a 30-km radius would have about 300 villages with a total population in excess of 300,000. If a fibre is available in a nearby town[3], we can provide one connection to each village using WiLL at a cost of Rs 12,000 per village. As demand in a village can be aggregated and serviced by an entrepreneur-driven kiosk located in the village, a connection becomes commercially viable within a few months. n-Logue Communications, a company incubated by IIT Madras is doing precisely this using CorDECT WiLL in many villages of India[4].

WiLL in rural areas is not just an intermediate solution, but is likely to remain the most cost-effective method to provide voice and Internet access for quite some time to come. Even as demand increases and fibre reaches the villages, connecting individual homes would be best done by WiLL. The speeds will increase and the form of WiLL may undergo a change. But connecting rural homes with any kind of cable is unlikely to be a meaningful option for a long time to come.

Conclusion

WiLL has already played a major role in India, in bringing down the per-line Capex of telecom—thus making telecom services widely available to people in India. For most areas, other than the densest parts of cities or towns, WiLL would remain the most cost-effective option to provide telephone and Internet to homes. For connecting rural areas, WiLL may be the only meaningful option, except in very sparse and hilly areas. Today, India has taken a technological lead in WiLL for emerging markets. With some effort, it can not only retain the lead but also become the driver for rural connectivity throughout the developing world.

Endnotes

[1] IS-95 or CDMA system is also referred to in India as WLL instead of mobile. While GSM standards evolved from Europe, IS-95 and its later version 3G-1X evolved as mobile communication system from USA. The fact that it is referred to as WLL has nothing to do with the technology, but more to do with policies. India had initially chosen GSM as its mobile platform and had awarded licenses to mobile operators. These with fixed line service license tried to cash on the popularity of mobility by declaring IS-95 as WLL, implying that it could be used by a fixed line Basic Services Operator to provide services. Since the two are competing mobile technologies providing equivalent mobile services, IS-95/3G1X will not be referred as WLL in this article.

[2] One often gets confused when the service is just introduced and one is amongst the first subscriber. The bit rate available to a user is immense. But as more and more people take the service, the bit rate falls drastically.

[3] As is the case in most parts of India where BSNL has a fibre terminating in almost every *taluka* (county) town.

[4] The story about applications driven by n-Logue is available at http://www.tenet.res.in/url/Day_in_Life.doc.

References

Jhunjhunwala, Ashok, Bhaskar Ramamurthi & Timothy A Gonsalves (1998). 'The Role of Technology in Telecom Expansion in India'. *IEEE Communication Magazine*, November, pp. 88–94.

Jhunjhunwala, Ashok & Bhaskar Ramamurthi (1995) 'Wireless in Local Loop', *Journal of IETE*, Vol. 12, Nos. 5&6, September–December, pp. 309–314.

DECT Reference Document (1991). ETSI Technical Report ETR015, European Telecommunications Standard Institute, March.

IEEE Standard release for 802.16. (2001). Air Interface for Fixed Broadband Wireless Access Systems. Part A: Systems between 2 and 11 GHz. IEEE 802.16ab-01/01r1, July.

6. Wireless and Mobile Impacts on News and Entertainment

The 'Third' Media Ecosystem

Madanmohan Rao

Editors' Note: This chapter provides an overview of key impacts of wireless channels on the news and entertainment sectors. From journalism tools and TV polling to music ringtones and mobile gaming, a vast new media ecosystem has been unleashed in the span of less than a decade. This chapter covers the opportunities (e.g., instant news alerts, political mobilisation) and challenges (e.g., IPR violation, information overload) that arise in the mobile media ecosystem. Case studies are drawn from around the world, with special emphasis on the advanced markets of Japan and South Korea, the vast SMS potential in India and China, and alternative news and political movements in Malaysia and the Philippines.

Introduction

Wireless technologies like Wi-Fi can have a visible impact in the design of workspaces for news organisations; handheld devices can improve information access and efficiency for journalists; mobile alerts can enhance the effectiveness of public relations agents; and mobile communication can open up a wide range of content and entertainment services for news consumers and revenue streams for operators and publishers. At the same time, challenges can arise via information overload, quality of information delivered wirelessly, and financial models for sustaining wireless ecosystems.

This spectrum of opportunities, challenges and learning points for the entire range of players in the news value chain is summarised in Table 6.1. While traditional media (print, broadcast) created the first major mass media ecosystem, the Internet created the second new media ecosystem with a combination of e-mail communication, Web publishing, searchable archives, and narratives like blogs. Wireless and mobile technologies have created the third media ecosystem, based on instant two-way communication, location-based services and dramatically improved mobility and portability. The three ecosystems intersect, of course, in interesting ways, and this chapter provides informative insights into the impact of wireless and mobile technologies on news, consumer power and entertainment.

News and Mobile Media

A growing number of reporters use mobile phones to stay in touch with sources, experts, event managers and colleagues; handheld organisers (sometimes built into smartphones) help manage contact information, calendaring and storing of company and people profiles in portable formats, and even early drafts of articles.

Editors and CIOs at news organisations will need to become well versed in the use of structuring languages such as XML for multi-purposed content, and familiarity with new devices like digital cameras and pocket PCs, according to Quinn (2002). Other tools and technologies useful for news professionals include Xybernaut wearable computer gear (which allows reporters to plug into workflows at all times) and tools for GIS and GPS analysis (e.g., for data mining and mapping of environmental stories).

Journalists need to structure content in such a way that it can be re-purposed easily-by deciding what the headline, blurb, body, related stories, related links and abstracts are; different versions may be needed for SMS delivery or for PDAs. Notable developments in the use of languages like XML for news include Reuters' NewsML and News Industry Text Format (NITF). Wireless devices are becoming important for consumer delivery of news as well as for newsroom workflow (via WLANs and handhelds). Developments on the technology front, of course, throw up complicated issues for news media ranging from speedy fact-checking to retaining a sense of context in the whirlwind of newsflow, Quinn cautions.

Photophones or cameraphones and even videophones are coming in handy as well. In Argentina, the radio programme 'La Cornisa' anchored by journalist Luis Majul features journalists carrying photo

Table 6.1: Wireless and mobile impact along the news value chain

	Opportunities	Challenges	The learning curve: Responses
News consumers	1. Any time, any where, any device access 2. Customisation, personalisation, forwarding 3. Chance to become 'citizen reporters,' participate in mobile text blogs and MMS photo galleries	1. Small size of display screens 2. Information overload 3. Weeding out fake rumours, scams	1. Tools, methodologies for managing information 2. Cultivating trust, researching news sources 3. Inspect online privacy policies before divulging personal information to operators and news sites
Journalists	1. Re-purposing story for multiple media 2. New narrative structuring (e.g., layered stories, mobile blogs) 3. Continuous access to colleagues, sources, experts	1. Information explosion 2. Loss of a sense of context, control over pace of industry 3. New legal risks (plagiarism, uncertainties over liability)	1. Tools, methodologies for managing information 2. Rigorous fact checking; advocacy in peer/industry associations 3. Sensitisation to cyberlaw issues
Editors	1. Meeting needs of different consumers 2. Extending shelf-life of editorial products 3. New forms of workflow, workspace design	1. Dealing with convergence 2. Evolving standards for structuring content 3. New legal risks (e.g., copyright of mobile content)	1. Co-location of different teams, roles 2. Joining consortia (e.g., for XML) 3. Legal counsel, signing formal agreements with content re-distributors
Business managers	1. Maximising editorial resources, content output 2. Meeting customer, advertiser at multiple access points	1. Sustainability, profitability of wireless operations 2. Structuring of new media units (e.g., Web, wireless)	1. Mixed models: some free, some fee-based content access 2. Incubate new media units separately, merge with parent company on maturity

(*Table Continued*)

	Opportunities	Challenges	The learning curve: Responses
	3. Expansion to m-commerce, alliancing	3. Creating a knowledge-focused culture	3. Conduct KM workshops, offer incentives, identify champions, invest in KM
Advertisers, marketers	1. Multiple targeting options: Web, e-mail, SMS 2. Demographic profiling 3. Permission marketing	1. User concerns over spam, invasion of privacy 2. Lack of standardisation of ad formats 3. Inconsistent reporting of traffic	1. Evolve new formats of advertising 2. Join industry consortia for interactive ad standards 3. Seek independent third-party traffic audits
PR agencies	1. Provide multimedia content for journalists: text, presentations, videos, e-mail newsletters, SMS alerts 2. Leverage archived press releases 3. Online reputation management services	1. Uneven familiarity with new technologies on the part of journalists 2. Unrealistic expectations fuelled by press releases 3. Fake reports, damaging claims, allegations about clients	1. Structure online news rooms for multiple access: e-mail, broadband, wireless 2. Insert legal disclaimers in press releases 3. Regularly poll journalists, clients; devise fire-fighting tactics
Alternative media, activists	1. Means of bypassing traditional gatekeepers 2. Networking with communities of interest globally 3. Scope for instant mobilisation, advocacy, fund-raising	1. Cost of wireless infrastructure, access 2. Lack of ICT-aware human resources 3. Censorship by authoritarian governments	1. Community networks, freeware, open source tools 2. Capacity building workshops, funds from donor agencies 3. Use of 'anonymous' proxies, mirror sites, encryption
Educators, academics	1. New areas of research 2. New ways of delivering courses and reference material	1. Rapid pace of change, fear of obsolescence	1. Leverage Web, wireless as a learning resource

(*Table Continued*)

	Opportunities	Challenges	The learning curve: Responses
	3. New research partnerships	2. Inadequate resources for digital labs 3. Lack of cooperation from industry	2. Seek partnerships with industry for resources, internships 3. Conduct joint studies, create centres of excellence
Government, national policymakers	1. Mobile dissemination of government content for media, businesses, citizens 2. Interactive, transactive m-government services 3. Regulation, initiatives promoting mobile industry, Wi-Fi, Wi-Max	1. Updating existing telecom regulations 2. Updating and enforcing laws regarding copyright, cyber crime, freedom of speech, surveillance 3. Devising spectrum allocation models (e.g., auction, beauty contest)	1. Set up ICT ministries, merge existing print/broadcast and telecom ministries 2. Lobby in existing international fora (e.g., ITU, WIPO) 3. Nurture collaboration between technology groups, standards, industry associations

Source: Author

phones and posting images on the web. The programme uses SMS to conduct listener surveys.

Mobile subscribers can now receive SMS alerts of news, business updates, sports and entertainment information. Subscription for SMS services for information as well as entertainment (e.g., ringtones) can be enabled via SMS, voice requests, SIM cards, operator kiosks, or the Website of the content provider. This applies not just to mainstream news but also to alternative media. For instance, Malaysia's alternative newssite Malaysiakini offers SMS alerts to subscribers who want to stay on top of political developments in the country.

News delivered via SMS in Russia could make the wireless channel 'rival' the Internet, according to a recent report in the *St. Petersburg Times*. Russia has about 5 million active Internet users and, by the end of 2003, about 32 million Russian citizens owned mobile phones. About 5 per cent of users subscribe to SMS lists to receive—voluntarily and at a cost to the subscriber—anecdotes, horoscopes, exchange rates, and news. News lists only began operating in September 2003.

Many wireless content disseminators have found billable SMS (or premium SMS alerts) to be revenue generating and even profit making. Chinese portals NetEase, Sina and Sohu began reporting increasing profits in 2002 thanks to billable SMS services for news alerts, information, cartoons, and games. At 1.5 yuan (20 US cents) per message, the huge volume of cellphone users and Internet traffic is enough to generate decent profits. In 2002, China Mobile, the country's biggest wireless operator, introduced a micropayment system that lets portals share in revenues for wireless content services. China Mobile users sent 80 billion SMS messages in 2002, up from 15.9 billion in 2001.

Traffic began to increase dramatically when China made its debut in the soccer World Cup in 2002. China's Sina.com portal recently acquired Crillion, a leading SMS valued-added service provider to temporary job seekers in China.

The Japanese market offers an excellent case study for other markets around the world on the importance of alliances between carriers and content providers to offer better services to consumers as well as increase ARPU. Challenges arise for content providers in furnishing information in various formats for the different proprietary standards of each carrier in Japan, and for re-purposing existing content for the smaller handheld device screen. All mainstream media are leveraging the wireless medium in Japan, and foreign media have staked out their turf in this medium as well.

Japanese content providers for information services like news have played a key role in boosting the wireless Internet as a medium in the country, thanks to strategic partnerships for delivery and billing with the wireless operators. Today, news is available on handsets in Japan from local and foreign news media like Reuters, Dow Jones, Bloomberg, CNN, *Asahi Shimbun*, *Mainichi Shimbun*, *Nihon Keizai Shimbun*, Jiji news agency, and Nippon Hoso Kyokai. Japan's major national TV networks offer a short video news programme consisting of the day's major stories.

'On the wireless Internet, few news and information services have been as successful as those in Japan. The news sites have played a prominent part in the growth of mobile Internet services in Japan ever since NTT DoCoMo Inc. launched its i-Mode service in February 1999,' according to Martyn Williams of IDG's Japan bureau (Rao, 2003).

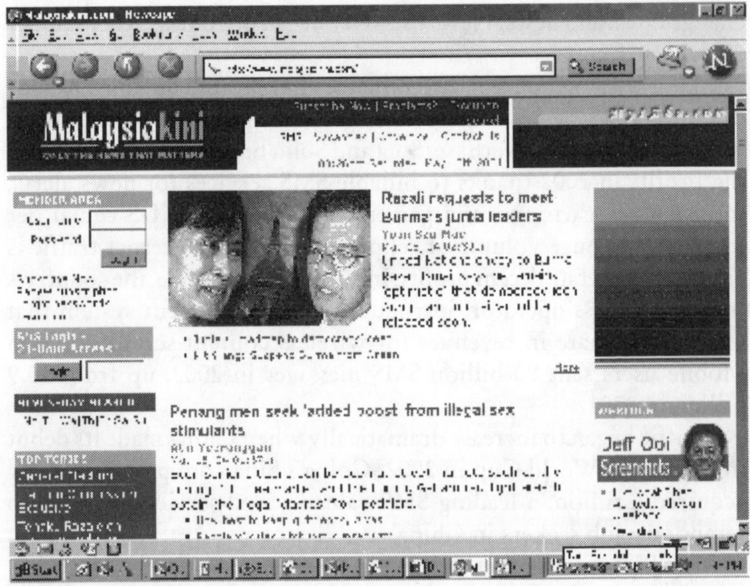

Figure 6.1: SMS services from the Malaysiakini Website

Cross-media Synergies: TV and SMS, MMS

Recognising the growing power of news consumers using the Internet and mobile phones, some news outlets have increased the channels of participation. This can include printing reader letters submitted via e-mail, SMS voting in TV and radio programmes, and posting of user-submitted photographs via e-mail and MMS on news Websites.

The 8888-branded SMS service of the IndiaTimes portal (part of the Times of India publishing group) reports a traffic of 30 million messages a month to users in 76 mobile circles across India, ranging from news and ringtones to horoscopes and stock updates. Its first interactive game was called Predikta; today it has promotional tieups with TV serials (e.g., Friends), TV channels (e.g., B4U, AXN), brand marketers (for Levi's, Temptations, Kwality Walls, Close up Lemon), FM radio stations (e.g., song dedications via SMS for 98.3 KKD) and movie houses (e.g., promotion for Chalte Chalte, Matrix, Jhankar Beats).

SMS is also pushing the envelope of 'participatory TV' beyond mere interactivity, said Mike Short, VP of technology for O_2 in UK. 'Every viewer is now a journalist,' added Joris van Heukelom, content director for MTV Netherlands. There are hundreds of thousands of digital video cameras in the Netherlands, and broadband connectivity in the country is also expanding, thus giving users more opportunities to share their interpretations and 'digital video diaries'. TV shows in Britain also let viewers vote yes/no via SMS on whether the UK should have participated in the Iraq war in 2003—and send in additional comments as well, observed Ashley Smith, senior analyst at research firm Van Dusseldorp & Partners.

In Europe, text message voting is considered a must for any reality TV show. Western European media companies and TV broadcasters generated $82 million in revenue from SMS voting in 2002. For the first time in MTV Europe Music Awards history, the winners of the best song category in 2003 were voted for live by SMS text messaging.

In UK, more than 1.5 million Big Brother fans used their cellphones to interact with the TV show in 2003, according to mobile company O_2; Big Brother sent out 40,000 picture-messaging news alerts during the programme, and received 30,000 requests for ringtones, logos, and Java games via SMS. In Australia, Telstra launched the first part of its Rugby World Cup marketing campaign with sports content on mobile phones such as moving pictures, scores, match summaries, news, and statistics via SMS, WAP, and MMS.

In some cases, news media (like broadcasters) may find polling via SMS to be more reliable and immediate than Internet polling. As Textually.org reports, BBC became the first British TV network to gauge public opinion by SMS on its Watchdog programme—recruiting a large pool of people willing to take part in text-message polling while watching TV. The survey project was conducted by MORI Research. While not yet the equal of traditional scientific polling, text-message polling looks a far sight better than self-selected Web surveys.

In St. Petersburg and Moscow in Russia, two local television stations have agreements with MegaFon, MTS, Bee Line GSM, and Infon to use chat exchanges on the TV-Chat-Infon overnight programme where subscribers can pay to see their messages on the television screen.

As more and more TV programmes start integrating participatory capabilities for cellphone users, a key challenge is devising technology platforms and programming features to allow real-time inclusion of SMS text messages on the TV screen (T2S). While this is already happening in many TV programmes in markets like Europe, the bigger challenge has been in smoothly integrating viewer-generated MMS (multimedia messaging services) content in the form of pictures (P2S).

SMS participation was successfully used to generate half a million Euros in revenue during a TV programme where viewers with mobile phones could vote to select Germany's contestant for the Eurovision song contest, according to Alexander Trommen, COO of the Minick Group, a mobile interactivity services group in Germany.

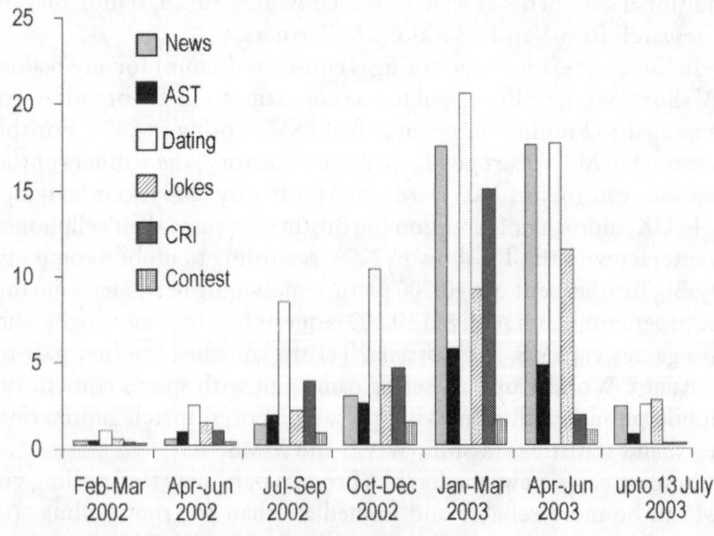

Source: *India Times*, 2003 (all figures in millions)

Figure 6.2: Popularity of categories at India Times' 8888 SMS service

Incorporating MMS pictures directly from viewers has not been that easy; instead, pre-stored pictures (usually thumbnail photos) can be uploaded from viewer Websites hosted by the telecom operator.

In terms of direct transmissions, many wireless operators have started broadcasting television clips (e.g., cricket in India) directly over their networks. The emerging Digital Video Broadcast-Handheld standard, which is designed to conserve battery and cope with moving receivers, is expected to take a few years to be commercialised and to pass some countries' regulatory hurdles.

Australian telco Optus has launched streaming TV feeds from channels ABC, SBS and CNN for users of its 2.5G wireless data service. Other offerings include video downloads of Channel 10 Sports Tonight, news from wire services, Channel Seven Finance News Updates, new Java games, and other content designed for multimedia mobile phones like the Nokia 3650.

Mobile Impacts on Public Relations

From SARS alerts and rapid changes in stock prices to new product launches and venue information, the wireless channel is becoming a crucial component of corporate and government public relations in Asia. While the Internet ushered in the era of 'paperless PR', the mobile ecosystem has created an era of 'instant PR'. The PR professionals, their clients, journalists, financial analysts, senior government officials and the citizen at large are increasingly becoming part of a 'wireless PR web' which seamlessly delivers information and alerts via fax, e-mail, voice and SMS.

Wireless communications helps marketers correct last-minute reporting errors, synchronise corporate news delivery across multiple stakeholders, and unleash new creativity, according to Kim Bayne, author of *Marketing without Wires*, who documents the successes of mobile marketing and PR in the US.

The world's leading press release aggregation and distribution service, PR Newswire, offers news distribution services to journalists' PDAs during events. It has signed up with mobile solutions firm Bluefish Wireless for events like PalmSource 2003 for Palm OS Developers, to deliver breaking corporate news and other event information on wireless devices to attendees at trade shows and conferences.

Issuing news to handheld devices at industry events is an extremely effective strategy to target audiences that are interested in the specific industry, and the wireless channel is the latest addition to online press kits and wire distribution to niche media. At PalmSource, Bluefish provided a comprehensive conference guide, including session information and speaker bios, to PDA users.

Auto giant General Motors has embraced wireless news dissemination during automotive industry events like the Detroit Auto Show. Tech marketing firm Hass Associates gives journalists a Sony Clie with pre-loaded press material during auto shows; future plans include using Bluetooth-enabled devices. PR 21 delivers wireless synopses of news reports on its US clients to their PDAs and mobile phones. Clients can sign up for such media tracking services on PR 21's password-protected Extranet.

Vocus PR (www.vocuspr.com) offers a variety of 'wireless office' solutions for the mobile PR professional. Sites like PrimAlert in the UK offer realtime financial alerts to clients via SMS. DigitalPaths allows PDA users to access micro-versions of Websites. 'SMS this headline' features are already available to news surfers on most major new sites in the world.

Media consultancy firms like Harvard PR in the UK offer their clients PR services ranging from lobbying and consumer phone-in services to Web communities and SMS alerts.

'Writing simply is extraordinarily difficult, particularly for those who have been rewarded for inflated verbiage through educational, bureaucratic and corporate promotion,' says Yvette Nielsen, a marketing and PR consultant in Brisbane (Rao, 2003).

Personal rapport and relationships between journalists and PR professionals can be strengthened, especially if managed with proper care and permission-based opt-in and opt-out features to avoid SMS spams.

PR executives and communications officers are increasingly signing up for not just e-mail news alerts from the popular portals and content services providers, but SMS alerts as well. Factiva—the news alerting and archiving service created by a joint venture between Dow Jones and Reuters—delivers top breaking news stories to communications officers and market researchers by using instant messaging (IM) technology as part of the Factiva Alerts news product.

The workflow of PR firms increasingly incorporates SMS services such as 'breakfast alerts' to notify PR professionals and clients of changes in market conditions, news coverage, client positioning and event updates. PR firms for technology clients in India already offer SMS alerts to clients by 9 am each morning with a brief capsule of which newspapers have coverage of their clients and on which page.

In the entertainment sector, the PR campaign for the movie 'Hollow Man' even included sending 50 reporters special Casio handhelds—loaded with a press release, clips of the soundtrack and movie, and information about where to buy tickets! Warner Brothers has also tied up with Casio to offer Web surfers the option of downloading movie trailers to their Cassiopeias.

Public relations strategies for leading firms increasingly include investor and financial analyst relations as well, particularly via wireless updates about stock price changes. Wireless PR is being effectively leveraged by government agencies as well, who are adding mobile services to e-government offerings. The SARS outbreak in countries like Hong Kong has also been accompanied, unfortunately, by fake rumour campaigns on SMS, and governments in the region are beginning to counter such false campaigns with more proactive SMS campaigns of their own.

Keeping direct contact with key media outlets via SMS alerts will become a must-have for corporate and government PR departments across the world. Wireless technologies are now firmly cementing themselves as the latest platform in PR agency operations.

Mobile Entertainment and Gaming

Japan was among the first countries to launch ring tone downloadable services in late 1999. Technology has rapidly advanced from simple 1–3 chords through 16-40 chords to the current 64 chords. The Japanese Society for Rights of Authors, Composers and Publishers (JASRAC) coordinates copyright deals between parties.

In South Korea, operators have embraced entertainment as a way to make money from their 2G and 3G networks. SK Telecom has its own artist, a band called Noel, whose material initially was available only on to SKT subscribers via their mobile phone. Mobile music revenue in South Korea matches that of offline music: In 2004, both markets will be worth around $150 million, according to Sony Music Entertainment (Korea). Approximately 30 per cent of the mobile music revenue comes from traditional song downloads, with the remainder generated by ringtones and ringback tones. (Ringback tones are music tracks which callers will hear—instead of the normal ringing tone—when they dial a number. These can be customised depending on the number called.)

In Europe, rich content is starting to emerge, e.g., football content on Hutchison-backed 3G operator 3. UK's Mobile Data Association predicted that the value of the ringtones market would exceed that for CD singles during 2003. Some UK-released songs already have generated greater revenue from ringtones than CD sales.

The popularity of ringtones has been further increased by record companies which now add downloadable ringtones to some of their releases. A few labels have even released the ringtone version of a single before the CD. The British Library has reportedly released animal sounds for mobile download, courtesy its vast sound archive, which include colobus monkey calls, a desert cobra attack, loons yodelling and cattle lowing!

A recent study by ARC showed that sales of cellphone ringtones worldwide jumped 40 per cent in 2003 to $3.5 billion. The worldwide music market is estimated to be $32.2 billion; more than 10 per cent of the overall music business is in ringtones.

In 2003, £65 million worth of singles were sold in the UK, compared with £75 million in ringtones for mobile phones. A growth rate of about 60 per cent will result in creating a £100 million industry in 2004. The European ringtone market alone is worth an estimated 1 billion euros annually and growing. Woolworths has created a top-10 chart for ringtones, displaying it alongside the singles chart in the hope that customers would buy the single and the ringtone at the same time!

MTV Networks Asia has launched a brand new business division, MTV Mobile, in nine countries across the region—China, Hong Kong, India, Indonesia, Malaysia, the Philippines, Singapore, Taiwan and Thailand. Youth audience services offered include MTV VJ and celebrity content.

The US accounts for about 5 per cent of the market for gaming in the global wireless industry (mobile as well as standalone), while Europe represents 15 per cent and Asia for over 75 per cent, according to Strategy Analytics. US gamers are reportedly far behind their Asian and European counterparts, with far fewer games being developed and less revenues being generated.

According to the Wireless World Forum (W2F), the mobile games market will be worth $1.9 billion in 2006, up from $520 million in 2003. Japan and Korea together currently represent 64 per cent of the total value of the global mobile games market, though the US is expected to become a major player as well in 2006.

Mobile gaming clocked over $1.5 billion in 2003, up from an almost negligible base in 2000; revenues could be worth $7.4 billion by 2007, according to the ARC Group. Mobile gaming encompasses cellphones (both online and voice-based), PDAs (which may also have phone capabilities) and mobile game platforms such as the GameBoy

Figure 6.3: Mobile entertainment and information services from Indian portal Rediff.com

Advanced. Almost all new phones can play games, and their CPUs have special capability for graphics and data compression.

Some interesting location-based games (LBGs) are emerging, especially in Europe and Asia, e.g., the Mogi a game that is about collecting and trading objects in the real and virtual world rather than fighting and maiming in the virtual world. Singapore's SingTel has launched Gunslingers, a multi-player game where participants find and challenge each other. The growth of GPS capabilities into mobile handsets will fuel LBGs.

For various reasons, this book will not address the impact of mobile technology on the adult entertainment industry, estimated to grow into a $1.2 billion-a-year industry in 2008. Another industry beyond the scope of this book is mobile gambling, estimated to cross $6 billion a year by 2008.

The Legal Frontier: Content Copyrights and Technology Licensing

Beneath the hype and hope of mass consumer market adoption and enterprise uptake of wireless technologies, major tectonic shifts are taking place on another front: Intellectual property rights (IPR), or the creation of technologies, content and applications that can be enforceably licensed for lucrative royalties at various points of the wireless ecosystem.

Entrepreneurs, big business, patent lawyers, standards bodies, policy institutes and industry forums have barely managed to realign themselves during the Internet tidal wave when the wireless juggernaut has thrown up a whole slew of opportunities—and challenges—for IP creators.

Wireless operators, chipset manufacturers, application developers, content aggregators and the entertainment industry are collaborating to try and figure out how to ensure that licensing of mobile content can build healthy revenue streams.

According to Jupiter Research, payments made with wireless phones, which will predominantly be for mobile content and applications, will total $812 million in revenue by 2006. Mobile entertainment revenues may actually cross $12 billion by 2005, says McKinsey.

Strand Consulting estimates that over one billion ringtones may have been downloaded in 2002; ringtone sales have already reached $1.5 billion in Europe and $300 million in Japan. Merrill Lynch predicts that the mobile phone content market will reach $25 billion globally by 2005, with a worldwide base of 1.6 billion wireless Internet users. Merrill Lynch estimates an average revenue of $1 per subscriber per month for content services. Firms like Probe Research predict that the handheld device market will be conducive to 'spur-of-the-moment purchases during periods of boredom' including music and games.

Ringtones, music clips, entire songs and videos are touted as popular download items for users of handphones, PDAs and portable MP3 players. Music publishers, according to industry estimates, received over $71 million by supplying handset ringtones in 2002. The challenge for the content industry is to exploit such new markets while also safeguarding intellectual property rights and avoiding 'mobile Napsterisation'. For instance, BMG, a supplier of music to several approved Web-based music download sites, is now getting into the business of supplying ringtones and music tracks for wireless operators including 3G mobile phone operator 3.

The youth market is being identified as a key market for such products. Research across the world shows that teens may be spending less on teen magazines and even chocolates so that they can spend more on cellphones and top-up cards! Legend Mobile licenses content from the website TeenHollywood.com for a new SMS service for mobile phone users.

Figure 6.4: BBC news Website lets users post photographs via e-mail or MMS

In addition to music content, other information like MTV chart listings are also valuable for the wireless entertainment market. Content banks like picture libraries and clipart databases for MMS are another emerging revenue opportunity for such intellectual property players.

On the legal enforcement front, one approach is to hard-wire or hard-code copyright protection mechanisms into consumer devices; another is to legally hammer out anti-piracy policies, standards and consumer rights regimes and only then dole out premium entertainment content. US policymakers are debating a new 'cellphone user bill of rights', jokes Adam Thierer, co-editor of 'Copy Fights: The Future of Intellectual Property in the Information Age'.

Players in the content rights management technology arena include RightsLine (www.RightsLine.com), a provider of solutions for automated administration, licensing and processing of media contracts, rights and transactions. Content assets need to be categorised by format, language, geography, usage and time, creating a complex model for international licensing and management. This also needs to happen in a secure, reliable and scalable manner.

BMI licenses content to more than 2,300 Websites as well as ringtone providers including Sprint PCS, Your Mobile, Zingy.com and Sonera

Zed. The US market is taking its cue from the performance of content rights players in the markets of western Europe and east Asia. Walt Disney Internet Group makes wireless content available through four major US wireless providers and 12 international carriers; content offerings include news and sports scores. AOL Time Warner also licenses Warner Bros. characters like Bugs Bunny for ringtones, screensavers, and wireless games directly to Motorola. Viacom licenses wireless-packaged content like games from Star Trek, Rug Rats Ratrace, Mission Impossible, and Mighty Mouse.

Revenues for such copyrighted content and services arise from a combination of upfront payment, download volumes, and even network usage fees. Handset manufacturers prefer to license content and games that can be pre-installed on their phones to give them a marketing edge.

nGame, an aggregator of entertainment content for mobile phones and digital interactive TV, has signed a worldwide agreement with Viacom to publish games, ringtones and graphics. For instance, the 'Britney Spears–Crossroads Trivia' SMS game, sweepstakes and ringtones for mobile phones were launched to coincide with the release of the movie.

Patents for standards, protocols, chipset designs and wireless applications are another major source of revenues on the intellectual property front. The challenge is to generate and protect IPR while also making it affordable for use by players at various points of the wireless value chain.

For instance, NTT DoCoMo, Ericsson, Nokia and Siemens are keeping W-CDMA royalty rates affordable (in the single digit range) via proportionate licensing arrangements. The companies together own a majority of W-CDMA IPR already used by about 110 operators worldwide. Forums like the 3G Patent Platform Partnership (3G3P) play an important role in the distribution and harmonisation of such IPR.

As for multimedia formats, Dolby Laboratories has formulated a licensing programme for the MPEG-4 audio format, the next generation of audio and video delivery specifically targeted at mobile devices. Its alliance partners include Apple, Cisco Systems, IBM, Philips and Sun Microsystems.

Licensing IPR abroad is a key component of market expansion for globally-minded players. For instance, Research in Motion (RIM) of Waterloo, Ontario, has a licensing programme called BlackBerry Connect for handset makers who want to use RIM's technology. Nokia has already licensed RIM's technology. RIM has also signed up agreements to bring BlackBerry to Vodafone's corporate customers in Germany.

The licensed IPR includes end-to-end security and push-based wireless applications. RIM and Symbian are working to bring RIM's BlackBerry wireless e-mail to Symbian-based cellphones.

On the legal front, complications over dispute claims and settlements can arise in a rapidly proliferating global market. Companies like Sun Microsystems and Qualcomm are already coming up with 'Priority Watch List' countries based on charges of unfair trade practices over the use of wireless patents and copyrights. Qualcomm, for example, has run into intellectual property controversy in South Korea and China. Bodies like the Wireless Internet Standardisation Forum will play a key role in negotiating such IPR disputes.

Filing patents for emering technology IP can be a costly and lengthy process, but may be well worth its rewards. Mobile business software maker @Road has secured a US patent for technology used in its GeoManager platform, which provides GPS applications for AT&T Wireless, Cingular Wireless, Nextel, and Verizon. 'We are very pleased that the investment we have made over the past several years in intellectual property protection and patent prosecution has begun to bear fruit,' says Krish Panu, CEO of @Road.

In Asia, China—the world's largest cellular market with over 180 million wireless subscribers—has a lot to gain by influencing existing wireless IP standards and even creating its own ones, such as TD-SCDMA, which is less costly than CDMA2000 and W-CDMA systems.

Another strategy for major technology vendors to leverage their IP is to create subsidiaries or spinoffs who can exploit special vertical niches in areas like broadband and wireless. For instance, Eastman Kodak plans to spin off a company focused on wireless imaging technologies like high-quality video, a strategy that will help expand the company's presence in the info-imaging market, estimated to be a $225 billion industry. The spinoff, called Appairent Technologies, will leverage patented wireless technologies from Kodak's research and development labs.

In sum, major opportunities are opening up for IP creators and infrastructure services companies on the content and technology front. A keen eye on market trends, killer applications, scalability and legal enforcement issues will ensure that potential players stay on the right track.

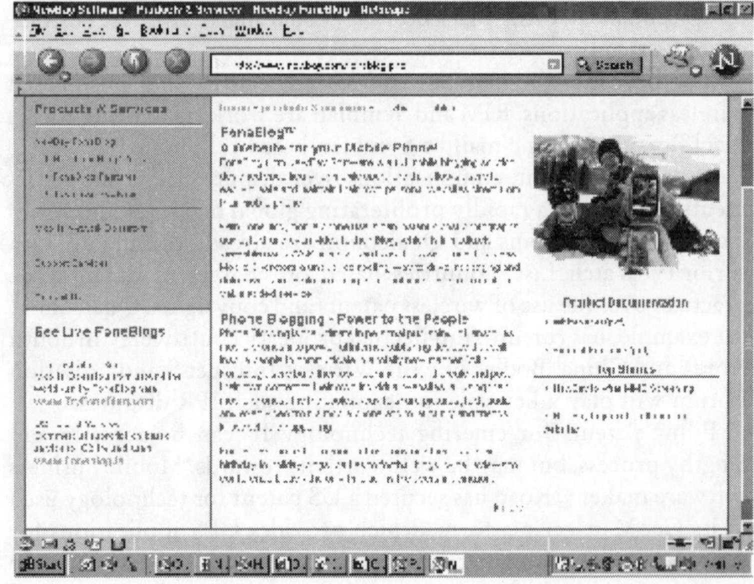

Figure 6.5: Newbay's applications allow blogging via cellphones

Political Ramifications: User Participation, International News Flows

From wireless videophones to user-submitted MMS photographs, wireless technologies definitely made an impact on coverage of the Iraq war.

News viewers tired of the cheerleading role played by US media in the war against Iraq in 2003 were able to turn to the Internet for better news sources, some of which included SMS delivery of news alerts. For instance, Arab TV broadcaster Al-Jazeera launched an SMS news service for wireless users in regional markets like Qatar and other markets further away via operator agreements. The service, available in English and Arabic, has been divided into two categories—breaking news and political news—and costs 15 riyals each per month. The service can be activated in two ways, via SMS or through a call centre. It is also available to subscribers while roaming abroad. Malaysian cellular operator Celcom signed a deal to distribute Al-Jazeera news via SMS. Celcom already offers an SMS news-alert service from Bernama, the national news agency of Malaysia.

The *Indian Express* also hooked up with the Arab news channel, Al-Jazeera, to provide breaking news from the US-Iraq war on SMS. Indian mobile phone users could get the latest updates by SMS-ing 'AJ' at 3636. Back-end technology has been provided by Active Media Tech. 'We came up with the idea as there is a keen interest among the public to know about the latest happenings from the ongoing war,' says Raj Singh of Active Media Tech.

Photophones: Birth of the Snooper Society?

Camera-equipped cellphones—cameraphones or photophones—already outstrip digital cameras in sales volumes, and soon more than half of all cellphones sold will have built-in cameras. They have been used as photography tools by journalists in a number of countries, and potentially turn every citizen into a photojournalist—or even a snooper or voyeur. On one hand, many news sites and even newspapers print photographs submitted by readers via MMS. But as concerns grow over unethical violations of privacy, national governments, local authorities and some businesses in countries around the world are starting to restrict the places these devices can be used.

The ministry of information and communication in South Korea has drawn up regulations aimed at protecting personal privacy, which is increasingly being threatened by the surreptitious taking of photographs in public areas, such as locker rooms and around swimming pools. Cellphone manufacturers will be required to design camera or camcorder-equipped handsets to make loud sounds when pictures are taken. The ministry has also drafted a new bill intended to punish those who take photographs of others with the phones without prior consent.

Samsung Electronics and LG Electronics have banned visitors from carrying camera-equipped phones to prevent corporate espionage. The staff at Seoul's Kyobo bookstore, the biggest in the nation, have begun keeping a watch to stop patrons from snapping pictures of books and magazines using the phones. Additionally, men have been arrested on suspicion of taking voyeuristic photos of women on the subways or streets.

In Japan, some fitness centres ban the use of camera phones and the Italian information commissioner has issued guidelines on where and how such phones can be used. In the UK, several councils have taken action to stop such phones from being used in schools, leisure centres and swimming pools.

A bill banning so-called up-skirt photos and other forms of voyeurism has made further progress through the political machinery in Washington. The taking of covert photos in places like locker rooms or bedrooms is likely to be judged a crime punishable by up to a year in prison and fines.

Source: *Korea Times, BBC*

Elections are also occasions around which a lot of SMS news and polling services can be launched by news media and independent analysts. Malaysia's independent online news daily, Malaysiakini, launched a comprehensive Website on the Malaysian general election in 2004; subscribers could get the results of polling via SMS. Cellphone campaigns via SMS also have been used in Iran to urge some voters to boycott recent polls.

'If there is a medium that has contributed most to make news run like hell during the last days of the strange and difficult Spanish election campaign in 2004, it is text messaging (SMS) through mobile phones,' says new media analyst Eva Dominguez. SMS was used to spread news among citizens as well as political parties. 'But the most impressive use happened the night before the elections. The spread of text messaging congregated some thousands of people in front of the political party running the country, Partido Popular, in just a couple of hours. The use of big media has not been as powerful as text messaging to spread the news and ask for the truth in the most intensive days of democracy in Spain,' observes Dominguez.

In 2003, during the protests leading up to the eventual launch of the war against in Iraq, the BBC news Website published pictures of protests taken from around the world, submitted by users via e-mail or MMS. SMS messages on cellphones were also being used to coordinate anti-war protests in China, Germany and Egypt. In Qatar, where US Gen. Tommy Franks directed the war in Iraq, locals have used SMS to unleash anti-American protests and a boycott of American and British products. In Denmark, demonstrators have used cellphones while riding bicycles to reconnoitre and update each other on police movements.

The 'blogosphere' of independent analysts and concerned citizens has been augmented via wireless channels for direct inputs to blogging sites in some markets. For instance, wireless operator O_2 Ireland has launched a service called FoneBlog that allows users to update weblogs from their mobile phones. The application is powered by Dublin-based company NewBay Software. Users can upload images to their blog from a camera phone or via SMS text, for a monthly fee of 2.50 euros.

But internal and external forces may undermine the 'we journalism' of blog and wireless communication, according to new media guru Howard Rheingold, author of *Smart Mobs*. Misinformation, disinformation, incredulity, and magical thinking all are problems on the supply side of these new reporting modes. Activists should now concentrate their efforts in technology-amplified collective action. Influencing elections and legislation is the sine qua non of effectiveness.

On January 20, 2001, President Joseph Estrada of the Philippines became the first head of state in history to lose power in a protest movement coordinated in part by agitated citizens using SMS to broadcast and receive messages. 'The Philippines might furnish early indicators of the way mobile communications could affect other countries where it is more cost-effective to jump directly to a wireless infrastructure,' Rheingold observes.

In the next few years, peer-to-peer, self-organised, citizen-centric movements enabled by smart mob media will either demonstrate real political influence, be successfully contained by those whose power they threaten, or recede as a utopian myth of days gone by, says Rheingold.

Conclusion

A complete survey and treatment of the impacts of wireless and mobile technologies on the news and entertainment industries could easily fill a whole book. This chapter has provided informative insights into a wide range of such impacts: news gathering by journalists, PR strategies, publisher revenue models, operator alliances, synergies between wireless and TV media, international news flows via SMS alliances, mobile user participation activities in traditional media, and activism in political environments. A wealth of research opportunities is opening up for the academic, industry, analyst, activist and policymaking sectors.

This chapter, in turn, opens the way for a number of further research questions in the 'third' media ecosystem. What are the challenges in conducting research with acceptable levels of precision (eg., was someone else using the subscriber's phone; was the SMS message really received; are results reported in focus groups about mobile behaviour accurate; what is the validity of SMS polls)? What are the ethical issues in conducting research on mobile users (e.g., children and early teens)? How can journalism and communication departments build capacity in design of mobile news and entertainment services? How can free versus fee-based models of content access be balanced? Will mobile media empower users in political movements, or open up new challenges of surveillance as in countries like China?

From a theoretical framework, this chapter has addressed three sets of fundamental research questions regarding wireless media.

1. What new and hybrid forms of media interactivity, user behaviour and organisational models can be created via wireless and mobile technologies?
2. What kinds of opportunities and concerns do publication and information accessibility via wireless channels open up regarding ownership rights and political economy of digital content, news media and entertainment industry?
3. Can mobile media empower citizens to bypass traditional gatekeepers of information and news, and create new kinds of interest groups, content hubs and political formations?

The findings and discussion in this chapter thus expand on prior inter-media, gatekeeper, international news, media organisation, global journalism, and global civil society studies.

Resources

Funk, Jeffrey Lee (2001). *The Mobile Internet: How Japan Dialed Up And The West Disconnected*, ISI Publications, Bermuda.

ITU (2002). *Asia-Pacific Telecommunications Indicators*, International Telecommunications Union, Geneva.

Quinn, Stephen (2002). *Knowledge Management in the Digital Newsroom*, Focal Press, Oxford.

Rao, Madanmohan (2003). *News Media and New Media: The Asia-Pacific Internet Handbook*, Times Publishing, Singapore.

Rao, Madanmohan (2002). *The Asia-Pacific Internet Handbook, Episode* IV: *Emerging Powerhouses*. Tata McGraw-Hill, New Delhi.

Rheingold, Howard (2003). *Smart Mobs: The Next Social Revolution*, Transforming Cultures and Communities in the Age of Instant Access. New York: Perseus Books.

Web Resources

Asian Media Information and Communication centre (www.amic.org.sg)
Online Journalism Review (www.ojr.org)
Pew Internet and American Life Project (www.pewinternet.org)
Poynter Institute (www.poynter.org)

7. M-commerce

Drivers, Challenges and Opportunities

Sunanda Sangwan

Editors' Note: Mobile commerce was once hyped as the 'killer app' for mobile devices but, as this chapter describes, there are significant challenges to be overcome with respect to size of the display screen, input characteristics, security and perceptions of trust. Still, the convenience and ubiquity of mobile devices makes for a compelling proposition, and m-commerce services are gaining acceptance in Hong Kong, Japan and South Korea. Content-to-person (C2P) sales of SMS and MMS data are proving to be lucrative, and trends to watch include the growth of P2P payments and the emergence of standards and consortia enabling mobile payment. The next chapter in this book covers m-business and media impacts of wireless technologies.

Introduction

One significant development of the network economy has been the growth of mobile devices. Different types of mobile devices have become the key access points to online consumer applications ranging from telephony to financial services. With a projected growth rate of 100 per cent by the end of 2003, business and end service users will increasingly use mobile devices as the highest entry point to the Internet (Cap Gemini Ernst & Young, 2004). Mobility can be defined as 'creating business value through mobile devices and wireless technology, by enabling communication, information access and business transactions.' (Cap Gemini Ernst & Young, 2004). Mobility uses the capabilities of mobile networks to connect computers, personal digital assistants (PDAs), telephones, cars, and household appliances (Kannan et al, 2001). Mobile commerce (m-commerce) involves the use of mobile computing devices in carrying out different types of economic transactions or enabling them to take place over space and time (Lyytinen, 2001). Mobility offerings of various organisations include everything from mobile strategy to development of server-and device-software. Mobile device manufacturers are introducing new types of hardware, wireless PDAs, communicators and smart phones in the market. Combined with mobile applications, these devices will characterise the future in making services ubiquitous, interactive and personalised (Kannan et al, 2001).

Telecom firms are also facilitating m-commerce by rolling out packet-oriented, wireless bearers such as general packet radio service (GPRS), enhanced data rates for global evolution (EDGE) and universal mobile telecommunications system (UMTS). These huge investments will be profitable only if a 'killer application' materialises, propelling an explosive growth in the demand for devices and bandwidth (Hui Lei, 2002). As the number of mobile device users increases and exceeds that of static terminals like personal computers, conducting business and services over these mobile devices is becoming attractive and is expected to drive the future development of e-commerce (Kannan et al, 2001).

For successful m-commerce activity, application providers, service providers, content providers and technology providers need to collaborate progressively. This integration is one of the fastest growing markets of e-commerce and it will involve the development and design of a host of new applications, services, business models and technological solutions (Lyytinen, 2001). E-commerce conducted via mobile devices is projected to become a $25 billion market worldwide by 2006, which will account for 15 per cent of the world's online commerce (Legard, 2002).

M-commerce offers many benefits to end users—including convenience for booking and paying for tickets, managing stock trading or financial transactions, and providing information at the point of sales and interactive commerce (Kannen et al, 2001). For merchants and banks, m-commerce reduces cash-handling costs, and an estimated 20 per cent increase in sales caused by customers being able to make payments more easily and conveniently. It enables them to get a feedback at the point of sales or usage, which has implications for customer services and personalisation.

M-commerce uses much smaller screens on mobile devices and it is important that the payment processing company provides a virtual terminal that can be viewed with a PDA or cellphone. The key distinction between m-commerce and e-commerce is the use of mobile or wireless devices to access the commerce Website, instead of using a wireline PC browser client (Turban et al, 2004). M-commerce enables users to access the Internet without needing to find a place to plug in a PC; it requires a multitude of wireless-related network, platform, and voice/data technologies—all technically different from today's more mature Web technologies (Kosiur, 1997). Wireless technologies are constantly evolving and, as a result, standardisation is needed on specific network platforms that often conflict. Competing standards make it costly for m-commerce vendors to develop stable applications on a number of ever-changing platforms.

Mobile Devices and Its Use

A mobile device, is a portable, generally small wireless device that can be used to access the Internet via a browser. All mobile devices are not created equal. There is a wide range of capabilities and functionality. In addition, the capability between the mobile device categories continues to be blurred. For example, some phone manufacturers are developing PDA-like functionality into mobile phones and providing larger screens and easier methods of input.

A mobile device may feature a scaled-down keyboard, a pen-like stylus, or both. Mobile devices are powered by batteries and need to be charged for use. They range in size from a stack of business cards to a paperback book. Mobile devices are basically mobile phones and PDAs. Mobile devices may be referred to as a pocket PC, handheld PC, palm, or palmtop. There is no standard nomenclature for mobile devices. Most Websites and store displays use the terms interchangeably.

Mobile devices will be small pocket-size devices, that are light weight with a long battery life and the ability to perform multiple functions. It is likely that users of mobile device will not surf the Web with their mobile device, instead they will demand personalised information such as priority e-mails, stock quotes and weather reports generated to match your exact location (University of Texas, 2002).

With a mobile device, you can update a task list, view your address book, check your calendar, and take notes, among other applications. With some devices, you can also run mobile versions of popular office software, play games, or listen to music. A mobile device is not a laptop, as it does not have the general-purpose hardware and software capabilities that a laptop does and it is not as large.

Some wireless phones can perform similar functions to mobile devices.

In Japan, NTT DoCoMo, the country's premier mobile communications company, launched a trial using mobile phones equipped with FeliCA, a contactless IC (integrated circuit) chip technology developed by Sony. The handsets supported public transport tolls, electronic money, personal information and other trial services.

Some mobile devices allow you to check your e-mail and surf the Web via a conventional or wireless modem, though some Internet features that are available to most desktop computer users may not be available on a mobile device. For example, you may not be able to view information in certain file formats, or use some multimedia programmes such as Shockwave or Real Player.

Most mobile devices allow you to synchronise data with a desktop or laptop computer. Additionally, some mobile devices can share data and software with each other through infrared transfer. You may have to purchase additional accessories to add these capabilities.

Mobile Device: Pros and Cons

Compared with a personal computer, the mobile device is small and portable. This portability is an advantage over the PC in usability, accessibility, and costs. The user can receive information anytime and anywhere by e-mail or a direct phone call. This enables people to receive information instantly. Mobile devices can also be switched on instantly. The user can operate a cellphone with a single hand, even with one finger or thumb. The size and weight of mobile devices and their ubiquity and accessibility make them an ideal tool for performing personal tasks. Lastly, mobile devices consume very little power and are relatively cheap in comparison to a PC browser client.

However, there are disadvantages when using a mobile device instead of a PC. The first is the quality of line connection. The connection speed to the Internet is relatively slow. The connection can sometimes be disconnected when a user is moving outside because of interference, being out of range, or outside the frequency.

Another important concern is security. If your mobile phone is lost or stolen while you are connected to a commerce site, the person with the phone can make purchases at your expense. Typically, the password is not visible when you enter it on a PC (* as shown). On a mobile phone, the four or six digits of numeric character are often used (lowered security).

Poor user interface is another drawback of mobile devices. It is difficult to enter characters from a standard keypad on a mobile device. In some mobile phones, the same key has to be pushed several times to represent different characters in the alphabet. The key layout is different from device to device and there is no mouse operation supported as in Windows. Other technical shortcomings include small display and cache, no attachment file support for e-mail (in many cellphones) and URL length limitations.

Mobile-Commerce Applications

Mobile Financial Services: Mobile financial applications are likely to be one of the most important components of m-Commerce. These could involve a variety of applications such as mobile banking and brokerage service, mobile money transfer, and mobile micro payments (Elaasar et al, 2003). These services could turn a mobile device into a business tool, replacing bank, ATM, and credit cards by letting a user conduct financial transactions with mobile money. Certainly more work is needed in providing transaction support in the applications and network infrastructure. Secure transactions are required before any of these applications are widely deployed. One interesting mobile financial application is micro payment involving small purchases such as vending and other items. A mobile device can communicate with a vending machine using a local wireless network to purchase desired items. Micro-payments can be implemented in a variety of ways. One way is that the user could make a call to a certain number where per-minute charges equal the cost of the vending item. This approach has been used by Sonera, a Finish wireless provider, in their famous Coke (and now Pepsi) machine service (Muller-Veerse, 2001). In effect, it collects money from the users and credits it to the vending providers. Another way to perform micro payments may be via using pre-paid numbers purchased from a service provider, bank, or credit-card company. To support financial transactions including micro payments, a mobile service provider could act as a bank, acquire a bank, or compete with a bank.

An excellent case study is that of Hong Kong, where most people use m-commerce just about every day (Nokia.com). The most popular form of m-commerce in Hong Kong is the long-established credit card, 'always on, anytime, anywhere'. For micro-payments the bank's credit card handling costs render them unprofitable, but people in Hong Kong use m-commerce for micro-payments most days, spending perhaps up to $20 a week on the metro, trains, ferries, buses and mini-buses, and occasionally to buy a cup of coffee at Starbucks or to purchase a carton of milk or a newspaper at a '7–11' store. The principal advantage is the convenience of not having to carry exact amounts of small money in coins and banknotes or indeed any small coins at all. This m-commerce is done in Hong Kong using the Octopus card.

The Octopus card costs HK$50 ($7) refundable deposit, and can store value of around $130. It is easily reloaded at customer service counters and add-value machines located in metro and train stations and any manned retail outlets that accept Octopus for payment and provides a means of payment 'anywhere, anytime' and is 'always on'. Soon it will become usable in Hong Kong taxis, and is gradually spreading to a wider range of shops. The card contains a smart 'chip' that is read by an Octopus card reader by means of low-powered radio frequency with license exemption. Communication between the card and the reader is secured by mutual authentication, but there is no security for cards that are lost or stolen.

In 2000, the Hong Kong government initiated an online service known as ESDLife (Electronic Services Delivery) using a portal built and operated by a joint venture between Hutchison and Compaq. The portal provides online access to a range of public and private services. In July 2002, ESDLife and Nokia announced two new initiatives. The first was to install Octopus card readers at ESDLife public booths that are located in metro stations, train stations, shopping malls, housing

Figure 7.1: M-commerce in Hong Kong with the Octopus card

areas and the like. Now an online transaction, such as booking cinema or concert tickets, can be conducted using the Octopus card. The second initiative is the incorporation of the Octopus smart chip into the plastic casing of a Nokia brand of mobile phone handset.

Mobile Shopping: Mobile creates new transaction opportunities. It is important to note that only a part of the purchasing process is conducted with the mobile terminal. The basic point is that you need to know exactly what you want before making a mobile purchase. It seems most likely that a shopping list might be created with a Web interface, which may then be executed from mobile.

Generally speaking, the quality of the image files displayed on a mobile device screen is relatively poor. On the other hand, items such as music CDs, games, books, and tickets are good examples of products that do not need high image quality for display on mobile devices. Repeat orders can be easily issued by referring to a previously purchased item from the commerce site. Auctions are also easily accessible via mobile devices. For example, your m-commerce application can notify you of an increase or decline of your bid (Elaasar et al, 2003). The mobile device can be used to identify the user accessing the Website. Registered users do not need to enter their personal information such as credit card number or phone number. In the near future, the mobile device itself will act as a personal identification device and be used as an electronic wallet.

Mobile Advertisement: Compared to the wired Internet, mobile Internet will not be dependent on advertisement revenue, because of its limited user interface. However, it is possible to mass customise a mobile phone for particular requirements which then in return would allow one-to-one marketing (Vyas et al, 2001). The condition for one-to-one marketing is ideal using mobile devices. The mobile operator has been able to build a data profile with a lot of information about that user's calling patterns, and by providing a mobile portal the network operator can get even more information needs, so that he/she will receive personalised and thus more valuable information.

Today, mobile advertising is carried out using SMS. Moving from simple mobile phones via WAP-enabled to GPRS and W-CDMA smartphones and communicators, the potential for advertising that includes audio, picture, and video clips on a colour screen is increasing. Monochrome pictures can be displayed and sent on a Nokia Communicator 9110, while Sumsung has integrated an MP3 player into one of their phones and Sanyo has already developed a mobile videophone.

Mobile Entertainment: With an increasingly mobile society, more and more people are on the move. Many interesting services can be offered to these people through mobile devices and wireless networks. This includes video-on-demand and other entertainment services, and information-oriented services. This might also include services like mobile gaming and mobile music (Elaasar et al, 2003). The technologies needed to support these services include: mobile devices with capabilities to match desired applications, suitable mobile middleware, and high bandwidth wireless networks and/or the use of video compression techniques. Continued connectivity is a real important issue as it may affect the perceived quality of service by a customer for entertainment/information services.

M-commerce Issues

Initially, m-commerce was expected to take off rapidly, but its growth has been slow. Nevertheless, the general feeling is that its time will eventually come. A usable interface to any application is critical. For example, properly designed Websites help ensure that users find what they are looking for and successfully perform online transactions. Given the uniqueness of the m-commerce environment, good interface design is especially difficult to achieve (Tarasewich, 2002). In the m-commerce environment, users and applications have to deal with a large diversity of devices (phones, handhelds) that continue to shrink in size and weight. While this achieves high device portability, usability of the devices can suffer. Mice and keyboards are being replaced with buttons and keypads. Smaller screens can be harder to read. Devices can be difficult to use with fewer than two hands. Changing environmental conditions, such as brightness, noise levels, weather can also affect the use of mobile devices. Difficulty in using devices can translate into wasted time, errors, and user frustration.

In a written summary released in 2000 of a 20-user field study conducted in London, users in the Nielsen study said they disliked a range of WAP wireless functions, including small black-and-white screens, the difficulty of navigating through small text with a phone keypad or a pen touching a screen, and an unreliable wireless data connection (Senn, 2000). The research shows that convenience and efficiency are the main advantages of mobile applications both for customers and company's employees. These are the two primary fundamental objectives that end-users expect mobile applications to provide.

Mobility is the basis for the values of mobile applications. The ability to be used anywhere, and at any place was highlighted by the interviewees as the main reason to adopt mobile applications. It is also the main advantage mobile applications have over traditional personal computers. In a way, it is a 'necessary condition' for mobile applications. Senn (2000) also found that the limitations of mobile devices being much smaller, mobile devices have small screens and complicated input mechanism, and are considered to be more difficult to use than personal computers. The low display resolution and small display screen have inhibited information to be displayed completely and clearly. The limited battery life has restricted the mobility of end-users, lasting only three to five days, depending on the amount of usage.

Slow adoption of m-commerce to the still voice-centric nature of the mobile device industry is because the users are turned off by slow connection speeds of 9.6 Kbps (compared with the 56 Kbps for a dial-up connection) and a lack of comfort and ease of use. The technical restrictions have posed a challenge for the wide diffusion of mobile applications (Senn, 2000). Low bandwidth, unstable connection, and limited coverage area are the main drawbacks of current mobile technology and services. These are listed as the main obstacles to overcome when using mobile applications.

Although it is believed that with the advancement of mobile technology, these problems would be alleviated or resolved in the near future, more research needs to be conducted at the current stage. The majority of mobile device users have a general lack of interest or perceived need to use their handsets for anything other than making phone calls. Most people have not come to the WAP system because they have not been persuaded that the journey is worthwhile. Most consumers are not yet ready to make Internet purchases with their mobile devices. While significant investment is being made throughout the world to bring mobile data services to the marketplace, consumer acceptance is lagging dramatically. To date, the promise of m-commerce has not been embedded in the buying habits of consumers.

Security is another significant challenge in the dynamic m-commerce environment (Ghosh and Swaminathan, 2001). There are potential benefits in storing sensitive data, including medical, personal, and financial information, on mobile devices for use by m-commerce applications. But the mobility of devices increases the risk of losing the device and its data. Furthermore, the risk of data access by unauthorised parties makes positive user identification a priority. Security is a major concern for mobile applications. Wireless transmission, in a way, biases end-users to perceive mobile applications to be more vulnerable and unsecured. Thus, more security enhancements and options must be provided to ensure the security of data and business transactions.

The problems hampering the growth of m-commerce can be summarised as below:

- The inherent limitations of the 'user interface', e.g., the screen display on mobile devices.
- Data transfer capacities (it generally takes much longer for text and pictures to be transmitted to mobile devices than to your desktop PC).
- Costs (accessing the Internet via a mobile phone usually involves higher charges for the access service and/or the time spent online than with a traditional PC connection).
- Different standards, in terms of compatibility between devices, services and systems.
- Issues like online privacy and online security are not as well developed as 'traditional' e-commerce.
- Lack of a dominant business model or approach—the 'browse and buy' approach—which you use when you sit at your PC and by a book from Amazon.com, for example, doesn't transfer well to mobile devices because of the limitations listed above.

It is necessary to overcome the physical limitations of mobile devices. In addition, flexibility of input and output modalities can allow adaptation to changing contexts and user needs. For example, text entry using a stylus or miniature keyboard on a PDA might be most efficient when both hands are free, but one-handed text entry using a thumbwheel (Tarasewich, 2003) might be better when the other hand is occupied. Speech input is a viable alternative for devices too small for buttons or for those without a screen, but may not be appropriate for noisy environments or where noise is restricted, example libraries.

Sound can also be used for output, taking the place of text or graphics. Holland and Morse, (2001) investigated an audio interface for a navigation system that leaves a user's eyes and hands free. The application used tones projected through headphones at locations relative to the user, example left, right, and forward, to indicate direction. A good user interface represents an intricate combination of multimedia and optimal terminal design adapted to the human physical and mental capabilities. The study of how security information should be handled in the user interface forms part of security usability. This field has mostly been ignored by researchers as well as application and hardware developers.

Effective security requires a different usability standard, and that it cannot be achieved through the user interface techniques appropriate to other types of consumer software. An example illustrating the subtleties of security is the padlock icon on Web browsers where an open padlock indicates insecure communication whereas a closed padlock indicates secure communication. This is seemingly a very neat and intuitive way of indicating that a Web server has been authenticated with SSL and that transmitted and received data are being encrypted. However, a closed padlock only tells the user that some Web server has been authenticated but not which Web server in particular. As long as the user does not do the extra mouse clicks to view the server certificate, he or she has not authenticated anything at all. Although it seems very simple, the padlock hides crucial aspects of security, without which authentication becomes meaningless (Turban et al, 2004).

Other security aspects lie in the physical loss of mobile devices due to its mobility, which can only be prevented by being extra careful on the part of device owners. Sensitive information stored on mobile devices for use in m-commerce can be protected with user identification and password before accessing devices.

Current Global and Regional m-commerce Trends

In recent times, many regions are finally getting all three key components that are needed for digital content and a better m-commerce experience: network capabilities, handset availability, and content availability. Mobile content services is still growing: ringtones, logos, and games have become popular in many markets. A number of commercial multimedia services offering video and streaming music have been launched in late 2002 and 2003. Use of premium SMS for delivering digital content and conducting marketing campaigns has grown enormously in Europe, and is picking up in North America. Markets are starting to experience movement from content used only as merchandising (the consumer purchases it) to include content as marketing.

Devices are driving data usage; a wide variety of cool handsets with colour screens at affordable prices seems to be driving not only handset purchases but also data usage on those handsets. The phones have features that enable new data services which definitely drive demand, particularly if the packaging highlights the new features and the user is educated about the functions of the device before leaving the retail store.

In a growing number of countries, users can add mobile Internet purchases or payments to their mobile phone bill, debit from prepaid mobile accounts, and charge to credit/debit cards. Examples of operators' m-payment solutions include DoCoMo's 'DoCommerce' service launched in May 2003, Vodafone's M-Pay bill and M-Pay card services, Telefónica's Mobipay, SK Telecom's 'NeMo' financial service, and KTF's K-merce.

Proximity payment services are also slowly expanding; commercial deployments of mobile point-of-sale (POS) payment solutions for use in retail stores are still limited but growing slowly. Currently, South Korea, Finland, and Spain are in the lead, with Japan close behind. Other proximity solutions for parking, public transportation and vending machines are somewhat more advanced than retail POS, but still not widespread. A growing number of mobile operators also offer merchants and service providers (such as deliveries, taxis, and restaurants) the ability to collect and process credit cards on mobile devices with a card reader in the device.

A recent report by Telecom Trends International (Muller, 2001) outlines the trends of m-commerce around the world. In Asia-Pacific operators are actively deploying a range of m-commerce services. With saturation in the mobile market, operators in Japan and South Korea have been searching for services to help them increase revenues. Digital content and mobile financial services are two main areas of focus. In Japan, KDDI sees the mobile phone as a 'personal gateway' without which consumers cannot leave home.

In Europe, pan-European operators are rapidly expanding services. First ringtones and graphics were the most popular mobile content in Europe, and now Java games and video clips are beginning to take off. User friendly content services with easy to use colour icons that bundle messaging and downloadables are becoming the norm. In February 2003, European operators Vodafone, Orange, Telefónica Móviles and T-Mobile launched the Mobile Payment Services Association (MPSA) to create an interoperable, commonly branded mobile payments solution. MPSA launched its new corporate brand, 'Simpay,' in June 2003.

As for the US, American operators are focusing on digital content: US carriers are focusing on services bundling downloadable applications and messaging. Services include Verizon Wireless' Get It Now, AT&T Wireless' mMode service, and Sprint PCS' Vision plan. Canadian operators are working toward inter-carrier short codes for premium SMS, as are their US counterparts. In Latin America, operators in more than 10 countries have launched or announced plans for digital content download services.

Conclusion: The Future of m-commerce

In the short-term, the real business for m-commerce will be what offers utility (including entertainment) and value to end-users. In the mid-term, the key to a sustainable business for mobile data and m-commerce will be building services based on what is unique about the mobile platform. In the longer-term, TTI anticipates a proliferation of types of devices and networks, as well as convergence of services and user profiles across networks, channels, and devices. As a carry-with-you device, mobile handsets will be ideal for handling transactions in this converged world, particularly as ease of use and security improve.

The range and type of transactions and goods purchased by consumers over mobile phones will also continue to expand. Digital content, although evolving in nature to include more multimedia, will retain a large percentage of total sales, while tangible goods (purchased remotely or at point-of-sale) and interactive-type transactions will continue to increase. Monetary exchange transactions will increasingly involve person-to-person (P2P) interactive transactions such as gambling, betting, auctions, classified-ad sales, and money transfers.

Content-to-person (C2P) has been touted as the next revenue opportunity for both premium SMS and multimedia message service (MMS). As the community of MMS users is built, there will be more opportunity for m-commerce applications, and MMS will enable a number of applications that cannot happen in SMS, such as purchase of video clips.

Brands are also likely to play key roles as retailers of recognised and trusted goods and services. Some big-brand companies will opt to become mobile virtual network operators (MVNOs) or co-branded mobile service providers. Branded content will gain more of a role in mobile data, both as purchased and free downloads, and exchanged within SMS and MMS P2P messages. Brands are becoming motivated to participate in m-commerce as they see opportunities to push their brand identity on mobile platforms with colour graphics and music.

Brands are being driven to use m-commerce as a sales and marketing channel as the size of the addressable market grows and offers more revenue potential.

Another evolving trend is the steady growth in m-payments. The market will move from today's early implementations and trials to agreement on and adoption of standards, which will ensure greater acceptance by operators and merchants. As the process of user education and early experimentation drives habit changes, m-payments will experience steady growth. The predominance of small purchases charged to the mobile bill may evolve into a mixture of small purchases and higher value transactions that consumers would prefer to have charged to their credit cards.

There is also expected to be more market segmentation and enterprise use. The evolution of target market segmentation will help drive up revenues and take m-commerce from the realm of early adopters to mass market. Operators are now getting a fair share of revenues from entertainment, and are looking at other m-commerce transactions and the enterprise market to evolve and grow their application download platforms. B2B auctions and reverse auctions, supply chain management, and the ability for field sales to conduct transactions remotely will lead the growth of m-commerce in the enterprise.

As the market for m-commerce progresses, the power of viral marketing—word-of-mouth marketing among users—will have a big impact on the uptake of m-commerce transactions particularly as younger age groups (that are more comfortable with electronic payments and more active consumers of digital content) gain purchasing power. In a few early segments, there is now better evidence that word-of-mouth marketing is beginning to drive the market.

After being rolled out in fits and starts since 2000, m-commerce services are poised for rapid growth. Globally, there will be 1.7 billion users by the end of 2008. At the same time, revenues from m-commerce will experience a CAGR of 141 per cent, growing from $6.8 billion in 2003 to over $554 billion in 2008. Sixty per cent of global mobile subscribers will use some form of m-commerce by 2008, which will be about one-fourth of the global population.

References

Brewin, B. (2001). 'Study: Mobile commerce loses luster', *Computerworld*, 2001.

Cap Gemini Ernst and Young, 'm-Commerce', http:www.cgey.com/solutions/adi/mcommerce

Elaasar, A., K. Lee, W. Green & M. Sakurai. (2003), 'Mobile Electronic Commerce', In *Proceedings of the 36th Annual Hawaii International Conference on System Sciences*, p. 87.

Ghosh, A. K. & T.M. Swaminathan, (2001). 'Software Security and Privacy Risks in Mobile E-Commerce', *Communications of the ACM* 44, pp. 51–57.

Hartung, F, & F. Ramme. (2000), 'Digital rights management and watermarking of multimedia content for m-commerce applications'; *Communications Magazine*, IEEE ,Volume: 38, Issue 11, November 2000, pp. 78–84.

Hamblen, M. (2000). 'WAP Gets Both Jeers, Cheers for Usability: Analysts, users divided over efficiency of Wireless Application Protocol, *Computerworld*.

Holland, S. & D.R. Morse. (2001). 'Audio GPD: Spatial Audio in a Minimal Attention Interface', In *Proceedings of Mobile HCI 2001: Third International Workshop on Human-Computer Interaction with Mobile Devices*, M. D. Dunlop and S. A. Brewster, (Eds).

Hui, Lei. (2002). 'Mobile Commerce: Vision and Challenges', in *Proceedings of 2002 Symposium on Applications and the Internet*, pp. 36.

infoMerchant.com, 'Mobile Commerce', http://www/informerchant.net.ecommerce/m-commerce.html

Kannan, P. K., Ai-Mei Chang & Andrew B. Whinston. (2001). 'Wireless Commerce: Marketing Issues and Possibilities, *HICCS* , 34th Annual Conference.

Kosiur, D. (1997). *Understanding Electronic Commerce*, Microsoft Press, Redmond, Washington.

Legard, D. (2002). 'M-commerce to hit $25B by 2006, study says', http://www.infoworld.com/cgi-bin/displayNewsArchive.pl?day=020322&week=

Luna, L. (2002). 'The m-commerce m-plosion: Is there still hope?' *Telephony*, Vol. 243, Issue 12, October 2002, pp. 48.

Lu, W.W. (2003). '4G Mobile research in Asia', *Communications Magazine*, IEEE,Vol. 41, Issue 3, March 2003, pp. 104–106.

Lyytinen, K. (2001). 'Mobile Commerce : A New Frontier for E-business', 34th *Annual Hawaii International Conference on Systems Sciences*, Vol. 9 .

May, P. (2001). *Mobile Commerce : Opportunities, Applications and Technologies of Wireless Business*, Cambridge University Press.

Muller-Veerse, F. (2001). 'Mobile Commerce Report', http://www.durlacher.com/downloads/mcomreport.pdf.

Schwiderski-Grosche, S. & H. Knospe. (2002). 'Secure mobile commerce', *Electronics and Communication Engineering Journal*, Vol. 14, Issue 5, October 2002, pp. 228–238.

Seager, A. (2003). 'M-commerce: An integrated approach', *Telecommunications—International Edition*, Vol. 37, Issue 2, February 2003, p. 36.

searchCIO.com Definitions of e-commerce, http://searchcio.techtarget.com/sdefinition/

Senn, J. A. (2000). 'The emergence of m-commerce', *Computer*, Vol. 33, Issue 12, December 2000, pp. 148–150.

Tarasewich, P. (2002). 'Wireless Devices for Mobile Commerce: User Interface Design and Usability', in *Mobile Commerce: Technology, Theory and Applications*, B.E. Mennecke and T.J. Strader, (Eds) Idea Group Publishing, Hershey, PA, 2002, pp. 26–50.

The University of Texas at Austin (2002). 'Mobile Computing Device Information', Central http://www.utexas.edu/computer/pda/

Turban, E., D. King, J. Lee & D. Viehland (2004). *Electronic commerce: A managerial perspective*, 3rd edition, Pearson/Prentice Hall.

Varshney, U., R. Vetter & R. Kalakota (2000). 'Mobile commerce: A New Frontier', *Computer: Special issue on e-commerce*, Vol. 33, Issue 10, October, pp. 32–38.

Vyas, A. & P. O. Grady (2001). 'Review of Mobile Commerce Technologies', *Futurist*, Vol. 36, Issue 6, November/December 2001, pp. 11–18.

8. Organisational Impact

Wireless in Industry, Government and Education

Madanmohan Rao

Editors' Note: This chapter covers wireless impact on business environments (e.g., sales force automation, supply chain management), government (e.g., use of SMS alerts for citizens, personal digital assistants [PDAs] by government officials) and education (e.g., Wi-Fi infrastructure on campuses, PDA usage for corporate learning). Wireless impacts on the media sector and commerce have been covered in earlier chapters. While wireless solutions (such as Wi-Fi and RFID for inventory tracking), handheld devices (e.g., PDAs for sales staff) and mobile communications (e.g., SMS alerts by health authorities) open up vast new opportunities for productivity enhancement, more effective learning environments and innovative citizen services, challenges can also arise due to information overload and inadequate security. Careful attention must be paid to metrics covering processes as well as employee impacts of wireless solutions, and the chapter concludes with a useful framework for classifying such metrics. Subsequent chapters cover wireless impact on organisational knowledge management, and classroom design.

Introduction

Information and communication 'anytime, anywhere, on any device'—this catchy slogan of the wireless era has had profound impacts on a variety of processes (e.g., information access for research or sales) in a wide spectrum of sectors (ranging from education and media to government and industry). Indeed, the intertwining of the Internet, Intranet and wireless ecosystems is being referred to as the EverNet, SupraNet, PermaNet or HyperNet.

Do not get distracted by the mass market hysteria of the wireless industry—there is enough action on the enterprise front, says Jaclyn Easton (2002), author of *Going Wireless*. While Europe and Asia lead the US in the consumer mobile markets (with sophisticated technology and services as in Japan and Korea, as well as widespread adoption of SMS in other Asian countries), it is in the enterprise wireless solutions market (especially based on PDAs and Wi-Fi-enabled laptops) that the US may have a notable lead, as the examples in this chapter indicate.

M-business: Mobile and Wireless Impact on Business Environments

We are witnessing a pattern of structural migration from 'tethered' PC-centric models of e-business towards mobile person-centric models, according to Kalakota and Robinson (2002). 'A substantial new creative cycle predicated on the marriage of the Internet, wireless technology and e-commerce is upon us. Businesses in the m-economy will have to deliver existing and next-generation services and applications with greater speed, intelligence, interactivity, and personalisation than ever before,' say the authors.

This section outlines key m-business impact on areas like supply chain management, sales force automation and payment; a sectoral case study of the pharmaceutical industry is presented (see Table 8.1).

Wireless and Supply Chain Management: 'Unleashing the Chain'

From satellite networks and RFID tags to sector-specific handheld devices and wireless security algorithms, mobile technologies are speeding up supply chains, increasing productivity, reducing costs and providing 'global dashboards' of distributed and shifting assets.

At the multi-enterprise collaborative supply chain level, mobile applications can improve speed, cost and accuracy considerations in procurement, supply chain execution (fulfilment), measurement (asset visibility), and service management (field force automation), according to Ravi Kalakota and Marcia Robinson (2002).

'Mobility isn't just a product or a technology. Mobility is a fundamental structural change in business. A company must plan as though it is confronting a moving target,' say Kalakota and Robinson.

In areas like fleet management, remote monitoring, and field communications, m-commerce adds greatly to the e-commerce

Table 8.1: Organisational and sectoral impact of wireless and mobile

	Opportunities	Challenges	The learning curve: Responses
Supply chain management	1. More speed and accuracy in tracking shipments and inventory 2. Providing 'global dashboards' of shifting assets 3. Visibility, security of assets	1. Securing 100 per cent coverage of wireless transmission 2. Systems integration 3. Lack of IT alignment with partners	1. Forming consortia with industry players 2. Joining standards bodies 3. Conducting systematic metrics studies
Sales staff	1. Information and/or connectivity even on the road 2. Less idle time during travel 3. Better data capture and faster decision-making	1. Designing documents and applications for display on small screens 2. Middleware integration 3. Standardisation of equipment, interfaces	1. Outsourcing to specialists 2. Capacity building, training 3. Standardisation of interfaces, security measures
CXOs	1. Organisational productivity 2. Better customer service 3. New markets	1. Strategies and tactics for coping with disruption 2. Ensuring security of information assets and relationships 3. Legal issues, compliance	1. Evolving RoI metrics 2. Promoting organisational innovation 3. Benchmarking with industry pioneers, competitors
Healthcare	1. Reduced errors, redundancy in data entry by pharma staff, doctors 2. Better coordination of medical product chains 3. Alerting services for emergencies, special patient needs	1. Design of usable interfaces for tablet/pocket PCs 2. Security of data, non-interference of transmissions 3. Compliance with government regulations	1. Forming industry consortia 2. Partnering with technology outsourcing vendors 3. Clear specification of legal issues governing wireless data and information services

(*Table Continued*)

	Opportunities	Challenges	The learning curve: Responses
Banking and finance	1. More contact points with customers 2. Real-time transactions and inquiry 3. Faster responsiveness to changes	1. Security 2. Privacy 3. Mobile device becomes competitor to credit/debit cards	1. Form alliances with mobile operators, payment consortia 2. Outsource to tech vendors 3. Conduct thorough system audits, security upgrades
Retail outlets	1. Real-time information access via WLANs, PDAs 2. Promotional campaigns (LBS, in-store) via SMS 3. Better user segmentation, customer servicing	1. Security of information 2. Customer education 3. Consumer distraction by instant comparison services	1. Enhanced data mining, pattern recognition 2. Better planning of promotional campaigns (e.g., competitions) 3. Delivering deeper value to preferred customers
Travel and tourism	1. Check-in services; CRM for delayed/ cancelled flights 2. Last-minute auctions 3. Travel advisory, local information services	1. Managing information in real time 2. Lack of standards 3. Cooperation between multiple players in multiple countries	1. Planning for data and communications integrity 2. Join industry groups, standards groups 3. Design services for best customers, most commonly occurring problems
Automotive	1. Navigation, traffic, shopping guides 2. Emergency services 3. Telematics devices	1. Designing telematics interfaces which do not distract drivers 2. Competition between multiple channels (e.g., dashboard devices, cellphones) 3. Complying with legal requirements	1. Form partnerships with tech, wireless, content players 2. Conduct focus groups, surveys of target audiences and desired services 3. Identifying pricing points, adapting available infrastructure and services

(*Table Continued*)

	Opportunities	Challenges	The learning curve: Responses
Government	1. Productivity, accuracy (e.g., during field surveys) 2. More responsive services for citizens 3. More outreach to rural areas (e.g., via Wi-Fi, WLL)	1. Coping with the speed of technology change 2. Tech/info integration across multiple departments 3. Cultural barriers	1. Forming partnerships with private sector IT players 2. Promoting e-government and m-government policies, benchmarks 3. Capacity building, change management
Education	1. Better information access on campus environments via Wi-Fi 2. Immediate access to educational content on handhelds for learners 3. New ways of repackaging content by educators, publishers	1. Designing educational content for small handheld devices 2. Security of infrastructure 3. Coping with technology obsolescence	1. Join e-learning consortia 2. Form partnerships with private sector IT players 3. Devise phased solutions for technology evolution

Source: Author

contribution to the logistics revolution, which has in itself been truly astonishing, say Peter Keen and Ron Mackintosh (2001).

And the evidence is growing. Singapore Airlines uses RFID tags to improve its cargo hub operations. The US Army uses Savi's RFID tags to track pallets in shipment. Australia's Woolworths, whose holdings include a grocer for perishable and gourmet items accessible via the Net, uses Descartes solutions for routing and scheduling.

Adidas Canada uses handheld terminals to speed up picking, checking and packing of goods. UK doctors use SMS messages to locate nurses at short notice. Mexican cement manufacturer Cemex has redesigned its business model so that its ready-mix trucks are on the road all day in a continuous flow, coordinated by despatch software and mobile devices.

Manufacturing outsourcing, JIT (Just in Time) and VMI (Vendor Managed Inventory) create visibility challenges for globally distributed supply chains. Supply chains with Internet and wireless enablement need to offer a mix of features like event monitoring, exception alerts (e-mail and SMS), user-defined queries and performance analysis. Shipment management means dealing with changing scenarios, and single routes created months in advance are not always the best guide.

Mobile technologies help coordinate dynamic pickup and delivery route management, along with the improved outbound and linehaul planning. Data can thus be moved 'ahead of freight' and coupled with graphical mapping support, thus enabling supervisors to view the relationship of the drivers and tasks within a route.

Transportation management solutions today cover everything from routing and ocean contract management to global inventory visibility and scheduling. Mobile Computing Corporation and Glentel have formed an alliance for wireless solutions in the service, maintenance, inspection and transportation markets. The range of solutions offered by the alliance includes a GPS-tracking application for small and medium enterprises (SMEs). MCC's m-LINX system provides users with the ability to rapidly retrieve virtually any field information that is required. The international wireless service provider Glentel has a vast network utilising two-way radio voice and data, MSAT mobile satellite services, VSAT and mobile data solutions.

Taxi company Taxco in Venezuela uses wireless technology for fleet management as well as credit card authorisation—and improved its turnover volume by 35 per cent in just three months, says Easton (2002).

On the security front, wireless asset tracking via location-sensitive chips can help reduce the annual $1 billion worth of heavy equipment theft in the US. In Europe, brewer Scottish Courage helps track unauthorised reselling of its kegs via RFID.

Aether Systems has a mobile logistics application, e-Mobile Delivery, which provides package tracking and proof of delivery as part of the company's suite of wireless supply chain management solutions. Trucking company TRL, which specialises in humidity-and temperature-controlled transportation, is converting its entire fleet of 600 trucks to Aether's MobileMax system for wireless tracking and messaging. The solutions includes on-board GPS devices, interactive display units, automated status messaging and proximity notifications. Aether's TrailerMax solution can even automatically report whether a trailer is tethered or untethered, loaded or unloaded, or when the doors are opened or closed.

The MobileMax mobile communications system lets fleet supervisors manage trucks on the road by knowing their locations and getting information to and from the driver. Data is automatically switched between land-based and satellite communications without driver or dispatcher intervention. MobileMax can even provide data on border crossings, vehicle fault codes and driver performance. Mobile communications has become the most vital management tool in the transportation industry today, says Aether sources.

Opportunities also arise here for software, portals and services enablers of SCO, such as i2, FreeMarkets, Mapics, Ryder Logistics, Nistevo, and BEA.

Technical challenges for CIOs and supply chain planners lie in the rapid shortening of product lifecycles as one move progressively down the line of operating systems, ERP databases, CRM software, and application plug-ins. Change is inevitable and CIOs must plan for periods of transition during transformation planning while balancing data velocity with accuracy.

FedEx UPS have been using special rugged handheld computers to manage their operations for decades now; other logistics companies are now leveraging the newer Wi-Fi standards and devices. Falling prices of RFID tagging systems in the coming years will undoubtedly boost their use in supply chain automation systems for companies like Wal-Mart, a tenth of whose goods originate in China.

Mobile Sales Force Automation (SFA)

Lack of crucial information on pricing, offerings and competitive positioning has often been a problem for sales staff on the road. The availability of reliable and timely business information in today's hyper-fast marketplace may make the key difference between winning or losing business by sales forces.

Sales forces on the road have been traditionally challenged by the lack of the same information and tools that their colleagues have on enterprise networks, but the emergence of mobile SFA (sales force automation) solutions may change all that.

The 'thousands of points of connectivity' such as wireless phones improve data flows and help bring everyone into the corporate loop, including those left out in the Internet Age because they did not have laptops, says Easton (2002).

SFA services company Salesnet offers its services on a range of devices include the RIM Blackberry. AvantGo's m-business server is used by sales professionals in companies like American Freightways, Avid (maker of film editing products), Brown Brothers Harriman (a private US bank), Future Electronics and Cisco; its clients in Asia include Swire Pacific in Hong Kong. Sales staff of Swire Coca-Cola track products at 'points of availability' (POA) in dozens of retail stores.

An estimated 30 per cent of the US workforce (or 36 million people) is out of the office travelling at any one time. The wireless channel will impact everyone in the sales and marketing chain, from the CEO and CIO to sales forces and customer service reps, according to Frederick Newell and Katherine Newell Lemon (2001).

The wireless channel will alter online and offline channels, move companies from 'tell and sell' to 'listen and learn' mode, and impact the 4Ps (product, price, promotion, place) as well as 4Cs (customer, company, competitors, collaborators) of marketing.

Mobile SFA applications can be purchased outright, or 'on tap.' Salesforce.com has a range of software-as-service applications for SFA. Vodafone and Telecom Mobile, along with mobile middleware vendors, also offer SFA solutions. Companies like HP offer SFA solutions as part of the wireless enterprise framework (WEF) for real-time information on order status, customer problems and fixes, stock levels, delivery times, pricing and product updates.

Toshiba's elevator manufacturing unit uses wireless SFA solutions to enable work order, repair history, and call reporting to be performed at the customer site via pocket PCs and Wi-Fi connections. In the first year of deployment, the company reportedly reaped a cost reduction of $1.5 million while cutting the response time for a new repair request in half.

End-user devices for mobile SFA solutions currently range from laptops and PDAs to cellphones and industry-specific tablet computers. For instance, companies like FedEx and UPS have been deploying tablet computers with digital pen signature facilities for over two decades now.

Wireless sales force automation at content aggregator Lexis-Nexis helps create personalised proposals and contracts with higher accuracy. Salespersons of the newspaper *Atlanta Journal-Constitution* use Palm devices to access data about potential subscribers during house calls. MSN's CarPoint has integrated wireless sales lead management into its DealerPoint network.

Companies assessing mobile SFA solutions should evaluate vendor offerings based on how they handle scheduling, order entry, order status, promotions, pricing requests, sales leads, appointment scheduling, product look-ups, invoicing, template creation, inventory tracking, territory management, sales history, credit checks, order fulfilment, competitive intelligence, 'dashboard' utilities (e.g., for win/loss analysis) and even e-learning and training modules. Challenges to address in mobile SFA include security, continuous availability, scalability and synchronisation across the enterprise.

Emerging applications include XML-based re-purposing of enterprise sales content, coupling with Web services, a tighter integration between supply chain management (SCM) and SFA, linking of SFA and CRM solutions, and the embedding of knowledge processes within workflow objects.

M-payment: The Tantalising Promise of One-stop, One-device, One-account Services

As with e-commerce in the last decade, the initial promise of m-commerce was overhyped—but slow and steady steps are now being taken to usher in a world of cashless and cardless payment with handheld devices.

According to research firm Ovum, wireless micro-payments—transactions of less than \$10—could reach \$200 billion worldwide by 2005, up from \$3.5 billion in 2000. Mobile payments could reach \$25 billion in 2006 worldwide, according to Frost & Sullivan. Tower Group Research predicts that 118 million western European and 145 million mobile users in Asia-Pacific will purchase low-cost mobile premium content by 2005.

As the epicentre of the world's wireless explosion, Asia will witness serious inroads by m-payments into retail transactions. For example, Singaporeans will soon be able to use their mobile phones to pay for parking, taxi fares and photo prints. Already in Suntec City, drivers can pay parking tickets via cellphones after having procured a designated PIN (personal information number) with the Telemoney service (www.telemoneyworld.com) of Systems@Work, a local m-payment services provider. Another cashless wireless service called Blink enables consumer purchases from department stores, cinemas and fast food outlets.

In societies where cellphones or PDAs are fast becoming pervasive, mobile payment solutions will inevitably increase in prominence as channels of transaction. The promises of m-payment include one-stop, one-device, one-account payment convenience for users, direct marketing, shortened transaction time, and increased sales during peak business hours for commercial establishments.

In the Asia-Pacific region, companies like Telstra are also targeting services which will collapse multiple credit card features into one single cellphone. In Melbourne and Sydney, mobile phone-equipped parking meters are widespread (thanks to features like meter expiry alerts via SMS), and sales at phone-equipped soft-drink machines are reportedly 10 per cent higher than at conventional machines.

Companies like USA Technologies are major electronic payment players in the vending, commercial laundry, hospitality and digital imaging industries. USA Technologies has developed over a dozen patents to tap into the $40 billion vending industry along the entire spectrum ranging from billing services to remote auditing of stock in vending machines. There are an estimated 30 million vending machines and 39 million point-of-sale locations worldwide.

Its partners include Dr Pepper/Seven Up, Kodak, ZiLOG (microprocessor manufacturer), CineMachine (which makes vending machines for DVDs and videogames), MEI (a manufacturer of machines which accept coins and currency notes), Wurlitzer (a manufacturer of jukeboxes), and even manufacturers of laundry machines for US college campuses. USA Technologies has acquired companies like Stitch Networks (www.stitchnetworks.com) for its wireless networking technology geared at PDAs and RFID devices.

The market for m-commerce in Europe is estimated to have crossed $23 billion in 2003. Europe and Japan have an 18-month head start over the US on the m-front. In the UK, Virgin Mobile customers can browse the company's Website via a cellphone and buy wine, CDs and appliances.

The star performer on the m-payment front is Japan's i-Mode, where millions of consumers are buying tickets, ordering books and getting news through cellphones. Partnerships have been struck with about 300 banks and brokers. I-Mode's 'I-vending' service lets cellphone users purchase drinks and accumulate points which can be redeemed for further drinks.

GartnerG2, a research service offered by Gartner, predicts that US consumers using m-banking services will cross 7 million by 2005. 'Wireless is the growth hormone for e-commerce,' says Gartner Group analyst Bob Egan.

Wireless solutions are already being aggressively marketed by the financial services sector, such as Schwab's PocketBroker m-trading feature on PDAs. M-payment is particularly ripe for countries with less PC-centric consumer infrastructure than the US, such as Asia and Europe.

'The best systems with the lowest transaction costs for small payments will also enlarge the economy and make it more efficient,' according to Peter Wayner, author of '*Digital Cash*.' However, challenges can arise in ensuring security, guaranteeing flexibility and ease of use, risk and liability allotment, and adequate but anonymous recordkeeping.

'Money is ultimately a political creation,' says Wayner. Thus, any designer of m-payment services must also address political challenges like international government controls, currency standards, accountability and taxation.

Opportunities along the m-payment value chain arise for consumer/vendor hardware, transaction software, back-end retail systems integration, real-time financial reconciliation services, and marketing/data mining analytics. The dotcom era saw the rise of numerous digital cash systems like CyberCash, CheckFree, DigiCash, NetBill and Magic Money—some of whom have since folded.

Cellphone companies are new but powerful entrants into the payments systems value chain—and also have the ultimate leverage to guarantee swift payments: the threat of turning off a customer's cellphone!

Challenges facing m-payment proponents include lack of standards for m-commerce between different financial institutions, inability of carriers to scale up and form partnerships with thousands of retailers, ensuring consumer loyalty, and proliferation of middleware channels for application developers.

Fortunately, a number of standards initiatives have been launched by players like Palm, Visa, CitiGroup, CyberCash, SNAZ, FreedomPay, Nokia, ProPay, and Paybox.

Booz Allen & Hamilton predict increasing joint venture activity between operators and financial institutions. For instance,

TelefonicaMoviles and BBVA are creating new mobile payments platforms across Europe, while the joint venture KPN/ABN Amro JV (called 'Money Planet') is aiming at e-commerce and e-banking.

KPMG predicts that when mobile phone technology converges with the Internet, the result will be a 'financial revolution' which will enable consumers to break free from existing business models to carry out complex transactions and make purchases from anyone, anytime, anywhere. But there will also be a 'major battle-ground' for m-payment between telcos and banks in the future.

The m-payment market in the US is gathering momentum as well. For instance, research by The Kelsey Group show that 25 per cent of all directory assistance requests made by a cellphone resulted in a sale.

Amazon and eBay already have wireless auctions. The restaurant chain Johnny Rockets lets customers use their cellphones to place and tally meal orders for pickup. Domino's Pizza has a similar wireless ordering service called PizzaCast. TellMe is a leading player in v-commerce, based on voice portals.

Hotels like Bass Hotels (owner of Holiday Inn, InterContinental and Crowne Plaza) and Choice Hotels allow for wireless booking and can leverage 100 per cent of their inventory. Employees at Boeing, DaimlerChrysler and Walt Disney can charge their purchases by using their smart ID cards. Handheld devices like Palm will increasingly be m-payment enabled via Bluetooth in future.

Taxi company Taxco in Venezuela uses wireless technology for fleet management as well as credit card authorisation. Famous Footwear uses WLANs, RFIDS and handhelds to cut down on price mismatches. Wireless companies like SkyGo have reinvented marketing, via strategies like interactive branding (e.g., Subway's games and digital coupons), sales alerts (e.g., ESPN's alerts on basketball goods), coupons (e,g., KFC's mobile coupons for subsequent redemption), incentive ads (e.g., CompUSA's deal for a handheld plus free case), and audio ads (e.g., links to recorded information).

Other huge m-payment markets are opening up in telematics and telemetry. GM's OnStar telematics service, launched in 1996, offers hands-free voice-activated services for cars, such as emergency notification, tracking stolen vehicles, remote diagnostics, and information services. Ford's competing service is called Wingcast. According to Questra Software, such 'automated-to-business' or A2B features will help create 'transparent commerce'.

On a humorous note, the adult and gambling industries too are at the cutting edge of m-commerce, as evinced by SinPalm.com (a site for risqué stories), PocketPersonals.com (with the slogan 'Love is in the air—everywhere'), Eurobet, and eGamingWire.com.

For the coming years, key moves on the m-commerce front will include determining appropriate financial models for m-payment, forging synergistic alliances, expanding the world of micro-payments, developing layered payments schemes for wireless content, standardisation of payment interfaces, and ensuring security.

In sum, while 'Web-enabling' may have been the clarion call of the Internet Age, 'handsizing' seems to be the slogan of the Wireless Age (Easton, 2002).

Wireless Solutions: Transforming the Pharmaceutical Industry, Medical Practices and Education

Wireless solutions are making a significant impact on pharmaceutical companies, healthcare institutions, and med-school campuses. Travelling salesmen instant updates of medical supply inventories; they ensure error-free data entry by physicians doing their hospital rounds, and enable roaming access to medical journals for students on campuses.

On the hardware front, new products like the Tablet PC products are also emerging which are targeted at physicians, salesmen and students. Better coordination between travelling salesmen, busy doctors and distant offices of pharmaceutical companies can be facilitated in novel ways via handheld devices in offline or wirelessly connected modes for SFA, ERP and CRM. Other options include voice portals for dealing with shortage of 'face time.'

According to a survey conducted in September and October 2002 by wireless applications company AvantGo and *Pharma Field* magazine, only 25 per cent of surveyed pharmaceutical sales managers and representatives feel that vital information is readily available to them when needed; many have only weekly access to SFA resources. Several have laptops, but PDA usage is now increasing. An average of 1-2 hours each day is spent completing administration and call reports—the equivalent of nearly a day a week.

'The more accurately and frequently data is reported, the better-equipped corporate headquarters is to plan effectively and maximise sales force efficiency,' according to the survey. The US reportedly has close to 100 million remote and mobile workers whose productivity is often hindered by lack of access to timely information and experts.

To meet this growing market, Safe Med, a medical communications and technology company, provides handheld units that allow doctors to order prescriptions through broadband wireless connections. In partnership with ViaFone, HP has also developed mobile solutions targeted at pharmaceutical sales reps. AvantGo Mobile Pharma has signature capture capabilities on handheld devices. Planar Systems has developed thin client Invitium devices that operate on Windows CE.net; built-in applications include alerts warning staff of potential adverse drug interactions in patients.

Enterasys Networks and Concentra Health Services have developed a wireless solution called RoamAbout which enables physicians to roam throughout a clinic visiting patients and recording patient care information into the ChartSource software on a handheld or a tablet PC. RoamAbout wireless connectivity has been implemented in approximately 190 of the 231 Concentra clinics in the US.

Ambulances in Sweden use WLANs and Symbol devices so that paramedics can record on-scene victim information, including vital signs, injury type, and administered medication, and then communicate with local hospitals about the status of incoming patients.

The learnings from e-business activities in the Internet era are now set to carry over into m-business in the wireless era, according to speakers at the EyeForPharma conference, 'e-Sales and Marketing 2002: From Mobile to Wireless'.

Wireless networks and PDAs can help reduce the cabling in hospitals, allow for flexible restructuring of bed and equipment arrangement, and enable entry and flow of crucial patient information to and from handheld devices.

According to research firm Cahners In-Stat Group, about 15 to 20 per cent of the 1.7 million WLAN hardware units shipped in the second half of 2000 in the US, were for healthcare facilities.

Wireless infrastructure company Avaya provides solutions for Canada's Queen's University Anesthesiology Informatics Laboratory (QUAIL). Healthcare professionals can get real-time access to the most accurate and current medical information at a patient's bedside on PDAs via Avaya Wireless and VPN Solutions, helping them make clinical decisions more effectively and with fewer errors.

Accessible information includes patient records, lab results, pharmacy data, and diagnostic images such as X-rays or CT scans from anywhere in the hospital. It incorporates encryption designed to meet the requirements specified in the Health Insurance Portability and Accountability Act (HIPAA), and the Personal Information Protection and Electronic Documents Act (PIPEDA).

'Mobility is proving to be an invaluable tool for professionals in the healthcare industry,' according to Patrick Lance, vice president of marketing, Avaya Canada. 3Com's AirConnect and Cisco's AiroNet are other products in this space.

Innumerable jokes—and tragedies—abound about the errors made in deciphering the sometimes illegible handwriting of doctors and physicians. Peter Waegemann, CEO of the Medical Records Institute, reports that hospitals spend as much every year on transcribing practitioner notes—$25 billion—as they do on computer hardware. Well-designed handheld systems can step in here and reduce data entry errors.

Challenges in wireless medical informatics still remain, though, in ensuring security of wireless networks, designing rugged but user-friendly devices, avoiding possible harmful side-effects of radiation from wireless devices, increasing awareness among medical professionals, and ensuring compliance with medical and legal standards. Legislation is also moving ahead in countries like the US on location-based services to track mobile users of 911 emergency calls who may be in need of urgent medical care.

M-Government: The 21st Century Face of E-government

M-Government represents the hi-tech new face of e-government: mobile government. From efficient citizen services to updated cyberlaws governing the wireless ecosystem, governments around the world are taking on an increasingly active role in the mobile world. These range from exam results and train ticket sales to courtroom alerts and public interest laws.

For instance, London's anti-congestion scheme to deal with the crippling traffic bottlenecks in the UK capital is garnering a lot of attention from other parts of the world as well. The 688 cameras at 203 sites scattered across the 8-square-mile anti-congestion area will photograph the licence plates of the 250,000 cars that traverse it each day, and fine offenders. Wireless technology is a key part of the solution: wireless links connect the cameras and the image store, and the customer has the option to pay fines by Web or SMS.

The Helsinki City Transport Company has reportedly sold over a million public transportation service tickets via SMS—thanks to Finnish mobile application provider Plusdial, which is in talks with other major European cities' public transit companies.

Thousands of students of government schools in Malta and millions in India now receive their examination results via SMS. Notifications of court deferrals in some European countries can be sent via SMS to clients and their lawyers. Renewal notifications for trade licences can be delivered to licence-holders via SMS. At the recent e-Europe 2005 seminar, the European Commission chose the 'm-Government' project submitted by the Maltese government for the e-Europe awards.

UK think-tank Demos has published a report which recommends that government should also provide SMS services like real-time public transport and traffic information and alerts to remind patients of appointments with government doctors.

Singapore's eGap II programme—e-Government Action Plan II—includes features ranging from an online donation portal to SMS payment services. South Korea, with more than 32 million mobile phone subscribers, is developing mobile government document issuance and related solutions. E-government services are already mature, such as land registry and tax payment. The e-government infrastructure is reportedly expected to save about 1.8 trillion won per year via issuance of various government-related documents and certificates, 3.2 trillion won via e-procurement, and 140 billion won by integrated tax payment.

Land registry services are available via WAP in the UK. In the US, the state of California already has wireless services like direct access to state lottery results, public notices, traffic alerts and energy notifications.

In the Middle East, government telco Oman Telecommunications Company (Omantel) has launched a wireless infrastructure project for government and telecommunications customer service offerings, based on Oracle9i. Oman's ministry of education already sends exam grades to over 200,000 high school students via SMS. Other ministries are planning their own interactive information services for SMS as well. Al-Jazeera, partly financed by the Qatar government, has launched a voice portal service in the Emirates through Etisalat.

Leveraging wireless channels to connect unwired citizens is as much a priority for governments as delivering services to IT-savvy citizens. The forthcoming World Summit on the Information Society in 2005 in Tunis features recommendations from many governments—especially developing countries—to promote wireless Internet access in rural areas.

The Wireless Internet Institute (W2i) and the UNICT Task Force recently hosted a conference focusing on 'the Wi-Fi opportunity for developing nations.' While technology problems can be addressed, obstacles remain in the form of rigid spectrum policies and protective regulatory environments.

Government telcos like France Telecom are extending the universal access vision to wireless users as well. The French carrier has launched a drive to bring high-speed Internet to most of France by 2005, including wireless broadband in rural and mountainous areas. Similar initiatives are underway in Japan and Canada.

The UK's Select Committee on Environment, Food and Rural Affairs recommends that specific funds be made available under the England Rural Development Programme to subsidise the cost of broadband in the most remote areas, with schemes that go by creative names like RABBIT (Remote Areas Broadband Inclusion Trial—www.rabbit-broadband.org.uk).

While expanding wireless access and m-government services, several government agencies are also cooperating on the tricky issues of governance of the mobile environment—most notably spam via SMS.

This is particularly significant in countries like Australia where companies now spend nearly double on direct marketing than they do on advertising. Marketers will no longer be allowed to send unsolicited messages to mobiles following the launch of a Code of Practice for SMS marketing by the Australian Direct Marketing Association (ADMA), in conjunction with the National Office for Information Economy (NOIE). Only those consumers who have previously indicated their interest will receive marketing promotions; the code also delivers increased protection for children, who are among the earliest adopters of new media.

The EU's Commissioner for Enterprise and the Information Society is also drafting measures to address the proliferation of unsolicited commercial e-mail and SMS. The Directive on Privacy and Electronic Communications is expected to be implemented across Europe by 2004.

Thanks to the global nature of wireless networks and inter-carrier roaming agreements, effective spam control will require multiple governments to cooperate on anti-spam legislation—and cracks already seem to be appearing between various governments.

Government regulators also need to update existing commercial laws to include SMS as a valid form of corporate record. In many countries, instant messages are subject to regulatory compliance that can be used as evidence in a court of law.

In democracies, the promise of online electorates for e-government—much vaunted during the years of the dotcom boom—are now finding echoes through 'm-voting' as well. Mid-year e-voting trials in the UK featured voting through Web, telephone, interactive TV and SMS, though these are still in the embryonic phase.

During the national elections in India in April–May 2004, the Confederation of Indian Industry (CII) launched a campaign in association with the Election Commission and the Cellular Operators Association of India (COAI) to request all operators to send messages to their approximately 35 million customers on the polling days urging them to exercise their right to vote. Polling reminders were sent at 9 am and 2 pm on the day of voting.

The Norway government offered phone or Internet options for tax filing in 1999, and the SMS option was introduced in 2003.

The former Prime Minister of Swaziland, Sibusiso Dlamini, was reportedly sacked by a text message in September 2003 following a political reshuffle by the southern African nation's King Mswati—causing a lot of embarrassment and controversy.

In SMS-savvy Manila, the Office of the President uses several mobile numbers that can be used by the public to reach the president's office. SMS-driven public query services have been launched by the Department of Health (e.g., for SARS queries), the Metro Manila Development Authority (e.g., traffic alerts) and the National Computer Center. The Philippines has 20 million mobile subscribers in a population of 80 million.

The department of the interior and local government in the Philippines plans to leverage SMS communication with citizens for government drug-enforcement agencies via creative campaigns like 'Text the Pusher, Save the User.'

In Australia, drug tipoffs can be sent by concerned citizens to law enforcers in New South Wales. Government agencies in European countries like France are already delivering SMS advice and encouragements to users who are struggling to give up smoking and drugs.

According to research analyst Michal Zalesak of EuropeMedia, new opportunities are emerging in standardised government-approved services like mobile identification cards. The handheld device can now be a 'mobile government service device on the move'.

Gathering of census data and healthcare information by government employees using PDAs are other promising areas of m-government.

Members of Malaysian ruling political party UMNO receive news about the party, its membership, and activities on their mobile phones via SMS. Wireless operator Celcom launched the information service, which also includes UMNO ringtones and logos of the Wanita,

Youth, and Puteri wings. Party directory, election results, UMNO calendar, and quotes by the Prime Minister will be available next.

In terms of maintaining public law and order, SMS can be a powerful channel for government authorities—but can also be the carrier of rumours and hoaxes. Tightly-controlled state media in China were circumvented by anxious citizens exchanging millions of SMS messages during the recent SARS crisis—prompting government agencies to scramble and deal with the misinformation that also flowed via SMS among the country's 221 million mobile phone users.

Some messages were reportedly re-sent over 40 million times a day, according to the Chinese official newspaper, *Southern Weekend*. Hong Kong authorities on one occasion had to send a blanket text message to about 6 million mobile phones to scotch hoax 'infected city' rumours during the SARS crisis.

On an equally serious note, the regional government in Gujarat, India, ordered the local mobile phone networks to freeze all SMS services during the daylight hours of the religious festival, Rath Yatra of Lord Jagannath. The government was apprehensive that SMS services 'might be misused for spreading rumours and panic among the people of our state,' according to a fax message delivered to wireless operators signed by JR Rajput, deputy secretary (law and order) in the home department.

The county government of Burlingame, California, wirelessly regulates its parking meters to stop meter theft and monitor coin deposits. The parking industry itself is estimated at over $500 billion. The US accounts for over 105 million parking spaces. Automobile-dense markets are perfect targets for m-payment services, which allow for advanced fee collection, decreased operational costs, fewer onsite employees, increased security and predictive parking services for customers. Companies like EximSoft have developed cashless m-payment solutions for parking.

The major challenge on the m-government front, however, lies within government agencies who need to be pro-active and agile in embracing m-channels. Government agencies should encourage their employees to use PDAs and other wireless devices as part of a long-term plan to incorporate more wireless applications into their e-government initiatives, according to the recent report 'Preparing for Wireless and Mobile Technologies in Government', sponsored by the IBM Endowment for The Business of Government.

The report recommends that government agencies should provide training classes; create peer support groups in which tech-savvy employees can walk their less-technical colleagues through wireless use; and start with simple applications like mobile e-mail. Care should be taken to address issues of security as well, the report cautions.

M-learning: The Next Frontier in E-learning

Campus Wi-Fi networks deliver course content to laptops and e-books and reference materials on PDAs. Audio lectures can be played back on portable MP3 devices. Assignment reminders are delivered via SMS. Corporate learning modules can be downloaded on laptops and PDAs by travelling sales staff. Music training modules are delivered to cellphones.

There are just some instances of 'm-learning,' the next logical step in the evolution of digital learning delivery, which started with standalone CD-ROM based systems and then Web-based training (WBT). E-learning itself was preceded by distance learning, which included paper-based delivery of instructional materials and assignments.

The growth of wireless networks and PDAs is also making the jobs of educators and students easier on medical school campuses and universities. 'PDAs will be as common as the stethoscope,' says Bob Trelease, associate professor of pathology and laboratory medicine, UCLA School of Medicine.

Texas A&M University's medical sciences library provides wireless access for laptop users. Stanford University's Mobile Med programme leverages PDAs for time management tools and medical calculators. The University of Louisville School of Medicine is reportedly the first medical school to provide all its medical students with mobile devices.

Microsoft's Tablet PC may also make inroads into the campus market.

'With students constantly on the move, medical schools must provide reliable information resources that fit their mobile lifestyle, and the appropriate set of tools to make the most of their time at each location. The learning environment has evolved with students increasingly studying in a variety of locations and situations outside the classroom,' according to a recent white paper called 'The Impact of Mobile Technology on Medical Education,' prepared by AvantGo and ArcStream.

According to Harvard Medical School, between 70–75 per cent of students entered medical school in 2003 with a PDA. Harris Interactive predicts that by 2004, half of the physicians in practice will be using handheld devices.

'The inevitability of mobile technology in medical higher education can't be ignored. PDA technology is such a natural in medicine that it is not a question of if, but when. And, when 25 per cent of our third-year students got together and purchased PDAs, the Office of Medical Computing recognised a trend that couldn't be ignored, and recognized the need to coordinate and assist the students' use of handheld devices,' says Ray Dannenhoffer, Director, Office of Medical Computing, University at Buffalo School of Medicine. At the University of Buffalo, the largest concentration of handheld devices can be found in the medical school.

PDAs can be leveraged in medical schools for polling students in classes and for mobile access to course textbooks, anatomy illustrations and drug interaction tables. The more mundane uses of PDAs in administration workflow can also be quite beneficial. For instance, Harvard Medical School has reportedly eliminated $150,000 in costs and realised a 50 per cent return on investment (RoI) simply by replacing paper-based course surveys with an automated, mobile solution.

Challenges remain, of course, in designing educational content that can be viewed and used consistently across multiple kinds of devices, and educating professors and lecturers on the use of such handheld devices in courseware.

For non-campus learners, interesting learning applications can be designed for cellphones and other mobile devices. Launched in 2002, on DoCoMo's i-Mode platform, The Melo-Dam School was the first mobile 'music school' service offered to subscribers in Japan. It offers an extensive collection of music-training applications designed for mobile phones in areas like voice training and beat training billed at 150 yen per month. Subscribers can see their own scores as well as top scores of other students.

On an administrative note, the Emirates Telecommunications Corporation of Abu Dhabi has launched a new School Connect Service that allows schools, colleges and other educational institutions throughout the country to communicate instantly with their audiences through mobile phones, using SMS messages. Arabic and English messages can be triggered by dates, where a pre-written communication can be programmed to broadcast on a particular day at a set time.

'The principal advantage of using a wireless device for e-learning is sheer convenience,' says John Setaro, director of research at Thinq Learning Solutions, which specialises in B2B corporate training. The other advantages are improved productivity gains from wireless information access. 'Wireless takes e-learning to the field, where the best hands-on learning takes place,' according to Robert Meinhardt, AvantGo's vice president of enterprise marketing.

In markets like the US, laptops and PDAs now outnumber desktop computers in the workplace. Effective training in this mobile environment will have to be dramatically different—solution-based, flexible and accessible anywhere and anytime, according to Diane Gayeski, author of *Learning Unplugged: Mobile Technologies for Organisational Training and Performance Improvement*.

Dominant trends identified on the e-learning front by market research firm Gartner include a growing market for e-learning content development and design, hosted e-learning infrastructure, wireless e-learning, corporate portals with knowledge management (KM) and e-learning capability, and use of gaming and simulation.

Cahners In-Stat Group predicts that there will be more than 1.5 billion data-enabled handsets, PDAs and Internet appliances equipped with wireless Web capabilities by the end of 2004. The mobile e-learning market will be worth over $5 billion by 2006.

'Learning should occur when and where people need it most and where it can most readily affect the success of their work,' says Gayeski. Europe and Asia have much more pervasive and advanced wireless infrastructures than the US. The m-learning medium is especially suited to developing countries, via blended realtime and stand-alone handheld solutions. Numerous notable successes have emerged of m-learning in action in corporate and university settings.

Blended e-learning solutions including m-learning can enhance management and utilisation of a company's knowledge assets, according to Larry Bielawski and David Metcalf, co-authors of *Blended e-Learning: Integrating Knowledge, Performance Support, and Online Learning*.

Canadian telco Bell Nexxia uses handheld computers for sales representatives to deliver courses on product knowledge. AdvanceWork International has created courses on English as a foreign language for pocket PC users.

Companies like Audible.com provide audio integration services to convey audio versions of research and instructional bulletins to users with MP3 players. 'Lecturettes' in the form of instructional PowerPoint presentations are available for download from numerous Websites and replay on PDAs. Antenna Audio and AvantGo offer instructional content on PDAs for museum visitors.

Generation 21 Learning Systems offers 'instant knowledge' solutions via information access and learning content on tablet PCs for engineering clients like PRI Automation. Centrax Corporation uses m-learning solutions called TrainingBYTES and PromoBYTES for technicians and sales staff.

InterWise has a product suite called Millennium which allows for the delivery of live and asynchronous courses over wireless connections, ideal for travelling executives or telecommuting professionals.

Docent, a provider of integrated software solutions proven to directly drive business performance through learning, has tied up with Queensland's Department of Emergency Services (DES) for its Online Campus Learning Management System which includes wireless delivery. The target audience is 7,000 emergency care officers and 40,000 volunteers who will be provided with the training necessary to help save lives and property, and protect valuable natural resources.

There is a growing library of e-books that can be displayed on handheld computers. Leading educational publishers like Reed Elsevier, Vivendi Universal, Pearson Education and Franklin Electronic Publishers are positioning mobile e-learning products in the academic sector.

The Handango Medical Student Suite offers a huge medical reference database with medical terms and textbook information on topics ranging from wounds and needles to sutures and surgical knots.

The SCATA logbook for Symbian devices is used by anaesthetics students to enter logs of their activities which can then be examined by supervisors. Massachusetts General Hospital uses AvantGo solutions to apply wireless e-learning in their department of neurology and create a 'living textbook' of latest research and cases.

Business school INSEAD has partnered with Singapore e-learning firm ICUS to offer courses that can even be accessed via WAP on Web-enabled cellphones. The most common uses are for multiple-choice questions, coach reminders, browsing content and voice communications. Specific applications range from assignment reminders and glossary information to quizzes and tips of the day.

Avaya has signed up for an initiative to upgrade the wireless network at Abu Dhabi Men's College (ADMC) of the Higher Colleges of Technology (HCT), to the new 802.11a standard giving 1,200 students faster and more convenient access to the Internet and extensive e-learning applications, deployed by the college over the last 12 months.

'This aligns with the college's vision of providing state-of-the-art infrastructure for its students to keep abreast of current technologies. Students will gain easy access to e-libraries, video conferencing, online courses, Web-based classes, course materials and much more, from anywhere on campus,' according to Dr. Tayeb Kamali, Vice-Chancellor and Director of Abu Dhabi Men's College, where laptops are becoming mandatory learning tools.

Laptops are also mandatory at the Indian Institute of Information Technology in Bangalore where students access learning resources from campus Wi-Fi networks.

East Carolina University has tied up with Ericsson to develop its Online Wireless Learning Solution (OWLS). University of Tennessee has a similar tieup with Dell and Avaya. Griffith University in Australia is also contemplating m-learning solutions.

Several efforts are underway to define learning objects as a medium for content and system delivery—especially the Advanced Distributed Learning initiative's Standard Courseware Object Reference Model (SCORM), according to Clark Quinn, director of Cognitive Systems at KnowledgePlanet.

IBM is actively promoting its wireless ThinkPads along with Cisco Aironet Wireless 802.11b technology for wireless classrooms. The University of Melbourne has tested its first prototype of wireless classrooms, where movable desks and wireless networks enable dynamic and creative configuration of classroom design, according to June Gassin, assistant dean of IT.

All this means there will be a call for a drastic re-design of learning content into 'learning bytes' or smaller packages of learning materials and job aids. Challenges arise, of course, in inadequate standardisation on the protocol and tools front and in squeezing instructional content into such small chunks for largely text-based environments.

Discussion and Analysis: Metrics for Success in the Wireless Ecosystem

The previous four sections of this chapter have addressed the wide array of organisational impacts of wireless and mobile technologies; indeed, each section could easily be the subject of an entire book. The impacts cited above range from anecdotal evidence to well-defined quantitative measures. In the post-dotcom era, CIOs and CFOs are becoming increasingly concerned with well-defined measures of IT impacts, or 'return on investment' on technologies like handheld devices and wireless communication. It would, therefore, be useful and important to carefully classify the different kinds of impacts of new media on organisational behaviour, and come up with measures for each of these categories of impact (Rao, 2003).

Metrics for wireless and mobile impacts can be classified into five key categories: technology, process, knowledge, people and economics. Taken together, they capture the full range of wireless projects and initiatives. Examples of these parameters are indicated in Table 8.2.

Table 8.2: Metrics for assessing wireless and mobile initiatives

Scope of metrics	Sample parameters
Technology/infrastructure metrics	Number of handheld or wireless-enabled devices, bandwidth, number of registered users, frequency of usage, multimedia nature of content (text/audio/video), number of messages exchanged, size of archived content base
Process metrics	Quicker access to information, faster response times to queries, improved service quality in specific verticals (e.g., telemedicine, retail), less idle time on the road, single-point data entry, better decision making
Knowledge metrics	Effective design of collaborative workspaces, more flexible and creative environments in the office, increased rate of innovation, better access to knowledge and learning assets in rural areas, mobilising of knowledge to staff on the road
People metrics	Feeling of being connected and in control, stronger sense of team spirit, customer empowerment, livelihood security, social inclusion and equity (e.g., in rural areas), economic upliftment
Economic metrics	Increase sales volume, growth in revenues generated, number of jobs saved/created/filled, new services offered, better customer satisfaction, new customers, new companies created, new products made, volume of exports

Source: Author

It is important to classify these metrics because far too often the wrong results are interpreted for the wrong reasons, e.g., decrease in SMS-based queries from customers in a CRM initiative may not actually mean that customers are satisfied and do not have questions to ask; they may have stopped sending queries only because the responses are inadequate and the customers have given up on further queries! Similarly, wireless communication may help employees be in constant touch with one another—but this may lead to increased levels of stress and concerns over invasion of privacy, unless company policies for wireless notification are clearly specified.

A news organisation may find a huge demand for SMS-based news alerts from readers, but unless it finds a good revenue model, it may find it uneconomical to provide this service. A Wi-Fi/Wi-Max initiative in a rural area may find that it has been able to improve information flow and educational services (process and knowledge metrics) but has favoured only the well-off and not bridged social inequities or included marginalised communities (people metrics).

Measurements of these parameters should also be taken on a regular basis in order to assess progress and the impact of wireless initiatives and take corrective action where necessary. From a methodology perspective, the parameters involved a mix of qualitative and quantitative analysis. A sample range of these parameters, including semi-quantitative parameters, is indicated in Table 8.3.

Table 8.3: Quantitative and qualitative metrics

Nature of metrics	Sample parameters
Quantitative	Reduced telecom costs, reduced idle time, speedier response times
Semi-quantitative	- Productivity (e.g., via speedier information access) - Employee satisfaction (e.g., improved morale, mood) - Knowledge assets (e.g., lessons learned/accessed)
Qualitative	Better innovation, empowered customers; stories, anecdotes

Source: Author

These metrics need to be devised by the planners (e.g., sales, customer service) as well as implementers (e.g., CTO) of the wireless projects; CFOs and CIOs would need these measures to assess performance and evolution of these projects. Discussions, meetings and focus groups should be held in the target communities (e.g., sales employees, customers, students) to conduct period content and service audits and find out what the community would really seek from a wireless information resource service.

Data must be systematically captured in the form of charts, tables, questionnaires, logbooks, Website visits, SMS traffic, network performance, peak usage patterns, and other statistics capturing the nature of information access, demographic profiles, user attitudes, and diffusion of skill-sets. Reports of the wireless and mobile projects should also be full of anecdotes, and even try to throw in touches of humour and actual quotes from the users and designers of the projects.

For this, specific training materials will need to be developed by the mobile project implementers so that metrics can be gathered and inferred in various forms (print, video, handwritten, site logs) and with differing frequency (e.g., weekly, daily, hourly).

In sum, paying attention to all parameters of the '8Cs' framework for the wireless ecosystem and devising comprehensive metrics and monitoring can help bring the fruits of the wireless-enabled productivity and knowledge revolution to a broad user base.

References

Bayne, Kim (2002). *Marketing without wires*, John Wiley, New York.

Easton, Jaclyn (2002). *Going Wireless: Transform Your Business With Mobile Technology*, Harper Business, New York.

Kalakota, Ravi & Marcia Robinson (2002). *M-Business: The Race to Mobility*, McGraw-Hill, New York.

Keen, Peter & Ron Mackintosh (2001). *The Freedom Economy: Gaining the m-Commerce Edge in the Era of the Wireless Internet*, McGraw-Hill, New York.

Newell, Frederick & Katherine Newell Lemon (2001). *Wireless Rules: New Marketing Strategies for Customer Relationship Management Anytime, Anywhere*, McGraw-Hill, New York.

Rao, Madanmohan (2003). *News Media and New Media: The Asia-Pacific Internet Handbook, Episode V*, Times Publishing, Singapore.

9. Knowledge Management and Wireless Ecosystems

Exploring Mutual Synergies

Madanmohan Rao

Editors' Note: This chapter identifies mutual synergies between knowledge management (KM), business intelligence (BI) and the wireless ecosystem. KM and BI are being used to improve efficiency and innovation by wireless players: operators, device manufacturers, network providers and software services companies. Wireless technologies are being used by a range of companies to enhance their existing KM strategies by 'mobilising' explicit knowledge and making it directly accessible to employees on the road. The chapter illustrates innovative practices on these fronts from Asia, such as NTT DoCoMo, LG Electronics, Samsung, Sun Microsystems Philippines, Hughes Software Services and Origin Exterminators (as well as examples from outside the region, such as Siemens and Ericsson Canada). The next chapter also illustrates the use of wireless technologies to transform classrooms and learning experiences in universities, with a case study from Australia.

Introduction

Knowledge management (KM) in the 21st century can be defined as the practice of cleverly using information and communication technologies, collaboration processes and conducive cultural environments to capture and leverage organisational learning. Initially emerging as a formally defined methodology about a decade ago, KM is becoming a key discipline in management, especially in large organisations and distributed networks—corporations, universities, multilateral agencies and NGO communities.

Organisational KM practices focus on explicit (codified) knowledge and tacit knowledge (embedded in employee expertise and teams) via best practice repositories, knowledge maps and taxonomies, communities of practice, data mining and knowledge discovery tools, collaborative platforms, processes for knowledge asset rating, and incentive schemes for knowledge sharing.

KM benefits documented over the years by practitioners, consultants and academics include cost savings, faster innovation, improved productivity, more effective training, knowledge retention, and focused conversion of intellectual capital into tangible market value.

The worlds of KM and wireless ecosystems intersect in two ways: using wireless technologies to enhance existing KM practices, and the use of KM practices by wireless companies to enhance their efficiency and innovation.

Origin Exterminators and Sun Microsystems Philippines are good examples of companies in Asia whose existing KM practices use mobile applications to enable employees on the road to tap into organisational knowledge assets. Wireless technology companies such as network and device players (e.g., Ericsson, Siemens), operators (e.g., Japan's NTT DoCoMo) and wireless software services providers (e.g., Hughes Software Services) use KM practices to stay at the cutting edge of innovation and efficient collaboration. Wireless operators (e.g., Singapore's StarHub and SingTel) also use business intelligence tools to 'mine' call records data and user profile databases for better segmentation of users and cross-selling of services.

Knowledge Mobilisation: Use of Wireless Media to Amplify Existing KM Practices

Access to knowledgeable experts as well as explicit knowledge can be 'mobilised' through the use of offline handheld devices like PDAs and real-time cellular connections like GSM handphones. Other kinds of wireless technologies also enable flexible workplace configurations to facilitate more creativity (e.g., Wi-Fi connections for the enterprise LAN).

While PCs and workstations have come under some criticism for 'tethering' knowledge workers to their desks, wireless technologies may be the perfect answer to 'mobilising' workforce knowledge by helping employees capture and harness key information and knowledge attributes wherever they are, whenever they want, and however they want. Strategies focused on knowledge mobilisation via handheld devices and wireless networks—ranging from Pocket PCs and cellphones to WLANs and RFID tags—can take knowledge management to an entirely new plane of performance, putting road warriors and field workers in the centre of the information and communications world via mobile portals and on-demand expert services.

'Wireless connectivity at the LAN level lets employees work creatively outside of office cubicles if they so desire. They can roam around and stay connected at the same time,' observes Jeanne Holm (2002), chief knowledge architect at NASA in the US.

'However, too much wireless interruption can be distracting at meetings as well,' she warns. Overall, wireless technologies improve employee productivity. 'Knowledge anywhere, anytime and on any device is critical in this day and age,' Holm says.

'Wireless solutions can help employees communicate easily in real time and function across boundaries of space and time,' says Robert Buckman of Buckman Laboratories, one of the pioneering practitioners and champions of KM and author of the bestseller *Building the Knowledge-driven Organisation*.

Such wireless solutions must be integrated well with the business culture; it is important to create a climate of continuity and trust so that employees may have pro-active knowledge sharing across an organisation's evolution, he advises. 'In addition to wireless connectivity, ubiquitous bandwidth is key. Otherwise, information and applications will have to be designed for too many different bandwidths and devices,' Buckman cautions. Asia is way ahead in wireless adoption, especially Korea, and the US market has much to learn from Asian countries in this regard. 'We Americans are enamoured of ourselves way too much,' Buckman jokes.

'Mobility-enabling technologies are emerging as drivers for the importance of knowledge,' says Manuella Mueller, director of knowledge sharing at Siemens Medical Solutions in Germany. Siemen's KM system, based on Livelink, deploys a knowledge mobilisation function called 'Med2Go' on Compaq's iPaq, which is used by hundreds of Siemens employees worldwide. As an incentive for employees who perform well in the company's 'share and succeed' KM initiative, the KM programme officers award prizes like the Siemens SL45 dual-band mobile phone with integrated MP3 player.

'Wireless will certainly bring great innovation to organisations, once stumbling blocks like inadequate standardisation are resolved,' says Paul Hearn, project officer at the European Commission's Information Society Technologies Programme.

It is important, however, for mobile content and applications to be workflow-oriented, and aligned with core business processes in order to take electronic networking and collaboration to a higher level of efficiency. Current challenges for KM practitioners include devising effective metrics to measure RoI (return on investment) in wireless KM solutions.

'The knowledge mobilisation opportunity using wireless technologies is so huge that no company can afford not to grab it fast and hard,' say Peter Keen and Ron Mackintosh (2001). This applies to external competitive/regulatory intelligence, structured internal knowledge, and especially informal internal knowledge.

For instance, Sun's field engineers are equipped with handhelds so that they have the collective knowledge of the company at easy disposal. Hotel giant Carlson's managers use iPaq handhelds with up-to-date information on room status and yield management. Boeing's engineers use laptops and WLANs to better access complex multimedia documents on the move. Bell Canada technicians use wearable computers and miniature cameras (Xybernaut Mobile Assistants) so that they can instantly tap office expertise even when they are on top of a pole.

Schlumberger reported first-year savings of $75 million through its KM initiative called InTouch, which improved operational efficiency by connecting technology centres and field workers via handheld devices and wireless communication. As a result, technical query resolution time fell by 95 per cent and engineering modifications update time was reduced by 75 per cent.

'Mobile accessible information puts knowledge to work right at the demand points. Mobile technologies enhance communication, information and collaboration, the three cornerstones of knowledge building and usage,' say Keen and Mackintosh.

Portable tools like digital cameras, pocket PCs and multimedia cellphones are transforming KM practices in the news industry, says Stephen Quinn (2002). Devices like Xybernaut wearable computer gear allow reporters to plug into workflows at all times. Savvy journalists and editors are structuring content in such a way that it can be re-purposed easily for print, broadcast, Web, Intranet, PDA and SMS delivery.

Case Study 1: Origin Exterminators

A rather innovative use of wireless technologies to conduct workforce research and leverage the learnings instantly comes from an insect and rodent control company called Origin Exterminators, headquartered in Singapore.

'Eighty per cent of our workforce is out in the field. Wireless technologies help us gather timely data which we can then harvest for information nuggets,' says Carl Baptista, head of R&D at Origin Exterminators.

'Wireless also helps us act on this knowledge immediately by contacting our field workforce in realtime via SMS alerts—for instance, about unusually large rodent activity,' says Baptista. The company also uses wireless networks to schedule dispatch times and route planning in an optimised manner.

Automatic wireless sensors and handheld devices operated by field technicians help the company sharpen its research on key issues: Which rat baits are working best? Is the population of roaches increasing? What traps are working best this week? What patterns can be detected in different seasons? How quickly can new extermination services be offered? Which chemicals are the most effective?

'It almost puts a new twist to the acronym CMS (content management system)—cockroach management system,' jokes Baptista.

Origin Exterminator's field force uses PDAs with applications built for the Symbol OS on Palm handhelds. 'Our rodent traps have bar coded tags, and our back-end solution leverages GPS technology to coordinate communication with our field force,' says Baptista. The company also deploys password-protected Web gateways for customers to access their data in realtime.

Case Study 2: Sun Microsystems Philippines

Digital KM platforms—including wireless media—have also shown their transformative power in Sun Microsystems Philippines (SMP), according to Margarita Alunan Abarquez, knowledge manager for business consulting services at SMP (Rao, 2003).

SMP, a joint venture between Sun Microsystems and erstwhile distributor Philippine Systems Products, was formed in 1999. Portal, content, mail and messaging solutions based on Sun technology were used to launch the SunPhil Corporate Portal and its KM system. Features and benefits of the KM practice include voting and feedback which allows users to assign importance rating to documents, profiling and filtered search, and collaborative authoring.

The time taken to prepare proposals and project documentation has been reduced tremendously, and innovative approaches have been explored to harness information mobilisation and real-time expert contact via PDAs and SMS (the Philippines, after all, is the world's SMS capital in terms of volume of mobile messaging per subscriber). SunPhil is even taking the KM message to the national level through its active support of the Knowledge Management Association of the Philippines (KMAP).

In pre-merger days, the company had relied heavily on traditional 'stand-alone' records management and distribution techniques. With users running on different platforms, access to information of different formats was also becoming increasingly difficult, and quick replies to queries, especially from field staff, just could not be provided.

The goal of the KM system was to enable the retention of valuable corporate documents and intellectual capital (best practices and methodology) and to facilitate retrieval via Web and wireless platforms. The completed system includes an electronic whiteboard called SunForum. SunPhil's Business Consulting Services' Java experts also developed J2EE-compliant software applications for KM called SunScope and Mercury.

SunScope is a software package that integrates KM with mobile devices and the Internet—thus making knowledge available to anyone from anywhere using any device. With the use of a web browser, SunScope allows users to access their company's knowledge base, whether they are within the company's premises, at home or out in the field. For example, corporate information published as FAQs (Frequently Asked Questions) may be searched using keywords.

When users are off-line, SunScope allows users to access the knowledge base anytime, anywhere using their mobile device. With SunScope, users have a greater degree of mobility and productivity by allowing them to carry a relevant subset of their company's knowledge base in their mobile device such as a Personal Digital Assistant (PDA).

Since mobile devices have limited storage capacity and data access speed, only significant information is stored. SunScope allows users to synchronise the information from their device with the company's knowledge base so they will always have up-to-date information, says Abarquez.

Mercury Messaging System is a software application package that offers seamless integration with existing business applications via SMS to send notification, alerts or messages between the user and the enterprise. From a single web application, enterprises may send SMS messages to clients, business partners, or staff.

'In the Information Age, how you manage information within your organisation and with partners and suppliers remains the biggest challenge to knowledge management,' says Cynthia R. Mamon, SunPhil President and Managing Director. Access to knowledge assets via mobile devices is thus an important component of SunPhil's KM practice.

Use of KM by Wireless Players

The other point of intersection between the worlds of KM and wireless is the use of KM practices by wireless players: operators, equipment manufacturers and software service providers. KM practices typically have three kinds of impact: improving efficiency (e.g., cutting costs, responding faster, performing tasks with fewer steps, using best practices), effectiveness (e.g., being able to do things better, like cross-selling and up-selling customised mobile services), and innovation (e.g., being able to do completely new things, like devise new business models).

Wireless Operators

The runaway success of Japanese operator NTT DoCoMo's i-Mode wireless Internet service has become the focus of a number of benchmarking studies and business model innovation books (see Part III of this book for reviews of three of these important publications). DoCoMo's business model regarding technology platform, content alliances, customer centricity and management styles are important sources of learning for organisations around the world, especially in the wireless sector.

DoCoMo's KM practice involves an intricate mix of inter-group, intra-organisational and cross-industry knowledge sharing environments. Japanese KM practitioners have evolved elaborate theories of knowledge nurturing in environments called *ba*, a combination of physical and virtual contexts along with agents and processes for knowledge emergence and alliancing (Takeuchi et al, 2004). The term *ba* refers to a space-time nexus, the physical and/or mental space shared by co-workers, whose nature defines the scale and scope of knowledge creation through its various phases like socialisation (originating *ba*), externalisation (dialoguing *ba*), combination (systematising *ba*) and internalisation (exercising *ba*).

DoCoMo was able to evolve entirely new knowledge models for its path-breaking wireless Internet service via a combination of a number of factors: bringing together employees with entirely different knowledge backgrounds, creating conducive knowledge emergent environments, shielding this environment from external conflicts, and opening the doors to external knowledge partners.

DoCoMo had a diverse management group: Keiichi Enoki (former branch manager, with an outspoken and non-conservative nature), Mari Matsunaga (editor-in-chief of a classified ad magazine for women), and Takeshi Natsuno (an Internet entrepreneur). Each brought to DoCoMo expertise in different areas: network externalities (Natsuno) and consumer focus (Matsunaga). Enoki helped shield or cocoon this group from the strong bureaucratic and conservative culture of the parent organisation, NTT, and played the role of knowledge activist by creating the appropriate tipping point. Groundbreaking concepts of the entirely new wireless Internet service i-Mode grew in this energising environment. This environment was also open to knowledge co-sensing and co-creation with value chain partners like content providers and banks, rather than conducting all the knowledge creation internally. These knowledge models enabled the i-Mode success story to spread its wings beyond the borders of Japan to other parts of Asia and to Europe and the US as well.

The discipline of KM also has a cousin called business intelligence. Business intelligence (focused on structured data) and knowledge management (focused on unstructured data as well as human teams and communities) can actually work in tandem in a very effective manner (Rao, Madanmohan, 2003).

Business intelligence (BI) addresses the processes, technologies and strategies for cleaning up transactional and operational data, analysing patterns in this data, discovering and extracting performance-oriented nuggets, reporting this information to a wide range of managers, and taking appropriate business action. Knowledge management (KM) addresses the cultures, tools and capacities for creating and harnessing knowledge stocks and flows in employee communities and their often 'unstructured' communication (e.g., emails) and documentation (e.g., presentations). KM is more of a socio-technical process than BI.

Both fields actually have a lot to share with and complement one another—especially in companies like wireless operators, who have high call-data flow and pace of innovation—and at higher strategy levels they can actually be planned and rolled out in tandem. In terms of implementation, there are parallels between the KM roadmap (e.g., KM audit, KM capacity building, KM task force, knowledge validation processes) and BI roadmap (e.g., business requirements discovery, dedicated BI team, tech/business metrics). Both call for a significant culture change in an organisation.

Domain experts and veterans in local market trends have helped companies like India's leading cellular operator Airtel exploit local consumer behaviour patterns along with the extended BI expertise of its overseas partner SingTel in Singapore, according to Rajeev Batra, enterprise architecture manager at Bharti Tele-Ventures, which manages the Airtel brand.

Singapore's wireless operator StarHub also uses tools like datawarehousing and mining to evolve across the stages of the BI path: analytical reporting (identifying 'what' happened), exploration (using behavioural segmentation to find out 'why' it happened) and prediction (forecasting 'what will' happen regarding customer behaviour).

BI tools from vendors like KnowledgeDynamics play an important role in StarHub's pre-campaign analysis, cross-selling campaigns, re-contracting activities, ARPU analysis, service quality metering, determining customer 'stickiness', direct mailing promotions, predictive churn management and cross-product correlations, according to Gilfred Chew, operations manager at StarHub.

With operations in mobile, cable TV and Internet services, BI plays an important role at StarHub—especially after M&A activity—in consolidating business requirements, datawarehouses and source systems for consumer and enterprise markets. Data sources for mobile usage include a myriad of contracts, price plans, billing transactions, inter-carrier roaming records, dealer channels, payments, pre-paid SIM cards, post-paid calling plans, voice and data services.

Wireless Equipment and Service Providers

Vendors like Hummingbird offer a range of compatible KM and business intelligence platforms with features like taxonomies, discovery, collaboration, data integration and unified views. Its tools power cellphone manufacturer LG Electronics' knowledge portal called LG-Net which has helped increase knowledge sharing particularly in departments like customer service and sales. The Korean electronics giant now has over 340 active communities of practice, and 63 per cent of its employees report satisfaction with access to valuable and frequently updated knowledge.

South Korean cellphone manufacturing giant Samsung leverages a variety of techniques like 'knowledge catch-up' mode (prevalent in many emerging economies), knowledge pioneering, knowledge exposure, co-opetition, absorptive capacity, migratory knowledge (from the Korean diaspora), M&As, and pro-actively creating internal crises to intensify the work pace and accelerate the process of knowledge conversion, according to Linsu Kim, author of *Imitation to*

Innovation: *The Dynamics of Korea's Technological Learning*. Other cultural factors—such as the extremely hard-working habits of the people of South Korea, its cramped and cold environment, memories of Japanese occupation, and a determination to out-compete Japan—may not be easy to duplicate in other countries, according to Kim (Nonaka and Nishigushi, 2001).

Mobile communications is a heavily innovation-based industry, and KM forms a key plank of R&D in companies like Ericsson Research Canada. Developing new products and services and moving these effectively and efficiently to market are main functions of the business operation. In the 1990s, the Montreal-based organisation—rated among top 10 companies in Canada with respect to R&D investment—grew rapidly and took on new product and technology mandates, recalls Anders Hemre, CKO (chief knowledge officer) of Ericsson Canada (Rao, 2004).

With many new products and technologies being developed during a relatively short time period (e.g., open systems, 3G wireless, mobile Internet), it was becoming increasingly obvious that successful market introduction of new products also involves effective transfer of knowledge, explains Hemre.

In 1999, two KM activities were conducted: an organisation and culture study and a technology assessment. A KM advisory board was formed in the summer of 1999 to oversee and guide the effort involved in moving the initiative forward. Members included the CKO, CIO, CTO, HR director and systems research director. A KM support team was also assembled with representatives from HR and IT to directly assist the CKO.

Early in 2000, the KM architecture was revamped with Orbital Software (now Sopheon) for the installation of their Organik product. In the summer of 2000, a formal initiative to harness communities of practice was launched, consisting of forums of experts in different areas of wireless technology. The first community of practice stalled shortly after launch in 2000 after the community leader relocated, recalls Hemre. This clearly demonstrated the crucial role of the community leader. It was decided to try shared community leadership with two individuals involved and later to use a small core group to plan community programmes and prepare meetings, says Hemre.

Early examples of value added included identification and use of a database for improved quality of business cases, identification of new business opportunities, design of a new security feature for a network product and finding a third-party technology provider for a new product offering. Other and softer benefits such as improved information sharing and stronger influence on methodology development were also stated, says Hemre.

Mobile device and services player Siemens has been rated as one of the top 10 KM-driven companies worldwide according to an international benchmarking exercise called MAKE (most admired knowledge enterprise), thanks to its comprehensive efforts at fostering, promoting and optimising knowledge utilisation.

With a diverse group of companies and almost half a million employees globally, Siemens is one of the world's oldest and most successful corporations which successfully adapted to the chaotic world of the Information Age to re-structure itself around its most valuable assets: its knowledge base and people.

In addition to KM tools, Siemens has focused on a culture of sharing, synergy, and customer focus, especially in markets and fast-moving technology areas like wireless where the customer needs are more for total business solutions and sector intelligence than mere technology components.

Its KM roadmap covers four phases: initiate, mobilise, institutionalise, and innovate (Davenport and Probst 2002). The company has developed various knowledge content domains such as best practices, customer knowledge, competitive intelligence, product knowledge, and financial knowledge.

There can be numerous barriers to sharing knowledge in a company: personal (lack of time or confidence), collective (in-house competition), structural (poor IT infrastructure), or political (lack of openness). Siemens' 'top' best practice sharing initiative try to overcome these by connecting people, incentivisation, designing a topic structure for relevant experience, providing content support for editing and structuring of experiences, and, finally, via a cascaded communication strategy via divisional workshops, posters, postcards, flyers and even matchboxes with the KM Intranet URL.

Transforming from a product seller (box mover!) to a solutions provider, Siemens Information and Communication Networks (ICN) devised a business development KM practice called ShareNet in 1999 to help share project knowledge across technologies and markets in different stages of maturity. Sales staff now find themselves playing the role of strategy-management consultants who have to be able to interpret trends and design new opportunities together with the customer.

Knowledge areas covered include financing, planning, engineering and operation. This helps sales staff devise customised telecom solutions using existing service packages, business plans and profitability paths. ShareNet helps tap and share local innovation in different parts of the world via project debriefings, manuals, codified databases, structured questionnaires, chat rooms, and hot lines.

Its success is due to leadership support (from the ShareNet Committee which includes local and global representatives), organisational support (global editor, regional contributors, training bootcamps), motivation system (via ShareNet 'shares' for contributions which can be exchanged for equipment or conference fees), organisational culture (promoting sharing via messages like 'Unlike in school, copying is not only allowed—it is required'), and quantifiable benefits (cost saving by re-use of tenders, increased revenues by competing faster, and alignment with customer needs by spotting worldwide trends).

Another innovative KM practice comes from Siemens ICN VD (German sales unit), called Knowledge Networking Service. The objective of knowledge networking at ICN VD is to 'create a living network of knowledge amongst all employees'. Key requisites for this are a mix of high tech and high touch networking, encouraging voluntary participation, and creating a mix of interdisciplinary backgrounds.

An editorial team helps manage a 'Knowledge and More' personal account statement for employees. Employees receive points for submitting business tips used by call centres and other service staff; these points can be tallied and converted into prizes. All staff are trained on how to use the KM Intranet via an 'Intranet Driving Licence.' They are also sensitised to use it regularly.

Courses on KM (like Knowledge Master) are offered as a joint partnership between Siemens Qualification and Training (SQT) and University of Munich, blending live case studies with academic trainers and business tutors. The Knowledge Web learning portal includes abstracts of relevant literature and a Web board for discussion.

Challenges faced by Siemens on the KM front, according to Davenport and Probst, include balancing energies, resources and rewards for local versus global KM initiatives on a daily basis, managing the knowledge-sharing tension between different business units, and nourishing KM during hard economic times.

Wireless Software Companies

KM is also being used successfully by companies like Hughes Software Systems (HSS), a major player in mobile network solutions. HSS spends about 14 per cent of its revenue in technical R&D, for which KM is a key driver; KM also extends to customer relationships, e-learning, and procurement (Rao, 2003). HSS has grown to 1,600 employees worldwide with offices in the US, UK, Germany, Finland and India. Through its two development centres (New Delhi and Bangalore) in India, HSS offers full-spectrum outsourcing services to its clients ranging from new product design and joint product development to sustenance engineering and testing services. These services are carried out in a variety of domains like wireless networks, network management, switching systems, voice over packet, convergent networks, broadband networks and digital signal processing. HSS also offers stand-alone software products for voice over packet and mobile data.

In addition to an evolving and closely-monitored KM architecture, HSS taps into external sources of knowledge such as industry consortia and collaborative research agreements with universities around the world. Top management support, quality certification groups, KM workshops and a day-long knowledge-sharing event help foster a culture of knowledge sharing.

In January 2000, HSS was assessed at SEI-CMM Level 4 for both of its development centres and all lines of business. HSS has been an ISO 9001 company since 1996 and has recently got certified as an ISO 9001: 2000 company. HSS protects its customers' intellectual property rights and has been certified as meeting BS7799 standard related to information security. Its customers include Alcatel, Cisco, Johnson Control, Lucent, Nokia, NEC, Motorola, Shanghai Bell, SK Telecom, Tektronics and Toshiba.

The company formally identifies process improvement opportunities, develops and deploys these processes, trains and shares knowledge among employees. Ranging from repositories of best practices to e-discussion on post-mortem reports, emphasis has been placed on creating and retaining knowledge through systemic processes rather than dependence on individuals.

HSS has organised a dedicated knowledge sharing day ('HSS tech day') to share knowledge among the entire company, according to Aloknath De (engineering director) and R. Sathyavageeswaran, senior manager, quality. HSS has a project repository, which contains information on all past and current projects at HSS, including the metrics from these projects. These are then used by other project managers and quality groups to quantify HSS' capability in terms of various parameters such as ability to meet planned dates for delivery, quality, and the like. This acts as a valuable source of information for better planning of future projects. HSS also has a Web-based application called Skill Inventory to search for any specific set of skills across the organisation; this is very useful while identifying the right set of people for a project.

KM processes are specifically designed to help finding existing knowledge (via internal forums and external consultants), creation of knowledge (via publication of technical papers), knowledge aggregation (via the library and best practice meets), and re-structuring of knowledge (via an Intranet and an active intellectual property management programme).

A repository of project post-mortem reports of all past projects is available from a central place to share the learning. A formal training programme (Oasis) is in place where buddies and mentors help new engineers do their first project that adheres to company processes. More than half of the training programmes (be it technical or soft skill development) are conducted by the internal faculty.

Every project plan contains a mandatory section on reuse, where the reusable components are identified along with the effect of reuse on the overall planned effort. Technical paper writing is encouraged to capture the managerial and technical challenges and innovations in every major engagement.

'Accuracy, integrity and security of data are important attributes of a good knowledge management system,' says De. KM has touched almost all functions and activities at HSS: CRM, procurement, e-learning and participation in industry initiatives. The company is now able to do more sophisticated forecasting and enhance post-sales support tremendously. The knowledge base on customer queries/problems also includes satisfaction surveys. E-learning is used to enable self-placed learning among globally dispersed branches, especially for emerging wireless technologies.

At an industry and academic level, HSS has been participating in many of the technical standard bodies such as Session Initiation Protocol (SIP) Forum, International Multimedia Telecommunication Consortium (IMTC), International Softswitch Consortium (ISC) in the area of packet-based, next-generation network. In the wireless space, HSS has been a member of European Telecommunications Standards Institute (ETSI), Universal Wireless Communications Consortium (UWCC) and Wireless World Research Forum (WWRF). HSS works with universities using different models of knowledge sharing for mutual benefit.

Lessons learned by HSS in their KM practice include the need to build the KM architecture in a modular and incremental manner, paying careful attention to change management, clearly spelling out accountability of knowledge assets, and measuring the KM system impacts.

Quality of knowledge assets is an important priority, cautions De, as well as governance and keeping knowledge initiatives in line with business goals. Upcoming challenges include promoting knowledge sharing across multiple languages like French, German, and Japanese.

Conclusion

The wireless ecosystem can deliver significant benefits to KM infrastructure and flows, targeting explicit knowledge assets, direct employee communications and creative workplace design. Wireless players themselves can use KM methodology to amplify their own efficiency, effectiveness and innovation.

Wireless technologies—handheld devices, cellular networks and Wi-Fi—can help infuse new blood into the field of knowledge management by providing employees with anytime, anywhere, any device access to explicit knowledge assets and to experts in different locations.

'How to outfit the frontline and manage applications across multiple client platforms will be a key strategy and operational focus area for the coming years,' says Ian Bertram, VP hardware platforms for Gartner Asia-Pacific (Rao, 2002).

Handheld devices and wireless communication, if properly harnessed, can help improve business processes, create new ones, facilitate knowledge mobilisation and even spur new business models. Challenges will arise, however, in managing standards for organisational information representation and flow across multiple devices in an environment of constant technology and network evolution.

As for mobile operators, KM and BI can indeed help manage structured and unstructured knowledge as well as bring out the best in all their teams and partners, but significant culture change and process investment will be required. The innovative case studies emerging from Asia set the trend for their counterparts in the region and elsewhere.

References

Bertram, Ian (2002). 'Gartner Summit: IT, KM and wireless (Mumbai, September) http://www.destinationkm.com/articles/default.asp?ArticleID=1007

Davenport, Thomas & Gilbert Prost (2002). *Knowledge Management Case Book: Siemens Best Practices.* John Wiley/Publicus Corporate Publishing, New York.

Holm, Jeanne (2002). 'Wireless Knowledge Management Technologies', lecture delivered at KM Asia, Singapore; http://www.destinationkm.com/articles/default.asp?ArticleID=973

Keen, Peter & Ron Mackintosh (2001). *The Freedom Economy: Gaining the m-commerce Edge in the Era of the Wireless Internet* McGraw-Hill, New York.

Nonaka, Ikujiro & Toshihiro Nishigushi (2001). *Knowledge Emergence: Social, Technical, and Evolutionary Dimensions of Knowledge Creation.* Oxford University Press, Oxford.

Quinn, Stephen (2002). *Knowledge Management in the Digital Newsroom*, Focal Press, Oxford.

Rao, Madanmohan (2003). *Leading with Knowledge: KM Practices in the Global InfoTech Sector*, Tata McGraw-Hill, New Delhi.

Rao, Madanmohan (2003). Knowledge Management and Business Intelligence http://www.destinationkm.com/articles/default.asp?ArticleID=1096

Rao, Madanmohan (2004). *Knowledge Management Tools: Practitioners Evaluate KM Tools and Techniques*, Butterworth-Heinemann, Boston.

Takeuchi, Hirotaka & Ikujiro Nonaka, (2004). *Hitotsubashi on Knowledge Management*, John Wiley and Sons (Asia), Singapore.

10. The Flexible Wireless Classroom

Impact on Teaching

June Gassin and Som Naidu

Editors' Note: The faculty of arts at the University of Melbourne has implemented three flexible wireless classrooms for teaching and learning in the Humanities. These classrooms are designed to support innovative teaching practices and enhance natural human interaction. This paper reports the results of a survey of the impact on teaching in these classrooms. Results indicate that the opportunities provided by these flexible wireless classrooms substantially influenced particular teaching practices. These included strategies for teaching, engagement of students with the subject matter, socialisation and interaction in class and providing feedback to learners. Attributes of the classrooms that had the greatest impact on teaching were the ability to configure space, move about in the classroom and use a variety of technologies.

Introduction

Wireless technology is rapidly being introduced throughout the world for education, business, and commerce. In the United States, universities such as Carnegie Mellon and Dartmouth have already implemented ubiquitous wireless computing throughout their campuses (http://www.cmu.edu/computing/wireless/; http://wwwdartmouth.edu/comp/wireless/using/index.html). Many other institutions around the world are adopting wireless technologies to varying degrees (Berger, 2001). Wireless technology, however, like any other technology, is not in itself a panacea (Clark, 1994). Reflection on how it can be used to support and encourage the teaching/learning process through the enhancement of interaction, socialisation and engagement is required (Bleed, 2001).

The faculty of arts at the University of Melbourne has implemented three flexible wireless classrooms for teaching and learning in the humanities. This paper reports the results of a survey of staff perceptions about their teaching practices in these innovative classrooms. The purpose of this survey has been to ascertain the extent to which the specific features of these flexible wireless environments may have made an impact on common teaching practices. This survey represents the first step of an ongoing investigation into teaching and learning in the wireless classroom. Follow-up interviews to the present survey will focus on the nature and description of actual teaching practice and how it is evolving over time. This topic will be addressed in a subsequent communication.

The Flexible Wireless Classroom

The faculty of arts pioneered the concept of the flexible wireless classroom within the university. The first such classroom, the VITAL (versatile, integrated, teaching and learning) environment, was implemented in 1999 and inaugurated for teaching in March 2000. A complete technical description of the facility and a virtual tour can be found at http://www.artsit.unimelb.edu.au/facilities/vital/

The development of the VITAL environment took place in the context of a large-scale integration of IT and multimedia into the arts curriculum. The project responded to a range of perceived needs: the need for a high quality electronic space for small group teaching in the humanities; the need to support and encourage innovative teaching practices involving technology and collaborative learning; and the need to conduct research into teaching and learning with digital technologies in the classroom setting.

The success of the VITAL environment gave rise to the development and implementation of two further wireless classrooms at Melbourne University in its Sidney Myer Asia Centre, which houses the Melbourne Institute of Asian Languages and Societies and AsiaLink. These new wireless classrooms are dedicated to language learning and teaching and have been especially designed and equipped for that purpose. They are unique in Australia and opened for teaching in July 2002. They comprise two 25-place wireless classrooms joined by a central multimedia control room.

Rationale

The rationale for developing the flexible wireless classrooms was based on both pedagogical and theoretical considerations. It is widely acknowledged that the quality of human interactions in the traditional computer laboratory is unsatisfactory. Everyone is familiar with the difficulty of establishing eye contact and encouraging student interactions when monitors, wires and machine noise combine to prevent natural human interactions. The physical set-up of the traditional lab with computers in rows and each student seated behind his/her computer is generally unsuited to collaborative or team work. Neither students nor teachers can move about freely. The traditional lab also makes a strong statement about what to expect from the teaching/learning situation: technology is clearly the focus of attention.

Theoretical reasons for the innovation mirrored changes in teaching philosophy and practice among academic staff. The extensive integration of IT and multimedia into the arts curriculum over the past five years has resulted in new modes of teaching, many of them based on constructivist learning theories and collaborative learning approaches. These innovative teaching approaches (Debski, 2003; Wallace, Jagose & Gunn, 2003) require new types of teaching spaces and generate new types of classroom dynamics. Socialisation, the very cornerstone of constructivist philosophy, remains a crucial aspect of both face-to-face and online learning.

Concept

The basic concept for the wireless classroom was to create an electronic space that was as flexible and versatile as possible. We were familiar with the Flexible Class-lab at Stanford University (http://acomp.stanford.edu/roomsflexLab.html) and used it as a model in the development of the VITAL environment.

We wished, however, to create an environment particularly conducive to natural human interactions and in which technology became almost invisible. The primacy of quality human interaction became an important aspect of the original concept. We believed that technology needed to be integrated into the teaching space in an unobtrusive way. It needed to support rather than direct learning; it needed to enhance and enrich the teaching/learning experience but not dominate it. At the same time, we wished to create a teaching space that was easily re-configurable and could be shaped to meet the changing needs of the learning situation. We conceived of the environment in an organic way, allowing movement of both students and teachers as the flow of activities required.

Design

The principal design features of the wireless classrooms are flexibility, mobility and wireless connectivity. Each classroom is equipped with lightweight modular furniture that can be easily re-configured. Tables are trapezoidal in shape to maximise flexibility. The VITAL environment has a movable data projection system installed on a trolley so that students or teachers may project their work onto any wall of the room. There is no fixed place for the teacher to sit or stand, thus creating a space that is less teacher-centred. An entire wall of whiteboard in the VITAL is a design feature. Fifteen Apple iBook laptop computers and the base stations for the Airport wireless technology are stored in a cupboard. The iBooks are easily accessed when required.

The classrooms in the Sidney Myer Asia Centre are larger and are equipped with standard overhead projection facilities, VHS, VCRs, audiotape and CD-ROM units. There are 50 laptop computers in all, 25 each of Apple iBooks and Toshiba portables. They are stored on trolleys that are wheeled into the classroom. Teachers, therefore, have the choice of using either Apple Macintosh or PC machines. The choice of platform is an important consideration as language teachers have different preferences in this regard. Wireless network connection is provided by Cisco Aironet technology.

Professional architects designed all the rooms. Colour schemes and lighting were chosen to create an impression of simplicity and unity. The classrooms also have cameras and microphones installed in the ceilings for security and research purposes.

Teacher Perceptions

Although we have obtained quite a lot of anecdotal information regarding the success of teaching in the wireless classrooms, we wished to begin the process of gathering data for a more formal evaluation. We were interested in finding out what effects, if any, the specific attributes of the flexible wireless spaces might have had on teaching practice. We decided to begin with a survey of teacher perceptions.

Method

A brief questionnaire was used to carry out this survey. General information sought from respondents comprised the following:

- Subjects they taught in the wireless classrooms;
- Length of experience with teaching in the wireless classrooms;
- Any experience with teaching in a traditional computer laboratory;
- If yes, how would they describe their teaching experience in the wireless classroom in comparison with teaching in a traditional computer laboratory; and
- How they would rate their overall teaching experience in the wireless classrooms.

The remainder of the questions in the instrument focused on the impact of the flexibility afforded by the wireless classrooms on specific aspects of their teaching. Flexibility afforded by the wireless classroom was specified in the following ways:

- Ability to reconfigure the physical space.
- Ability to move around in the classroom.
- Ability to use a variety of teaching technologies.
- Ability to connect wirelessly to the Internet.
- Ability to move laptops around in the classroom.

We wanted to ascertain the extent to which the foregoing affordances of the flexible wireless classrooms influenced the following attributes of teaching:

- Selection of *teaching strategies*.
- Planning for the *engagement of students* with the subject matter.
- Organisation of *socialisation and interaction activities* in class.
- Providing *feedback* to learners.

These attributes were drawn from an analysis of the fundamental principles of teaching outlined by Naidu (2003). They seemed particularly relevant to the current project on wireless classrooms.

Data Gathering

The questionnaire was sent out in a paper-based format to 20 faculty members who had taught in the wireless classrooms. We asked them for their names, telephone numbers and e-mail address in order to be able contact them for further information if necessary. Fourteen out of 20 questionnaires were returned. That is a 70 per cent response rate.

Results

Subjects taught in the wireless classrooms included:

Cyber Society, Critical Thinking, Latin Paleography, Exeter Anthology, Introduction to Computer-Assisted Language Learning (CALL), Japanese (various levels), Multimedia Japanese, Introduction to Translation (Japanese), Advanced Grammar (Japanese), Reading Contemporary Texts (Japanese), Advanced ESL-IT, Project-based Communication Skills, Language and Media, CALL Software Evaluation, Online Language Learning Media, CALL Software Design and Implementation, Structural Editing and Editorial English, Business and Professional Communication, and Exploring Style and Meaning in Language (ESL).

Five of the 14 respondents had taught in the wireless classroom for one semester while nine of them had taught in it for two or more semesters (three had taught for two semesters, five of them had taught for three semesters, and one for more than three semesters). Eight of the 14 had taught the same subject in a traditional computer laboratory, while the other six had not. Seven with experience in both settings (wireless classroom and traditional computer labs) described their experience teaching in the wireless classrooms as more effective while one respondent was unsure. Moreover, all 14 rated their overall experience in the wireless classroom as positive (six Positive, and eight Very Positive).

On the Selection of Teaching Strategies

One of the most important aspects of the teaching process that we sought to observe was the selection of teaching strategies in the wireless classroom. We were interested in the extent to which these were influenced by the flexibility afforded by these innovative teaching spaces.

Figure 10.1 shows the extent to which specific affordances of the wireless classroom influenced teaching strategies in the classroom.

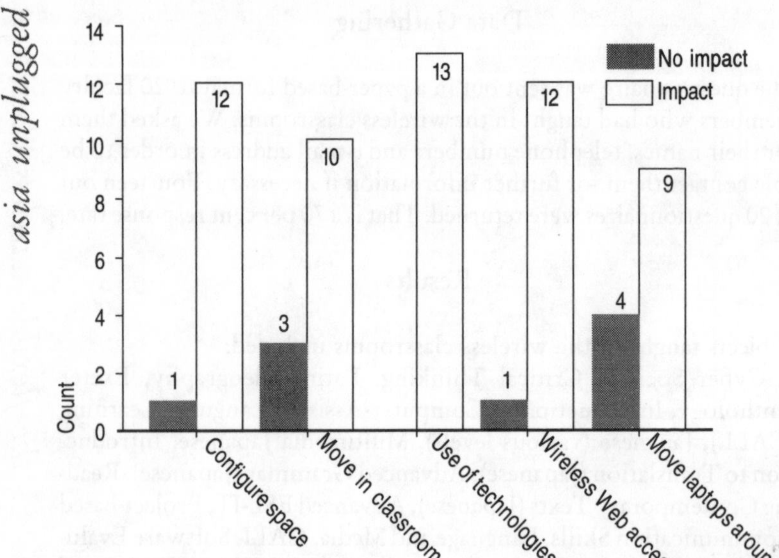

Figure 10.1: Impact on the selection of your teaching strategies

These included the ability to reconfigure the physical space, move about in the classroom, use a variety of teaching technologies, connect wirelessly to the Internet, and move laptops around in the classroom. Clearly, the majority of the teachers reported that these capabilities influenced their teaching strategies in the classroom. No impacts were reported in a few cases such as in the cases of the ability to move about in the classroom and move laptops around in the classroom.

Planning Activation (i.e., Engagement of Students) in Learning

A second fundamental principle of teaching that we were interested in observing was the activation of the learning process, which is the engagement of students with the subject matter content. We were interested in observing the extent to which this had been influenced by various opportunities afforded by the wireless classroom.

Figure 10.2 shows the extent to which various affordances of the wireless classroom influenced teachers' ability to activate learning by engaging students with their subject matter content. Clearly, the majority of the respondents felt that the various attributes of the

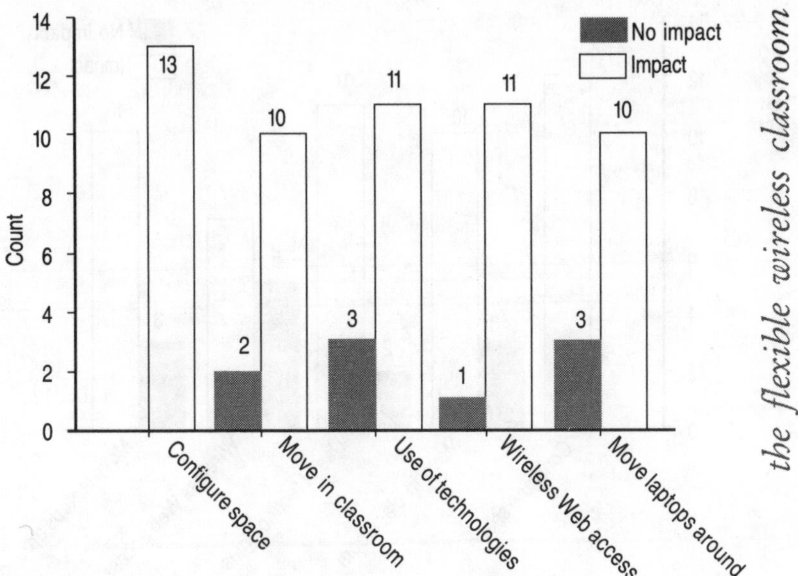

Figure 10.2: Impact on the engagement of students with subject matter

classroom had an impact on how they went about this aspect of their teaching. Note, especially, the impact of their ability to reconfigure the physical space, move about in the classroom, use a variety of technologies, and connect wirelessly to the Internet.

Organisation of Socialisation and Interaction in Class

Another fundamental principle of teaching that was the focus of our attention was the organisation of socialisation and interaction activities in class. We were interested in seeing if the use of these activities as part of one's teaching were being influenced by the capabilities afforded by the wireless learning space.

Figure 10.3 shows the extent to which specific affordances of the wireless learning space is reported to have influenced the organisation of socialisation and interaction activities in the classroom. Most noticeable among the affordances, in the extent of its impact on this aspect of teaching, is the ability to reconfigure the physical space. Except for the ability to connect wirelessly to the Internet, other critical opportunities had similar sorts of influences.

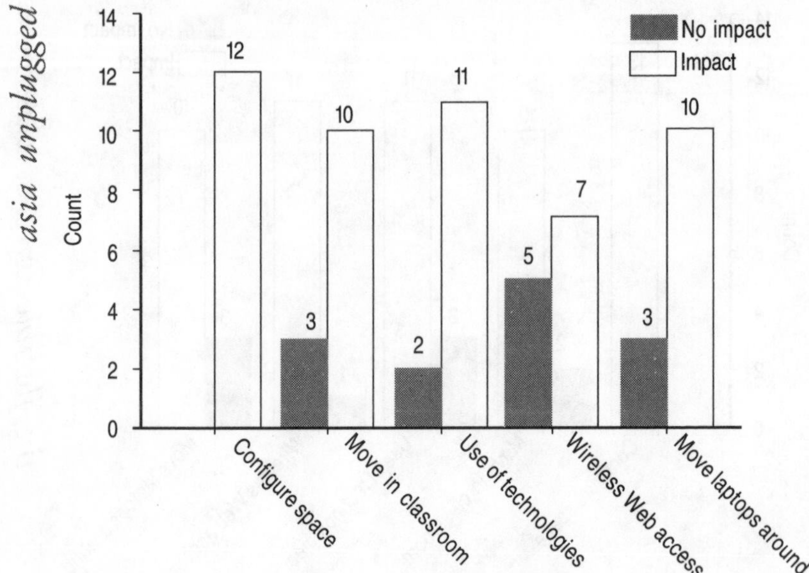

Figure 10.3: Impact on socialisation and interaction in class

Providing Feedback to Your Learners

Finally, not of any less interest to us were strategies that lecturers used to provide feedback to the learners as part of the teaching and learning process. We were keen to see if the flexibility that the wireless classroom afforded, had any impact on how the teachers went about providing feedback to their learners.

Figure 10.4 shows the results of this analysis. The two factors that seemed to have had substantial impact on this aspect of any teaching activity are the ability to reconfigure the physical space and move about in the classroom. The other factors seemed to have influenced the feedback strategies adopted by fewer numbers of the respondents. The reasons for this observation could be several, and probably have little to do with the opportunities afforded by the classroom itself.

Discussion and Conclusions

This survey did not set out to uncover any association between attributes of the flexible wireless learning space and learning achievement. Rather it sought to see if the specific opportunities afforded by the flexible wireless classroom had an impact on teaching practices.

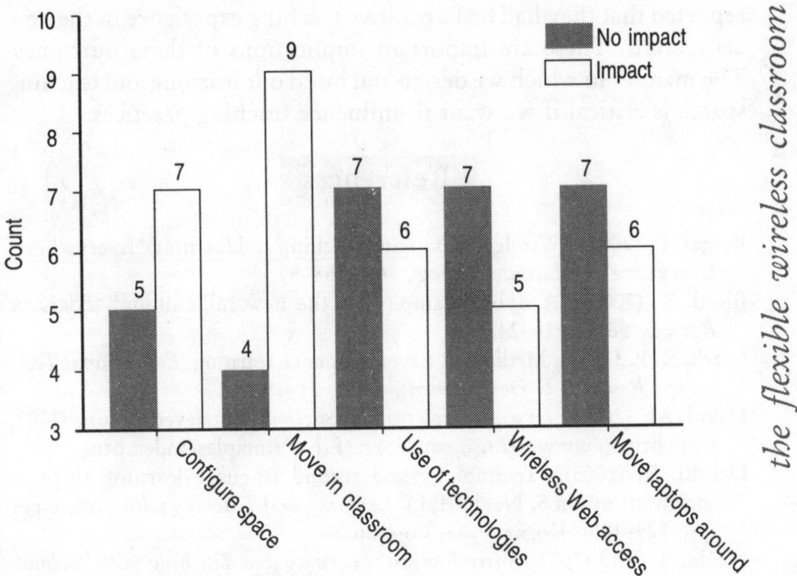

Figure 10.4: Impact on providing feedback to learners

Our results show that the particular attributes of the flexible wireless classroom have substantially influenced teachers' approaches to teaching. The attributes of the classrooms significantly affected their choice of teaching strategies, how they activated student learning and the way they organised socialisation and interaction. The attributes of the classrooms also affected, although to a much lesser extent, how teachers went about providing feedback to learners. Nine respondents felt that moving about the classroom had an impact on providing feedback.

In some cases, teachers reported no impact of the physical space on their approach to teaching. From this survey, we could not ascertain the reasons for this response. Was it because of the subject matter they taught, their teaching style or philosophy? This is something we intend to explore in the follow-up to this survey.

What we can confirm from the results of this preliminary investigation is that the flexible wireless environment has had a considerable impact on the choices teachers made about some of the fundamental processes of teaching. In addition, seven out of eight of those teachers who had taught their subjects in both a traditional lab and in the wireless classroom reported that they felt their teaching had been more effective in the flexible space. All 14 respondents reported that they had had a positive teaching experience in the new classrooms. There are important implications of these outcomes. The manner in which we design and build our learning and teaching spaces is critical if we want to influence teaching practices.

References

Berger, C. (2001). 'Wireless: Changing teaching and learning "Everywhere, Everytime"', *Educause Review, 36*(1), 58–59.

Bleed, R. (2001). 'A hybrid campus for the new millennium', *Educause Review, 36* (1), 16–24.

Clark, R. E. (1994). Media will never influence learning. *Educational Technology Research & Development, 42*(2), 21–19.

David, M. (2000). *First principles of instruction*. Retrieved 6 June 2003, from http://www.id2.usu.edu/5Star/FirstPrinciples/index.htm

Debski, R. (2003). 'Technology and second language learning through socialisation'. In S. Naidu (Ed.), *Learning and Teaching with Technology* (pp. 129–145), Kogan Page, London.

Naidu, S. (Ed.) (2003). Introduction. *Learning and Teaching with Technology: Principles and Practices* (pp. 1–9), Kogan Page: London.

Wallace, L., A. Jagose & C. Gunn, (2003). 'Virtual learning in cultural studies: Matching subject content and instructional delivery,' in S. Naidu (Ed.), *Learning and Teaching with Technology: Principles and Practices*. (pp. 97–112), Kogan Page, London.

11. Wireless in Asia

Investor/Equity Dynamics

Loudon Owen and Paul Gardner

Editors' Note: This chapter provides a sweeping overview of venture capital dynamics in Asia's wireless sector, with insightful comparisons with the US and Europe. Topics and case studies covered include industry lifecycles, management dynamics, regional/global opportunities, and alliance strategies—featuring giant established telcos and nimble startups along the entire wireless ecosystem. Though overall wireless penetration of Asia lags Europe, it is in areas of concentrated adoption that Asia is moving into the vanguard of the wireless industry, as in Korea and Japan. After disastrous forays of 3G in Europe, Asian innovators and consumers may save the technology by proving it is commercially viable. Asian companies are rivalling Scandinavia for technological leadership in the wireless world. Areas ripe for investors include cutting-edge emerging technologies in the advanced markets as well as newer markets in developing countries where tried and true mobile technology can be diffused. Expansion in the North American and European markets presents an opportunity for Asian-based players with matured and proven solutions.

Introduction

This chapter reviews how venture capital investors seek opportunities, what has happened and will happen in the wireless sector in Asia—and gives a framework for the analysis of wireless investments, or from a different perspective for those parties seeking capital.

Venture Capital: Foundations

As the major source of financing for the entrepreneurial community and their high-growth opportunities, venture capital (VC) serves as the key engine of growth for the wireless industry focusing on providing funding and expertise to early-stage companies looking to exploit new technologies and services. Venture capital is risk capital, an alternate source of financing for high-risk high-reward companies versus bank loans or other sources of debt-related financing.

The United States' National Venture Capital Association defines venture capital as 'money provided by professionals who invest alongside management in young, rapidly growing companies that have the potential to develop into significant economic contributors'. In exchange for receiving equity in a company for their financial investment, venture capitalists support their portfolio of companies by providing strategic advice, building strong management teams, and opening doors leading to product sales, business partnerships, follow-on financing and exit opportunities including public offerings, mergers and acquisitions. With this vantage point, the venture capitalist has a unique perspective on industry trends and directions of the wireless industry.

The first venture capital firm was American Research and Development—formed in 1946 by Karl Compton of the Massachusetts Institute of Technology (MIT) and Doriot, professor of Harvard Business School (HBS) and founder of the leading France and Singapore-based business school INSEAD—to commercialise MIT technologies after World War II. According to VentureWire, the US venture capital market exploded in 2000 with $106 billion of capital deployed, more than the total funding available in all the 15 years before that. With the recent softened global economy, this figure flattened in 2002 to $21 billion. As a comparison in scope and scale, there was $1.9 billion in venture capital investments disbursed into Asia-Pacific companies in 2002. Such wild swings were not only caused by macro-economic factors but are also indicative of a relatively young and maturing industry.

The venture capitalist is the general partner of a group of limited, passive, partners including institutional investors such as private and public pension funds, endowment funds, corporations, wealthy individuals and foundations. The active venture capitalist's goal is to generate returns to the limited partners and share in the success. Venture capital funds generally have a fixed lifespan ranging from eight to 13 years. They are typically compensated with an annual management fee plus a variable compensation on the fund's profits—on average 20 per cent of the fund's profits. The target for a venture firm is to average 1-2 home runs for every 10 investments with the rest providing an average return or failing. Over time, successful venture capital firms will generate a 20+ per cent annual rate of return.

> **Korean Cash-free Suburb Likely to Ring the Cash Machine**
>
> Seongnam is a bustling, middle-class Seoul suburb. But what sets it apart is the ambitious goal of its municipal officials plan to transform the town of 930,000 into the world's first digital city. No cash, no credit cards, just digital handshakes. The technology platform is relatively mundane—digital cellular phones. This is just one example of the leadership across Korea as the nation is determined to be a digital pathfinder. For example, in Korea today, more than 50 per cent of the households have broadband connections. Over 60 per cent of the population have cellular phones, and it is the stomping ground for test marketing 3G services. Seventy per cent of all share trades are already done online.

As venture capital firms invest for equity in the business at a negotiated valuation, their ability to realise a return on their investment is through a liquidity event. Venture capitalists eventually seek an exit within five to seven years of their initial investment in a company. Their liquidity options range from an initial public offering (IPO), a merger or acquisition, or a redemption of their shares through management buyouts (MBO).

Venture capitalists expect a company to demonstrate that it possesses a significant growth opportunity and that it has a sustainable competitive advantage that will lead to measurable and substantial returns. The market will not fund mediocrity. Investors exact premiums in the form of high equity requirements if a company cannot mitigate risk through a superior management team, detailed strategies, tactics and contingency plans. An exhaustive due-diligence process will focus not only on the company's technology, market opportunity and business model, but most importantly the company's management experience, ability to execute, leadership capabilities and track record.

Depending on the stage of a company's growth cycle, there are different forms of financing available (see Figure 11.1). In addition to placing the company appropriately within the growth cycle, key issues that need to be considered include the company's appetite for dilution of ownership and control, the need for value-added services from the investor beyond capital, and future financing requirements.

Venture capitalists range from generalists to specialist investors. They may invest globally (3i Asia Pacific PLC) or prefer to have

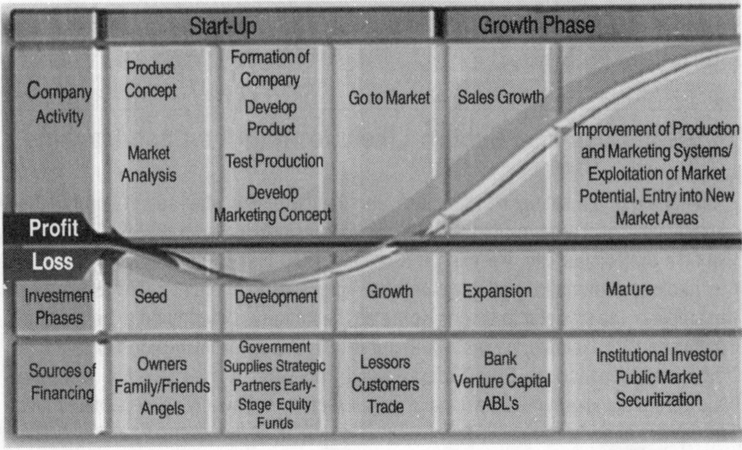

Figure 11.1: The funding lifecycle

their portfolio regionally (Shenzhen Capital Group). Not all invest at all stages of the business cycle. Some focus on startups providing small amounts of capital, generally less than $1 million, in seed financing. Others prefer to wait until the company is already generating sales and is looking for expansion capital (The Carlyle Group Asia Technology Venture Fund). The trade-off for the venture firm investing at a later stage of the growth cycle is lower risk for a lower rate of return on their investment.

The consistent denominator for all types of venture capital firms is their active involvement to guide, lead and grow their investment. Most venture firms syndicate investments between two or more firms in order to mitigate risk and bring another value-added player with complementary expertise to the table. The most common structure for a VC firm is independent with no affiliations with any other financial institution (Taiwan-based Fortune). Venture firms can also be subsidiaries of other financial institutions such as commercial banks, investment banks or insurance companies (Morgan Stanley Private Equity Asia), making investments on behalf of outside investors, the firm's clients, or the parent's capital (GE Asia Pacific Technology Fund).

Another form of venture investing is corporate venturing (Acer Technology Ventures, or groups such as Vertex Venture Holdings which was originally the VC arm of Singapore Technologies Group). These divisions use the company's own corporate capital to invest in opportunities that are synergistic with the parent company's

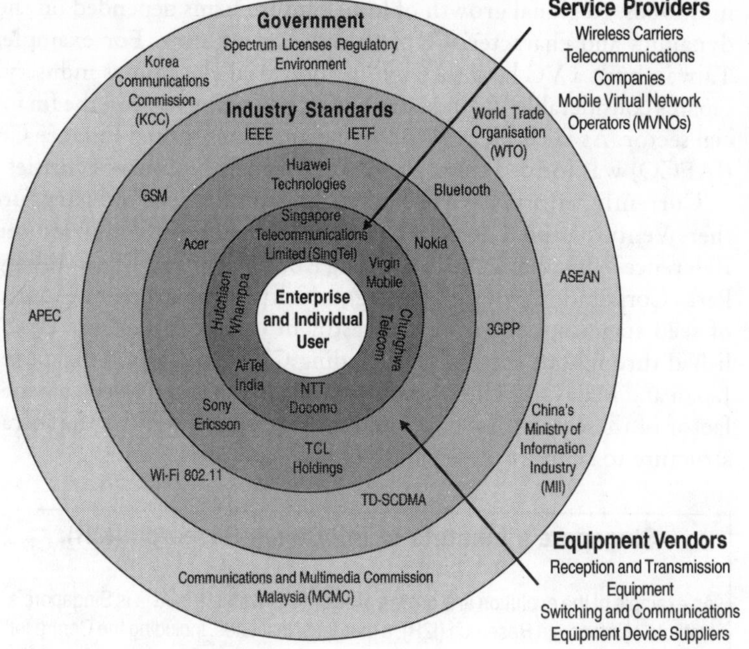

Figure 11.2: Wireless ecosystem

strategic plan, be it complementary technology to enhance their product line or reduce their operating costs (SingTel). Corporate venture capital has been of particular importance to the development of the wireless industry as it provides strategic alliances to early stage companies to navigate the complexities of the wireless supply chain (Figure 11.2).

Venture Capital in Asia

Venture capital in Asia started in the 1970s. In a region of the world where funding was often private and based upon pre-existing relationships, it was a concept imported from North America. Among the early and dynamic Asian venture capitalists were entrepreneurs such at Mr Samuel C. Hong of Hotung, and Keen Whye Lee who initially brought Rothschilds Ventures to Singapore and is now managing director of Strategic Alliance Capital. Perhaps best known is Lip Bu Tan, founder and chairman of Walden International. Other pioneers migrated from the US to Asia, such as Advent International and Hambrecht and Quist, establishing Southeast Asian bases in the mid 1980s. Regional growth of local venture firms depended on the dynamics and characteristic of their home country. For example, Taiwan built a VC base through its industrial electronics industry. Japan's considerable VC industry had its beginnings within the financial sector. As far back as 1973, the Japanese Associated Finance Co (JAFCO) was formed through brokerage giant Nomura Securities.

Currently, efforts have been made to stimulate the industry further. Venture capital associations now exist in most countries (see References). Incubators such as Hong Kong Science and Technology Parks Corporation (www.hkstp.org), which provide an ongoing source of seed stage capital for new investment opportunities, are established throughout the region including China, Taiwan, Singapore, Japan and Malaysia. The wireless industry has not only been a benefactor of these scientific parks and institutes but provides the infrastructure to support these endeavours.

Singapore's Institute of InfoComm Research (I2R)

An example of the evolution and growth of scientific institutes in Asia is Singapore's Institute of Infocomm Research (I2R), a merger of institutes, including the Centre for Wireless Communication, to capitalise on the necessary economies of scale to integrate wireless technologies within a portfolio of converging technologies. Under the care of A*Star (Agency for Science, Technology and Research) in Singapore, and with 350 researchers and more than 80 support staff, it is by far the one of largest Asian research centres.

One of the most significant moves to develop the venture capital industry in the region occurred in 1999 when the Singapore government launched the most ambitious programme in the history of a government to stimulate venture capital. The $1 billion Technopreneurship Investment Fund (TIF) was launched as part of a number of key efforts to attract the commitment of major venture capital firms both to either expand in Asia if they were already present, or to establish a presence if did not yet have operations in the region. This was part of the Technopreneurship 21 initiative to promote a viable and thriving sector of high growth technology-oriented companies in Singapore. Venture capital funds that were financed in part by this fund-of-funds represent a well diversified group of venture capital managers including North American Funds (Venrock Management, Sequoia Capital, and McLean Watson Capital), existing Asian funds (India's JumpStartUp Fund Advisors and

Fortune Financial from Taiwan), and representatives of leading venture firms from around the globe (Germany's TVM TechnoVenture, Israel's Giza Venture Capital, France's Viventures and Sweden's Argnor Wireless Ventures).

With the commitment and co-operation of the public and private sector, the Asian venture capital community has the depth, experience and diversity to serve as a catalyst for growth in the wireless industry. Exploiting a very strong regional demand for anything wireless, the ecosystem of institutes, incubators and venture capital firms promise to cultivate ideas and technology that deserve funding.

Chronology

The wireless industry has been one of the fastest growing industries in the past decade and a half. Ten years ago, all that subscribers could do was talk on their wireless phone. Today, one can check e-mail, access the web, or access a company's database application. Innovation continues to proceed at full speed. Convergence of voice and data communications will be ubiquitous. Users will be totally wired in, but wireless, from anywhere in the world.

The wireless industry took off in the late 1980s with the commercialisation of cellular phone technology. With more and more people purchasing cellular phones, the initial emphasis within the industry was on expansion and growth rather than quality, service, or technology. Most customers were business people who travelled substantially and were reimbursed by their companies for the cellular service. Price was not an issue. Cellular technology was also in the early stages of the learning curve.

When the Personal Communications Industry Association (PCIA) asked the US Federal Communications Commission (FCC) for permission to introduce digital cellular phones (otherwise known as PCS) in 1989, the cellular phone industry entered a phase of hyper growth. With an avalanche of competition, new entrants introduced new technology and business models, leading to lower prices, better quality, and better service. These qualities made wireless access more appealing to everybody. By the mid-1990s consumers began purchasing cellphones for everyday use. The market became saturated and leading equipment vendors, such as Nokia, and service providers led by NTT DoCoMo in Asia began to introduce more and more new products at a frantic pace.

Wireless devices today can use third party application software such as e-mail programmes, Internet browsers, and enterprise

Inmarsat

One interesting area of investment which has been highly sought after is telematics. Inmarsat was the world's first global mobile satellite communications operator and is still the only one to offer a mature range of modern communications services to maritime, land-mobile, aeronautical and other users.

Formed as a maritime-focused intergovernmental organisation over 20 years ago, Inmarsat has been a limited company since 1999, serving a broad range of markets. Starting with a user base of 900 ships in the early 1980s, it now supports links for phone, fax and data communications at up to 64 Kbit/s to more than 250,000 ship, vehicle, aircraft and portable terminals. That number is growing at several thousand a month.

Inmarsat Ltd is a subsidiary of the Inmarsat Ventures plc holding company. It operates a constellation of geostationary satellites designed to extend phone, fax and data communications all over the world. The constellation comprises five third-generation satellites backed up by four earlier spacecraft.

The satellites are controlled from Inmarsat's headquarters in London, which is also home to Inmarsat Ventures as well as the small Intergovernmental Treaty Organisation (IGO) created to supervise the company's public-service duties for the maritime community (global maritime distress and safety system) and aviation (air traffic control communications). Inmarsat has regional offices in Dubai, Singapore and India.

Today's Inmarsat system is used by independent service providers to offer a range of voice and multimedia communications. Users include ship owners and managers, journalists and broadcasters, health and disaster-relief workers, land transport fleet operators, airlines, airline passengers and air traffic controllers, government workers, national emergency and civil defence agencies, and peacekeeping forces.

The Inmarsat business strategy is to pursue a range of new opportunities at the convergence of information technology, telecoms and mobility while continuing to serve traditional maritime, aeronautical, land-mobile and remote-area markets.

software. They can send or receive e-mail, browse the internet, trade stocks, play video games, check your bank balances, pay your bills, get news and information, and send a fax—all from a handheld device weighing a couple of ounces. Recent features include built-in digital cameras, spreadsheet software, GPS location services and stereo-quality music features.

Formerly wired office devices—such as computers, printers, photocopiers, and telephones—are also now being networked through wireless means with new technologies entering the market one after another. Short-distance Bluetooth personal area networks (PANs) are being supplemented by Ethernet wire-speed wireless local area networks (WLANs). New technologies continue to go through standards committees such as ultra wide band (UWB) in a never-ending goal to transform the traditional wired office into a historical relic.

Wireless pioneers have been equally dynamic as their industry. While many of the major players are established companies—DoCoMo's parent NTT was founded in 1869, SingTel is 120 years old, and Ericsson launched in 1876—they have all adjusted dramatically over the past 20 years to ride the wave of the wireless industry. These companies launched their wireless initiatives off the backs of their legacy businesses because the capital requirements were so large to deploy wireless infrastructure, products and services. Moreover, the products and services they offered often built consumer awareness of their brands. Divisions were spun off or re-branded to attack the wireless markets, with the funding coming from public offerings. At the same time, an entire category of venture-backed telecommunication firms were built. The most successful was US-based Cisco Systems, founded in 1984 by a small group of computer scientists from Stanford University. More recently, niche technologies, such as Singapore-based Ace Fusion and their wireless asset tracking devices (acquired in 2001 by Savi Technologies) have ridden the venture-backed wireless wave.

The adoption of wireless technology in Asia-Pacific has seen an unprecedented acceleration in the recent 18 months across several countries. There have been many reasons offered for this growth, but the most plausible has been the aggressive moves by various carriers to enhance their ARPU by getting their subscribers to use more than just voice calls with their cellphones. In 2003, DoCoMo announced it breached the one million subscriber level for their high-bandwidth 3G technology service in Japan. All indicators point to a much higher adoption rate with richer applications being deployed on the platform. Another clear indicator is the runaway success of gaming companies in Korea, Japan and China that prompted leading equipment giant Nokia to launch its gaming-specific N-gage cellphone platform. Coupled with the phenomenal use of text messaging, the cellphone handset is transforming the technology landscape in more ways than we ever imagined. This insatiable consumer and business demand continues to make the wireless industry one of the hottest spaces for venture capitalists to invest in (see Table 11.1). As one company grows, clusters of technology expertise sprout and regions develop into industry powerhouses.

At a fundamental level, growth in the Asian wireless market has been a function of:

- A need to connect dispersed islands of populations and geo-political fragmentation.

Table 11.1: Wireless players and investors in Asia

Country	VC/Corporates investing in Wireless	Wireless Companies
Japan	• ITX Inc. • H&Q Asia Pacific • Softbank	• UI Evolution Inc. • Access C Ltd.
Korea	• Jafco • Softbank	• COM2US • NCSoft
Singapore	• 3i • Walden International • McLean Watson Capital	• Skywave • MobiApps
Taiwan	• Fortune Venture • Via Technologies • Acer Technology Ventures	• CC&C Technologies • NextComm
China	• GIC • Softbank • Intel Capital • Walden	• Airoam • United Platform Technologies

Source: Authors

- Poorly implemented legacy-wired infrastructure that could not support next generation products and services, especially in developing countries.
- An established industrial base such as the Taiwanese semiconductor and electronic device industry, to build and support wireless products and services.

While the overall penetration of wireless phones in Asia lags Europe, it is in areas of concentrated adoption that Asia is moving into the vanguard of the wireless industry. Penetration rates are highest in South Korea at approximately 60 per cent and are close to 50 per cent in Japan. In 2002, China surpassed the US in the number of individual cellphone users. The number of wireless users is projected to continue to grow at impressive rates for at least the next five years. The worldwide leader and pioneer of wireless Internet services, NTT DoCoMo of Japan, claims over 47 million customers.

As a result, Asian companies are rivalling historical Scandinavia, home of Nokia and Ericsson, for technological leadership in the wireless world. NTT DoCoMo, for example, has set the standard for wireless Net access with its i-Mode service. The company is providing the only commercial network in the world today that allows subscribers constant access to the Net from their mobile phones. They can play video games, swap digital pictures, browse web pages, reserve tickets, and download music. The service has 39 million subscribers.

The Asian mobile market will present formidable global competition. SK Telecom in Korea already has a tight grip on its market with at least 50 per cent of the country's mobile phone market. Its next leap is wireless Internet services. SingTel, NTT DoCoMo, China Telecom and SK Telecom are all looking to expand beyond their country borders through partnerships, acquisitions and launches of independent subsidiaries.

Investor Offerings

There is a passion in Asia both for wireless technology and for investing in its leading companies. It is often considered the most promising wireless market in the world—thanks to its 3.5 billion people, who represent approximately 60 per cent of the world's population. Penetration rates vary enormously, allowing technology and management skills transfer to neighbouring nations. While the most enticing long-term market is China simply due to its size, there are a host of areas and vertical markets that hold promise.

Venture capitalists are avid consumers. In an industry with no borders where they are constantly on the go, they are often early adopters of every mobile device available—mobile phone, two-way e-mail pagers, and wireless enabled notebook computers. It is, thus, no surprise that VCs have been attracted to the wireless world seeking investment opportunities. However, looking for success is more easily stated than accomplished. The wireless space has its equal share of false starts and failures. No VC firm has been spared the general downdraft in the last three years. Within the wireless industry, prominent early players have come and gone after receiving substantial venture investments.

Australia's Davnet launched high-profile deployment in Singapore to provide in-building ethernet service provisioning, with intentions of moving onto localised public high-bandwidth wireless hotspots offerings.

Unfortunately, commercial success fell short of expectations and they have since shutdown their Singapore operations. In Malaysia, there are currently a dozen hotspot players providing location-based services with no proven revenue model. No one really knows how long the players can keep up the infrastructure going if the hoped for revenues does not kick in. When WAP (Wireless Application Protocol) was all the rave in 1999 for the delivery of wireless multimedia services, companies like Singapore-based Edgematrix received enormous funding from leading VCs like 3i to dominate as the wireless systems integrator for Asia. Needless to say, the company could not carry on business after the dotcom crash.

What is the key driver for future growth? Wireless adoption! Is the space a disaster zone? No. As with any evolving industry it requires care, analysis, and some luck to succeed. There are successful stories of companies who manoeuvred the murky waters carefully and came out winners. Take Koei Corp of Japan, a gaming player listed on the Tokyo Stock Exchange, with four straight years of record net profits since 1999. In Korea, companies like COM2US have helped carriers in Asia generate revenues for their mobile games. With the slew of new handsets supporting Microsoft, Symbian and Brew operating systems, the proliferation of games means big bucks for the developers and carriers. Even little niche players providing downloads for ringtones and images are clocking real revenues.

The continuing enthusiasm of investors is guided by the meeting points between the state of the industry (early vs. late, slow vs. fast), the positioning of the company within that industry or market (leader vs. follower), and the quality of the leadership running the company. Instead of focusing on the idea, the investment is driven by the opportunity.

Which wireless technology will attract venture funding? One can superimpose a framework in analysing potential new investments.

Opportunities Often have the Following Characteristics:

- Management. Business is people, and management is always the number one issue. Investors are now looking at investing in regions that have the supporting management development infrastructure to supplement the entrepreneurial teams they invest within. A good example is the leading international business school INSEAD, which opened a campus in Singapore parallel to its campus in France.
- Defensible part of the food chain. Is there a need, a willingness to pay, and an ability to deliver, regardless of whether the technology or service fills a gap or creates a new market?
- Reasonable time period. Results must be achieved without perpetual financial losses, hopefully within 18–36 months. Time frames for risk assessment have become far shorter since the recent technology downturn.

- Manageable cost to achieve profitability. Growth cannot come at any price. It must be measurable.
- Market size. Even though every billion dollar market will attract competition, early movers can establish a beachhead before the space is saturated.
- Local Talent. Success breeds clusters of expertise. New ventures find their genesis in existing companies. Not only can they launch disruptive technologies that represent industry discontinuities, but they can also sustain a competitive advantage.
- Good business models. Cash management is the fundamental measurement of an early stage company's ability to survive.

Fundamental Industry Drivers must Exist, Including:

- Need to be mobile, based on job or personal requirements
- Leapfrogging competition, either on a persona/corporate basis, or on a national basis as in China
- Lower cost, higher productivity
- Entertainment: games and other personal wishes
- An existing infrastructure to support new services and technologies. There are always interdependencies, especially in the wireless industry (see Figure 11.2). An example is the telematics industry in Asia where it is well positioned based on the number of satellites. Depending on political imperatives, there are a host of satellites available (Japan, China, Malaysia and India); therefore satellite-related businesses can choose and have great flexibility.

In the wireless arena, it is both challenging and enticing to invest in early-stage companies. Clearly, professional investors avoid early-stage companies that are inherently weak or that seek to position into large, competitive and capital-intensive markets. The allure of wireless investing is that success often results in enormous growth because of the virtually global appeal of wireless solutions and their fast rate of adoption relative to most other sectors. At any point in the wireless cycle, the capital intensive and commoditised segments are the domain of the giants—the carriers, and massive service providers who are able to leverage both debt and equity markets. They are the giant standard bearers who provide distribution. Most venture firms target points in the wireless good chain that can deliver exceptional growth, but where they are less likely to be commoditised quickly.

Areas That Scare Away Venture Capital Firms

- Weak management no matter how good the technology is.
- An over-dependence on unproven emerging standards.
- Huge capital risk to deploy the technology product or service (e.g., 3G licences).
- Local only—the wireless industry is not identical everywhere, but it is global.
- Market saturation from competition.
- Technology risk that cannot be measured.
- Exits only through acquisition. VCs seek companies that can achieve liquidity through initial public offering. In the wireless industry, trade sales are often the only logical exit as minnows are swallowed by whales. This reduces both the VC's enthusiasm because it limits the number of potential exits, and leaves portfolio companies dependent on third party players for liquidity.
- Complex dynamics—sometimes the food chain is simply too complicated!
- Too early, too late—timing is everything!

Industry Ecosystem

Because of its youth, the wireless ecosystem has not yet constructed a hardened governance structure. The supply chain (see Figure 11.2) continues to evolve with lines blurring between government involvement, industry standards, equipment vendors and service providers. Instead of splitting the industry into consumer-driven segments, the following outlines the key constituents/factors in addition to consumer user base.

The Governments

Spectrum is finite in nature, yet it must accommodate future radio and telecommunications needs. Governments must allocate spectrum because wireless operators must have access to frequencies, and because only one operator can conduct wireless operations at a certain frequency in an area. Further, spectrum management is international in character and cannot be dealt with solely on a domestic basis. Similarly, the regulation of different product standards and competition in such a capital intensive wireless industry must reflect a government and private sector consensus as to what services are technically possible, economically sound, spectrally efficient and likely to benefit the public. Allowing unfettered corporate-driven agendas will lead to fractured markets, increased equipment costs, delayed research and product development, reduced time to market, and increased potential for interference among users. Cooperation through global organisations, such as the WTO and ITU, and regional bodies such as the Asia Pacific Forum for Economic Cooperation (APEC) are attempting to keep the borders open to competition and coordinate internal regulatory matters including product and service competition and spectrum allocation.

The Standards

As the number of wireless users and services increases, it becomes more important to ensure interoperability between different nations and regions. The current standards (CDMA, GPRS, GSM) in place were initially developed by equipment makers, the components of the infrastructure required for operation of the industry. However, the service providers have established an important role in the development of new standards including 3G and 802.11. The need for convergence of digital service and development of a wireless Internet standard has led to this increased activity.

The Equipment Vendors

Equipment vendors, including software firm providers, work very closely with service providers to build consumer demand. Wanting to unleash the wireless tornado market as the next stage of network market growth, innovation is constantly making previous competitors, such as Sony and Ericsson, into partners. Such linkages in this area are likely to create value-added products and services, especially with the convergence of the physical and wireless infrastructure. Equipment vendors are also targeting the enterprise and individual end-user directly with their Nokia smart phones, CDMA-enabled Palm Pilots and WLAN-ready Acer laptops.

The Service Providers

The service providers' ultimate goal is to provide simple, unified multi-service packages (data, voice and video) to their customer base, either providing all the pieces themselves or in partnership with other providers. They are continually looking for new technologies and opportunities to integrate into their service offerings such as gaming for individuals or wireless broadband access for their enterprise clients. Until recent years, telecommunications companies like China Telecom were national. With deregulation, companies such as

Hutchison Whampoa (with their 3G network rollout) are expanding beyond domestic borders. Liberalised markets, privatised firms, and the expansion of merger activity are all catalysts for this change.

Alliances and Partnerships

There are many kinds of alliances that populate the wireless landscape: technology, standards setting, sales, product and service integration, market development, and financial. Shifting alliances dominate the wireless industry.

DoCoMo is typical of the giants who know they need a global footprint. In addition to wholly-owned subsidiaries in Europe and North and South America, the company is expanding its global reach through strategic alliances with mobile and multimedia service providers in Asia-Pacific, Europe and North and South America. NTT DoCoMo is listed on the Tokyo (stock code 9437), London (stock symbol: NDCM), and New York (stock symbol: DCM) stock exchanges.

SkyWave & JaburSat: Telematics in action

How do you know where your trucks are once they leave the city? What if the driver is asleep in his motel, and someone steals the spare tyre? What if someone steals the whole truck? Every day, new wireless applications are rolling out to protect trucking assets and, of course, the drivers themselves. With wireless technology, companies like JaburSat (a systems integrator for trucking companies) have developed solutions that are a 24-hour passenger-riding shotgun alongside the main driver. The system never sleeps, and works worldwide. It can shut down an engine, lock a cab, lock or open a spare tyre contained, or anything else it is programmed to do. Companies who own moveable, valuable property want to know where it is, and its condition. They need inexpensive, fast, global solutions – and you can imagine how impossible it is for wires to be used. JaburSat uses the SkyWave telematics terminals and services—the world's leading manufacturer of low cost telematics terminals and a manufacturer/operator of the Inmarsat D+ network. Inmarsat D+ is the only truly global telematics system in the world today, and it is poised to grow dramatically.

The SkyWave/Inmarsat D+ system is widely used in Asia in the maritime and transportation industry (fishing, International Maritime Organisation applications such as piracy, location and condition reporting, pipeline and forest fire monitoring). Anywhere you are in the world, SkyWave feeds you the information you need virtually in real time via its relatively inexpensive mobile devices. Small devices were designed to function in remote and multi-jurisdiction locations (e.g., the Strait of Malacca, waters between islands in Indonesia) and regions that have little or no telematics infrastructures (e.g., Afghanistan, Pakistan, and western China).

Alliances are formed by business needs, standards initiatives and public-private partnerships. Industry giants like IBM, Intel, Motorola, Nokia and Ericsson not only partner with each other but bring young upstart companies into the marketplace to validate their new technologies. On the standards front, recent initiatives such as 802.11 and ultra wide band (UWB) bring together competing firms to establish ground rules and developments for new technologies. The Bluetooth working group's key members include 3Com, Agere, Ericsson, IBM, Intel, Microsoft, Motorola, Nokia, and Toshiba.

The public-private initiatives between semiconductor firm Intel and Infocomm Development Authority of Singapore (IDA) are developing a framework for seamless roaming for mobile telephones, based on Wi-Fi network technology. Asia has become the primary testing ground for new wireless technologies and initiatives.

Partnerships are also used by wireless companies to cross borders and enter new markets. In September 2003, Germany's Deutsche Telekom unit T-Mobile and British Telecom's firm BT group Plc joined an Asian partnership launched in 2003 by Singapore's StarHub, South Korea's KT Corp, China's Netcom Corp, Malaysia's Maxis Communications Bhd and Australia's Telstra Corp to promote wireless broadband technology and allow subscribers to roam on other networks when travelling. A Japanese partner is expected shortly. The alliance has coverage of 13,000 locations or 'hotspots' worldwide, and plans to double this by the 2004. The partnership has a combined subscriber base of over 48 million broadband and mobile users.

Revenues

As with all sectors of the technology industry, profitability tracks the life cycle of the industry and the companies within it. At one end of the spectrum, the development/early stage results in material investments and losses. At the other end, services and products become commodities, and margins are squeezed. The key for investors is the time to achieve good margins and the period of time during which they will be sustainable—through either an attractive business model and/or continued innovation. While software players can expect 20–25 per cent pre-tax earnings when they hit this performance sweet spot, the capital requirements of network providers and equipment manufacturers leads to far smaller proportionate returns.

Investments/Listing

In the volatile telecommunications market, the wireless sector is clearly a great opportunity in Asia. Both 3G and mobile Internet are on the horizon, and drive the need and investment in improved infrastructure, software and applications.

Whereas the wireless industry in regions like Europe went through a period of excessive enthusiasm (too much invested, at too high valuations for wireless portals and wireless Internet enablers), the general consensus is that the superior Asian plays are poised for excellent economic performance. For example, after disastrous forays into 3G in Europe, Asian innovators and consumers may save the technology by proving that it is commercially viable. Asian investment in 3G applications and services continues to grow at a relatively swift rate with NTT DoCoMo already achieving a one million user subscriber base.

Wireless can be segmented from an investment perspective into industrial/commercial and consumer. Within those categories, it is segmented into applications (products, devices, software, solutions and services), content and entertainment, consulting and professional services, infrastructure (networking equipment and software, components and semiconductors), and service providers (Internet access and mobile communications).

As one would expect, the major players in any of these segments have either corporate arms or affiliated investment groups that are very important. Some, such as Ericsson, have their own venture capital subsidiaries. Others, such as IBM, have partnered with independent venture capital firms. The subset of venture capital firms that have focused on wireless investing in Asia is large (e.g., 3i, H&Q, Venrock). At the current time, the key areas of interest include Bluetooth, wireless semiconductors, enterprise applications, rich media and 802.11. While 802.11 remains somewhat overfunded, prospects for Bluetooth and fixed wireless appear promising.

Arguably the most successful wireless investments in the Asia-Pacific rim have been VC-backed with participation from strategic investors. One example is Access C Ltd of Japan, which received VC funding from H&Q Asia Pacific, followed by subsequent rounds with investors like NTT DoCoMo, Sony, Motorola and Tokyo Electric Power; the company is publicly held in Japan. Other notably active investors include 3i, Asiatech Ventures, PCCW, Softbank, and SingTel Ventures.

Regulatory Issues

There are inevitable regulatory challenges and issues given the complexity of the wireless industry (see Figure 11.2). Some situations are driven by nationalistic and political motivations. For example, in contrast to the inefficient pan-national European satellite consortium, many countries in Asia have launched their own satellites. This includes India, China, Malaysia, Korea and Japan. This diversity offers both excellent infrastructure due to the supply of satellite services, but also forces business models that are consistent with these fragmented satellite markets.

Another type of regulatory issue that arises are national decisions that lead to standards wars such as the decision by Korea to adopt CDMA instead of GSM phones or China's decision to develop a unique CDMA standard quite apart from that of North America. Interoperability of the mobile Internet is also an example of competing commercial standards with arguably seven operating systems and three development languages vying to become the default standard for mobile Internet.

As usual, there are countries in Asia where the governments have imposed tight regulation over the content being delivered through wireless technologies. While pornographic magazines are not available off-the-shelf in many places, the Internet provides a host of options for determined consumers. The same principle applies to many wireless-related service offerings, be it casino gaming or adult-oriented content. Suffice to say that the floodgates have been opened, and try as the authorities might to stem them, it is likely to be driven by demand. If service offerings cannot be available openly, they tend to go underground usually at higher prices and often providing superior margins.

Management

Are entrepreneurs taught or born—or both? This question has never really been successfully answered. There is no foolproof way for assessing and selecting management in any area. However, in the wireless industry there are characteristics that successful managers inevitably share. Whilst focused on their own business, they have an open mind and carefully track major industry changes. In this industry, fossilised thinking is terminal. The top managers and entrepreneurs are also able to blend business discipline with the desire to achieve great product or technology success. With tight product delivery timelines and the need to work within a complex ecosystem of suppliers, partners, customers and investors, disciplined execution skills are fundamental. As an added factor, many of the entrepreneurs are serial company builders who are veterans of other startups and bring the credibility and affiliations that improve the odds of success.

Learnings and Recommendations

The fundamentals of investing in wireless technology are similar to what Roman generals faced thousands of years ago: Position top people into areas of the globe where you can win. What is a critical piece of information in this commercial battle is that Asia offers a host of lucrative and interesting opportunities. The opportunities are dramatic ranging from cutting-edge technology development in partnership with leading companies and institutes, to simply bringing tried and true mobile technology to newly-developing regions. While China naturally represents some of the most enticing growth rates and speed of development ever experienced in any industry, there are also plenty of attractive opportunities throughout the region where consumers and businesses have an insatiable demand for anything wireless.

Management, timing, discipline, positioning and lady luck all will play their part in choosing winners and losers!

The Future

The opportunity for wireless-related investments remain bright, and we will certainly see more successful companies emerge from Asia in this space. Several of the early players like COM2US now have a firm footing in Europe and North America, and the global market opportunity will clearly draw more entrepreneurs.

Asia provides a host of attractive and unique backdrops. Korea is booming with early adopters, whilst Japan continues to blaze the trail for many 3G applications. The giants of the industry may dominate their segments, but there is plenty of space along the food chain for well-positioned and well-managed companies.

The hot markets remain Japan, Korea and, of course, the giant China. But large markets like India, Indonesia and the Philippines will potentially be experiencing strong growth over the coming five years. Even in Thailand, there have been a number of successes with text messaging and mobile games developers, which is a good sign given that Thailand has traditionally not been a hotbed of technology development. Investors who can identify players with differentiated offerings (either content or platform software) in this space should be well rewarded. The greatest appeal for this space, as far as VCs are concerned, is the convergence of messaging and wireless protocols both at the development and deployment end—even the text messaging-shy populace of North America is slowly warming up to this usage (interestingly, the Blackberry offering by RIM is a clear sign that even a well designed text-based service can achieve reasonable traction, which will indirectly prime the consumers in general for SMS-related service offerings). This development in the North American market presents an opportunity for Asian-based players with matured and proven solutions.

There are also excellent opportunities in purely industrial applications, such as telematics, that provide coverage via satellite across disparate and fragmented islands and locations.

Resources

Wireless Trends

www.wirelessweek.com
www.americasnetwork.com
www.unstrung.com
http://www.telecomweb.com/wirelessdata/
http://www.nwfusion.com/topics/wireless.html
http://www.wirelessnow.com/
http://www.telecoms-mag.com/default.asp?func=channel&channel=Wireless&Id=74
http://www.internetworld.com/wireless/
http://wireless.itworld.com/News/
http://www.palowireless.com/wireless/news.asp
http://www.gsmworld.com/index.shtml

VC News

Wireless.VentureReporter.net
Singapore Venture Capital Association (www.svca.org.sg)
Hong Kong Venture Capital Association Ltd (www.hkvca.com.hk)
Thai Venture Capital Association (www.venturecapital.or.th)
Malaysian Venture Capital Association (www.mvca.org.my)(http://www.mvca.org.my)
Taiwan Venture Capital Association (www.tvca.org.tw)(http:/www./wwwhttp://www.tvca.org.tw/.tvca.org.tw)

12. SingTel Group

Multiple Services, Multiple Markets—Case Study

Michael Sainsbury

Editors' Note: This chapter provides a detailed case study of the SingTel Group, a diversified regional telecommunications services group with offerings in mobile telephony, fixed line and wireless Internet. It has accumulated minority interests in other mobile companies in the region—Globe in the Philippines, AIS in Thailand, Bharti in India, and Telkomsel in Indonesia—and also acquired Optus, Australia's number two telecommunications carrier and number two mobile network. The case study covers content plans, tariffs, user adoption, growth rates, 3G trials, aggregated alliance savings, churn rates, roaming services, ARPU, and regulatory issues like mobile number portability—issues of interest to media analysts, investors, regulators, technology players and other mobile operators.

Company History

SingTel, the incumbent former monopoly, was incorporated in March 1992 and became a public company in October 1993 although the Singapore government still owns majority share in it. It has made vague promises to sell this down at some time in the future, but no clear timetable has yet been laid out. SingTel was listed on the Singapore Exchange in November 1993 and on the Australian Stock Exchange in September 2001. SingTel's market capitalisation is about S$25 billion. SingTel's first overseas office was set up in May 1993 in New York City. Mobile communications revenue continued to be the top revenue stream, contributing 38 per cent of group operating revenue in 2003, up from 34 per cent a year ago.

Mobile: Driving International Strategy

After attempting to run a regional Asian strategy by setting up branch offices offering services to larger corporations, the company began to consider a regional mobile strategy—sensing that mobile was the best way forward. Since unlike fixed line services, mobile services continued to be a growth industry and next generation networks were on the horizon.

Meanwhile, it accumulated minority interests in regional mobile companies—Globe in the Philippines, AIS in Thailand, Bharti in India, and Telkomsel in Indonesia.

In October 2001, SingTel made its biggest purchase by buying Optus, Australia's number two telecommunications carrier and number two (of four) mobile networks in Australia. Bought when the depth of the global telecommunications crash was becoming evident, observers at the time believe SingTel had paid too much (indeed, when Optus' accounts were changed to use Singapore accounting standards, it was making a loss of almost $400 million!). SingTel bought Optus for its business and mobile divisions and was originally planning to jettison its loss making consumer and multimedia unit.

But in the past two years, Optus has turned around, moving into profit and providing SingTel with a growth engine as competition and poor economic environment hit the company in its home market.

Content, Marketing and Tariffs

SingTel introduced new price plans and value-added services in the second half of 2003 to counter growing competition. The company introduced new features such as connecting ringtone personalisation and IDEAS e-library. Mobile data services continued to grow steadily, contributing to 16 per cent of cellular revenue from 13 per cent a year ago. SMS usage increased by 17 per cent. Usage of other data services such as GPRS and MMS also grew significantly by about five times from a low base.

Optus has long used wildlife and nature videos to advertise its services. Late in 2003, the company took this a step further branding its new mobile content portal as Zoo. SingTel and Optus have also combined their mobile data platforms and they are now run out of Optus headquarters in Sydney.

In December 2003, SingTel put about 100 of its mobile customers on trial for selected 3G services including video calls, video-streaming for faster downloads of video-clips and movie trailers, and high-speed Internet access. Charges for voice calls, SMS, MMS and value-added services will apply based on their existing SingTel price plans.

Mobile Karaoke

SingTel's mobile karaoke service enables users to playback full-length songs on Java-supported mobile handsets with synchronised lyrics karaoke-style that prompt users to sing along. The service is available to SingTel's post-paid mobile subscribers at S$4.16 (GST included) for each download—covering the song's licence fee. Some Christmas songs were included nearer to the festive period. More than 100 songs were available by 2004, including English and Mandarin hits from artistes such as Celine Dion, John Mayer, Wong Lee Hom and Harlem Yu. Twenty songs are expected to be added to the list every month thereafter.

Two of SingTel's regional mobile associates have also launched the mobile karaoke service. Advanced Info Service (Thailand) introduced the service in February 2003, while Telkomsel (Indonesia) launched theirs in late 2003, witnessing an average of about 1,700 downloads per day.

TV on Mobile

Six months after its rivals joined TV services on mobile phones, Australia's second-biggest mobile company joined the chase for more revenues from data services. It uses re-broadcast services from ABC, SBS and CNN of the US as a lure in a mobile content strategy called Zoo. But question marks remain over demand. Mobile revenues across the industry in Australia hit more than A$9 billion in 2003 but only a tiny proportion are from non-voice or SMS sales. Australia's mobile networks have been trying to sell customers mobile data services since late 1999 when WAP (wireless application protocol) flopped. The picture version of SMS, known as MMS, has been slow to take off.

User Base

By September 30, 2003, the group's aggregate mobile subscriber base in the region hit 40.8 million, representing a 41 per cent year-on-year increase in the total number of mobile subscribers of SingTel, SingTel

Optus and SingTel's four regional associates, compared with the base of 28.9 million customers a year before. On a proportionate basis, SingTel's mobile subscriber base in the six markets increased 34 per cent to 16.1 million as of September 30, 2003.

SingTel's mobile associates in the region—Advanced Info Service, Bharti Group, Globe Telecom and Telkomsel—expanded their total subscriber base by 49 per cent from 2002 to 34.2 million as on September 30, 2003.

SingTel's strategy is to combine the strength of the six companies' total subscriber bases. The company claims that this had enabled them to derive synergy and savings from a number of areas, including discounts from bulk purchases of SIM cards and handsets, as well as from joint negotiation for infrastructure and other network expenditure. During the financial year ended March 31, 2003, the grouping claimed combined savings of $55 million, up from $20 million the previous year.

The post-paid churn rate has not been negatively impacted by the introduction of free mobile number retention this quarter. The average monthly churn rate dropped to a low of 1.2 per cent with the implementation of a customer retention programme.

The average acquisition cost per subscriber also fell to S$130 from S$212 a quarter ago. During the quarter, SingTel also introduced new price plans and value-added services, e.g., connecting ringtone personalisation and IDEAS e-library, to its range of offerings.

Mobile data services continued to grow steadily, contributing to 16 per cent of cellular revenue from 13 per cent a year ago. SMS usage increased by 17 per cent. Usage of other data services such as GPRS and MMS also grew significantly by about five times, albeit from a low base.

SingTel's wholly-owned subsidiary Optus continued to enjoy strong growth despite intense competition in the Australian mobile market where it competes for customers with five other networks (two owned by its largest competitor and former incumbent Telstra, two by Hong Kong-owned Hutchison, and one by UK-based Vodafone).

For the first time, Optus crossed the 5 million mark to register 5.07 million mobile customers as of September 30, 2003, an increase of 17 per cent from last year. Optus has about 34 per cent share of the Australian mobile market. Revenue from Australia accounted for 65 per cent of the SingTel Group's total operating revenue, compared to 52 per cent a year ago due to a strengthening Australian dollar and strong revenue growth of 20 per cent by Optus.

Table 12.1: Subscriber base of SingTel Group (as on September 30, 2003)

	Aggregate Subscriber Base (million)		YOY change*	SingTel's Proportionate Subscriber Base (million)		YOY change*
	Sept 03	Sept 02		Sept 03	Sept 02	
Optus	5.07	4.34	16.8%	5.07	4.34	16.8%
SingTel	1.52	1.51	0.66%	1.52	1.51	0.66%
SingTel's regional associates	34.20	23.0	48.7%	9.47	6.18	53.2%
Total	40.79	28.85	41.4%	16.06	12.03	33.5%

Note: *YOY = Year on Year
Source: Author

Alliances

There is some divergence regarding the choice of alliance partners between SingTel and Optus, which the management began to address during 2003. The key difference is in the network providers: SingTel had had a long-term alliance with Ericsson whereas Optus has an alliance with its rival Nokia!

In 2001, Optus signed a deal worth almost A$1 billion for a third generation network (W-CDMA) with Nokia. Trials for the service began in 2003 but Optus Mobiles managing director Allen Lew says the company will not begin 3G services until mid-2005. SingTel's customers can access the Internet at more than 2,000 Wi-Fi locations in 14 countries through its partnership with leading global roaming carrier, GRIC Communications. They will be able to access Grace's Wi-Fi service in places such as international airports, hotels and convention centres when they roam overseas. As part of the partnership, SingTel will also open its 200 over wireless surf zones in Singapore to the GRIC TierOne Network which comprises more than 300 international top-tier service providers.

Revenues

In 2003, mobile communications exceeded international telephony to be the second highest revenue stream at SingTel accounting for a 21 per cent share, up from 18 per cent a year earlier. SingTel's group revenue exceeded S$10 billion for the year ended March 31, 2003. As part of its drive towards privatisation, SingTel sold 69 per cent of its shares in Singapore Post (SingPost) in an initial public offering in

May 2003. SingPost was listed on the Singapore Exchange as a separate entity for the first time on May 13, 2003.

The mobile communications business of the SingTel Group is well diversified with investments in regional mobile markets that are at different stages of development, according to Lee Hsien Yang, SingTel's President and CEO. In Australia and Singapore, the group aims to preserve the value of its customer base, and manage the business for profitability and higher ARPUs by promoting use of data services.

Though the mobile subscriber base declined marginally, ARPU increased following the recovery in roaming-related revenue as the effects of the SARS outbreak subsided. On a year-on-year basis, mobile revenue recorded a small decline of 2.1 per cent caused mainly by lower minutes of use as consumer spending remained weak. The number of subscribers, however, rose to 1.518 million as of September 30, 2003 from 1.514 million the previous year.

SingTel

For the year ending September 2003, mobile communications exceeded international telephony to be the second highest revenue stream at SingTel, accounting for 21 per cent (up from 18 per cent a year earlier). Compared with the preceding quarter, mobile revenue rose 2.9 per cent to S$206 million. Although the mobile subscriber base declined marginally, ARPU increased 5.7 per cent following the recovery in roaming-related revenue as the effects of the SARS outbreak subsided.

On a year-on-year basis, mobile revenue recorded a small decline of 2.1 per cent caused mainly by lower minutes of use as consumer spending remained weak. The number of subscribers, however, rose to 1.518 million on September 30, 2003 from 1.514 million the previous year.

Optus

Optus Mobile continues to deliver profitable growth, with revenues up 21 per cent and EBITDA margins of 38 per cent. This reflects a continuing successful focus on higher value customers and reductions in unit subscriber acquisition and retention costs, which fell 19 per cent compared with the same quarter last year. Mobile service revenues, which exclude sale of equipment, grew 19 per cent. Total customers increased 17 per cent to over 5 million, with post-paid churn remaining low at 1.4 per cent. Post-paid ARPU increased by 7 per cent for the quarter, reflecting 5 per cent increase in minutes of use and a substantial increase in data revenues, which reached 14 per cent of service revenues. Data revenues continue to be driven by SMS growth, with the number of messages per subscriber increasing by 33 per cent compared with the same quarter last year.

Mobile cash capital expenditure for the half year was A$129 million, 36 per cent of the Optus total and down 3.2 per cent from A$133 million in the last corresponding half year. A base station rollout programme is under way to significantly improve coverage in targeted regional areas of Australia. Under Singapore GAAP, Optus expenses acquisition and retention costs as they occur. This can result in a sequential margin decline.

Regulatory Issues

In the Singapore market, subscription-free Mobile Number Portability (MNP) was mandated by the industry regulator in August 2003. This development had little impact on SingTel. 'The company's range of price plans, which come with features such as free incoming calls, extra bundled talktime and calls to a home number from just one cent a minute, means that our customers do not need to look elsewhere,' says Lucas Chow, CEO of SingTel Mobile.

But evidence shows, even from SingTel's Australian susbidiary Optus, that MNP takes about 12 months to really make its mark. More valuable customers—those from corporations, small businesses and high-end consumer users—tend to be on contract. It is a point Optus has often made.

'Compared to July 2003, the number of SingTel's post-paid subscribers rose in August and again in September despite market concerns about the impact of MNP. In fact, SingTel recorded its lowest post-paid churn rates ever with 1.1 per cent for the month of August and a similarly impressive 1.2 per cent churn in September 2003,' counters Chow.

13. Legal Issues in the Internet and Mobile Arenas

Asia-Pacific Roundup

Tan Min-Liang

Editors' Note: The 21st century information society calls for a converged regulatory approach to traditional telecommunications, mobile telephony and the Internet. This chapter provides a 'regional scorecard' of the diverse Asia-Pacific regulatory and legal scenario, ranging from Japan and Australia to India and Singapore. Concerns about security, management of competition, and convergence continue to dominate the discourse. Topics covered include IPR of content for mobile transmission, content filtering, electronic transactions, cyberstalking, wiretapping, database privacy, electronic evidence, Wi-Fi and VoIP. (A case study of cellular regulation in India is also presented in a later chapter.)

There is no other continent in the world that is as diverse and unique as Asia—in the old and new economies. Asia-Pacific now has the world's largest Internet and mobile telephony user base. Yet, the region still presents the conundrum of having the widest spectrum of level of sophistication of Internet users in the world today, with pioneers as well as laggards.

For instance, South Korea boasts of having one of the most developed Internet infrastructures in the world[1] and their passion for digital entertainment has propelled computer gaming to a semi-professional status[2], an unheard of phenomenon. Other countries in Asia-Pacific have not lagged behind; the renowned expertise of the Indian software developer and the relocation of most manufacturing hubs worldwide to China are testimonies of the growing influence that Asia-Pacific has on the burgeoning digital economy. Singapore,

Taiwan and Japan have unveiled public initiatives to establish their countries as intellectual property hubs in the region.

Whilst Asia-Pacific has embraced the Internet and mobile telephony in the digital millennium, technology has challenged the hallowed policy, legal and regulatory frameworks built brick by brick over the years. Instead of a homogenous landmass with a uniform government lattice or central body of government, Asia-Pacific is a diverse hodgepodge of economic, social and political establishments.

As such, Asia-Pacific has several governance issues unique to the region. Countries have widely differing levels of governance and technology legislation. Singapore and Malaysia already have in place extensive policies, legislation and regulations to govern the growth of ICT in their countries. Yet others are still in the infancy stages of developing their own national legislation.

Even those countries that have readily jumped on the ICT bandwagon with corresponding panache in providing for legislation are now re-looking at amending their legislation with the benefit of hindsight. Legislative trends have changed tremendously since the 1990s and more countries are taking steps to provide for regulations with a native flair, instead of reproducing the legislation of their western counterparts or taking the cue from international bodies. Legacy legislations that adopted a technology specific approach now see greater wisdom in taking a neutral stance in light of constant change.

The recent global events such as September 11 and the Bali bomb incident have also changed the outlook of legislature in providing for overarching powers for the executive in areas such as surveillance. With many countries already lacking in general laws for privacy and data protection, this trend toward policing citizens has also thrown up a gamut of legal issues for the Asia-Pacific.

Other recent developments have also thrown the digital millennium into disarray. The rise of Free & Open Source Software (FOSS), led by the poster penguin Linux as an alternative operating system, has challenged the traditional mindset of the use of propriety software in the industry. The traditional proprietary licensing model measured against the FOSS model has created a maelstrom of controversy. Governments worldwide have had to take a step back to assess the impact of FOSS in the industry; whilst some have actively taken policy decision to encourage the use of FOSS, others have taken a decidedly non-interventionist stance. Singapore is a notable example of a country that is treading the path of caution, despite having taken a pro-innovation slant in the past. The development of FOSS in Asia-

Pacific is nonetheless unabated given the acerbity of adoption by the developing countries in the region.

Despite the diversity of the Asia-Pacific experience, some general trends remain the same: Concerns about security, internal and otherwise; the role of the government to teeter on a thin line between a *laissez faire* type regulation and over-regulation; and the convergence of information technology, telecommunications and the media. Such threads bind the policy makers in an inextricable web that present a challenge in each unique environment.

This chapter is a macroscopic perspective of the various policies, legislative and regulatory regimes in Asia-Pacific from the ICT perspective and the challenges that they face. Specific examples will be gleaned from the region to provide a broad overview of the spectrum of regulatory perspectives and their efficacy—or lack thereof.

Content and Regulation

Internet News and Information The interplay between the instinctive need of the government to control the media and the call for freedom of the press is a theme which is omnipresent throughout Asia-Pacific. Western critics have often cast aspersions on the lack of editorial integrity and the almost draconian grip that certain Asia-Pacific governments have on the media, yet many of the same critics do not appreciate the fine cultural sensitivities that underlie the respective societies. Needless to say, the advent of the Internet has created a new world order in respect of freedom of the media in these countries, the citizens of which are now subject to a constant barrage of information, regulated and otherwise.

The Internet has been dubbed, by one of the authors, as the Fifth Estate[3], the most powerful of them all. Indubitably, this is recognised, though not openly, by various governments in the Asia-Pacific.

China has no qualms in managing content on the Internet by way of a host of laws and regulations, including the Measures for the Administration of Internet Information Services[4], Regulations for the Administration of Internet Electronic Messaging Services[5] and Provisional Regulations for the Administration of Engagement by Internet Sites in the Business of News Publication[6]. China has also taken a number of pre-emptive measures to filter out objectionable content to Chinese users, with some of these measures having had far reaching effects, which have invited the critique of Western observers who have likened it to the 'Great Firewall of China'[7]. The immensely popular Google Website was off-limits to Chinese surfers for two weeks as the authorities reportedly banned it for 'inappropriate content'. Whilst this ban has since been lifted, several key search words are reportedly still blocked. It has also been reported that only after the popular search engine Yahoo! agreed to filter their content, was the search engine allowed to be accessed by the general Chinese public.

Singapore has also taken a similar stance in regulating content for their citizens. The Media Development Authority[8] (MDA) of Singapore is entrusted with the task to ensure that only appropriate content is filtered through the pipes. The regulation is done by way of excluding objectionable content through mandatory proxies that Internet Service Providers (ISPs) designate for their clients. Pornographic content is banned though the proxies are unable to provide an all-encompassing filter against all undesirable content.[9]

Likewise, Malaysia has enacted the Communications and Multimedia Act[10] that defines the parameters of which users may harness the communications and multimedia infrastructure. The Act also sets out the broad aims and objectives for the national ICT initiatives. Further, the Act effectively provides for a convergence of telecommunications and the media, being the successor of the Telecommunication Act and the Broadcasting Act.

Content regulation has since seen a paradigm shift in some countries. Bhutan, which used to have strict content regulations, has since relaxed its rigid stance. Previously, Internet content was filtered by way of proxies through the county's single ISP, Druknet[11]. Today, the regulations have since been removed though there is still no single cohesive digital policy in place to govern new media.

Australia has a unique co-regulatory scheme for Internet content, the inception of which was premised upon the *Broadcasting Services Act*[12], which was recently amended in 1999. Co-regulation provides for a non-intrusive regulatory regime that sets out a broad framework of best practices. Instead of having a draconian top down approach with the regulatory body wielding the whip, Australia has opted to allow the public and interest groups to police the Internet. In cases where the codes of practice have been breached, these groups would then revert to the regulatory body, which would take the necessary action. This self-regulatory policy has served Australia well as this takes the burden off the regulatory body that often has limited resources at its disposal and puts the onus on the public to scrutinise available content.

As the Internet community matures, there is a slight yet perceptible shift toward a lesser government role in regulating the Internet content. This is also due to practical issues such as the difficulty in regulating content on the Internet given its multi-jurisdictional nature, the myriad of technological advances and the dynamic nature of content generation. Even content on the Internet has seen significant changes with the rise of new media platforms such as weblogs where regulatory and legal issues abound[13]. The difficulty of providing for a singular content regulation policy is apparent from the range of regulation policies that the Asia-Pacific governments have attempted to impose on their users. Further, the gamut of policies underscores the cultural differences of each country. Undoubtedly, such policies will gradually diminish the role of a single regulator to a more mature, user-centric self-regulatory policy that Australia has undertaken.

New Media and Technology

In light of convergence of information technology and telecommunications, regulatory bodies have sought to find the balance in regulating and de-regulating.

New industries such as ringtone and logo repositories have excited the ICT market, jumpstarting new industries such as mobile commerce, though such emerging markets are similarly hampered by concerns over intellectual property rights (IPRs) infringements.

Most countries in the Asia-Pacific region have sought to transpose existing legislation to address concerns over new media.

P2P networks that provide for the exchange of audio, video and software files have also recently come under fire[14]. Whilst associations such as the Recording Industry Association of America[15] have taken the active stance in combating filesharing in the United States, such action is decidedly muted in Asia-Pacific—though recent reports have indicated that the Recording Industry Association of Singapore[16] has also commenced with legal initiatives targeted at the end-user to protect the IPRs of their artistes, by working with the ISPs in Singapore. It was also reported recently that countries such as Korea and Taiwan have taken similar steps to address this stem of digital piracy.[17]

E-commerce and E-business

Governments in Asia-Pacific have recognised the rise of e-commerce and e-business as a purveyor of trade and commerce for homegrown companies seeking a foothold in the global arena. Astute governments have also recognised that a stable political and legal environment is required to encourage foreign trade and investment. To this end, governments in the region have sought to establish an e-commerce friendly legal environment where local parties can trade with their foreign counterparts via the Internet. As such, the need for a definitive e-commerce legal structure and provision for virtual trade has become imperative for most countries. There is no doubt that countries in the region have already taken significant steps in this area to provide for electronic transactions legislation.

Singapore has taken the bold first step in adopting legislation to govern e-commerce and e-businesses. In Singapore, legal effect has been given to the legality of electronic contracts and the liability of network service providers have been limited in circumstances where they 'merely provide access'. The Electronic Transactions Act[18] takes a leaf out of the Illinois electronic signature legislation and the UNICITRAL Draft Articles. The Act incorporates semi-technology specific digital signature legislation coupled with extensive provisions by way of subsidiary legislation, the Electronic Transactions (Certification Authority) Regulations, governing third party Certification Authorities. Similar laws based on the Singaporean experience have been enacted in Brunei, being the Electronic Transaction Order[19].

Hong Kong has enacted the Electronic Transactions Ordinance[20] (ETO) in April 2000 to provide a clear legal structure for electronic transactions and digital signatures. Similar to their Singaporean counterparts, the ETO provides for legal recognition of electronic records and contracts.

South Korea has similar legislation by way of the E-Commerce Act and the Digital Signature Act which provides for legal effect to online transactions and the like.

Other countries such as Australia have established extensive consumer laws, which are in turn transposed for e-commerce. The Trade Practices Act 1974[21] is administered by the Australian Competition and Consumer Commission[22], which has a separate section for e-commerce and a list of best practices.

Whilst Asia-Pacific countries are quickly establishing their foothold in the e-commerce world map, Vietnam and Mongolia have lagged behind in putting laws and regulations in place for e-commerce. However, these countries are already in the process of enacting a legislation to address these issues. Vietnam, for instance, has issued a number of decrees and decisions governing specific areas of technology law and is currently in the process of putting in place a specific e-commerce legislation.

More often than not, legislation for e-commerce and e-businesses is based on existing legislation in the United States or on the Model Law on Electronic Commerce[23] issued by the United Nations Commission on International Trade Law[24] (UNCITRAL). In 2001, the UNCITRAL adopted a Model Law on Electronic Signatures[25]. To date, only Thailand[26] has enacted digital signature legislation based on the aforementioned though it is expected that more countries will follow suit. Singapore, whose ETA is based on the Draft Articles of the UNCITRAL, has been criticised for being dated and is expected to overhaul the existing electronic signature legislation in light of recent developments.

Computer Crime, Civil Liabilities and Electronic Evidence

Computer Crime and Civil Liabilities No other area of cyberlaw has captured the interest of the public as much as computer crime. The notion of hackers entering into virtual bank vaults and cleaning them out from the comfort of their home has aroused the imaginations of Hollywood and would-be hackers. Yet this seemingly cinematesque scenario is closer to home than many would imagine.

In 2002, the Development Bank of Singapore discovered that 21 electronic bank accounts were compromised by a hacker who had made away with about $35,000 before the transgressions were discovered. Whilst it was reported that the electronic bank was not compromised though the customer accounts were, the Singaporean public was shocked to hear of the online heist. The authors have similarly addressed these concerns in light of the Internet bank hack.[27]

Other cyber crimes have been regularly reported throughout Asia-Pacific, ranging from Denial of Service attacks to website defacement. Other online civil misdemeanors have also surfaced, from Internet defamation to cyber stalking. In this respect, governments have painstakingly hammered out statutory laws to combat this growing threat.

Malaysia recognised the need for cyber crime legislation in the mid 1990s and enacted the Computer Crimes Act[28]. Likewise, Singapore enacted the Computer Misuse Act[29], which provided for a specific piece of legislation outlining stringent sanctions for 'unauthorized access' into computer systems. Brunei has similar legislation based on the Singapore Computer Misuse Act.[30] The Thai Computer Crime Law also provides for criminal sanctions for crimes of electronic nature.

Other jurisdictions have incorporated cyber crime legislation within an all-encompassing electronic statute. For example, the Philippines provides for provisions on cyber crime, specifically hacking and online piracy, in the e-commerce Act. Similarly, existing legislation has been used to address cyber crime in Hong Kong where the Crimes Ordinance addresses the need for cyber crime legislation.

Other countries in the Asia-Pacific who are lacking in cyber crime legislation are in the process of enacting such legislation to address this festering problem. For example, Thailand is in the midst of enacting the Computer Crime Law, which is expected to be in force anytime soon.

To date, some jurisdictions have already utilised cyber crime legislation to stem the flow of computer crime in their countries. Singapore has a number of decided cases of website defacement and unauthorised access where the judiciary has taken a stern stance against cyber crime perpetrators by rigidly enforcing the laws and passing a deterrent sentence despite the tender age of some of the cyber criminals.

Civil action for instances such as Internet defamation have already become run of the mill in most jurisdictions as the use of ICT becomes prevalent in the workplace and at home. The use of electronic mail, instant messaging tools such as ICQ or MSN Messenger, and SMS have created a host of social issues. Cyber stalking has become the new social ill that is plaguing most developed societies. The prevalence of ICT devices and the ease of access have created new avenues to harass and stalk individuals.

Electronic Evidence

Whilst legislature gives legal credence to electronic transactions, the issue of computer evidence has also given rise to some controversy, given that electronic evidence is not expressly provided for in most legacy legislations.

Certain legislatures have sought to address the issue of electronic evident by way of express amendments to their current legacy statutes. As an illustration, in 1997, Singapore amended the Evidence Act to provide for the general admissibility of computer evidence in court. Further, specific subsidiary legislation, being the Evidence (Computer Output) Regulations, has been introduced to establish the criteria for certain archival systems to recognise the archived documents as good evidence in court.

Given the constant state of change of technology, the amendments to the *Evidence Act*, have been reviewed recently in light of the criticisms to it being outmoded and overly technology specific[31]. To this end, the Singapore Academy of Law has issued a consultation paper and has embarked on a review of the current state of law in relation to electronic evidence. Accordingly, it is expected that most technology specific cyber laws will suffer the same fate in light of recent advances in modern technology.

Privacy and Data Protection

Despite the progression of technology and the advent of the electronic database that provides for ease of access and retrieval, countries in Asia-Pacific generally suffer from the lack of a cohesive data protection initiative, and many countries do not have any privacy or data protection legislation at all. On the other hand, certain countries that have adopted a pro-liberal stance have entrenched in specific legislation, the right to privacy and unauthorised intrusion.

Since 1994, South Korea has had a *Privacy Act* that circumscribes the right to privacy of an individual. The Act provides for extensive criteria in relation to the collection, processing, access and retrieval of personal information and makes specific references to the rights of the owner of the information with regard to the disclosure and modification of the same. Further, the country has instituted a Public Record Management of Public Institutions Act that provides the parameters for collecting and maintaining public records in the government.

Hong Kong enacted the Personal Data (Privacy) Ordinance[32] in 1995, which sets out clear boundaries for the access and use of personal data. Likewise, New Zealand enacted the *Privacy Act*[33] of 1993 that protects the personal information of its citizens sans international entities when dealing with domestic companies.

The Philippines has taken a unique turn to enact the Anti-Wire Tapping Act which criminalises unauthorised tapping of any wire or cable to intercept or eavesdrop on any communications or oral discussions. This is also in line with the constitutional right to privacy that every Filipino citizen is entitled to.

On the other hand, Singapore that has led the forefront of enacting legislation to address ICT developments is surprisingly absent from the roll of countries with an established privacy or data protection legislation. Whilst Singapore's *Banking Act*[34] provides for data protection of bank records and the interception of telecommunications transmissions is unlawful, there is no specific legislation that addresses the right to privacy. The only feeble attempt in this direction is the Model Data Protection Code for the Private Sector[35] released by the National Internet Advisory Committee. The code, though a laudable step in this direction, has been criticised for lacking legislative teeth, as adherence to the code is purely voluntary.

Further, many of the developing countries in Asia do not have any specific data protection or privacy legislation despite the gradual trend toward the recognition of the rights of the individual over and above constitutionally entrenched rights to privacy—if the respective constitutions actually provide for the same in the first place.

The general outlook from the Asia-Pacific perspective is that privacy and data protection do not hold an exalted position with respect to the rights of the citizens. The governments are generally more preoccupied with addressing bread and butter concerns such as trade and commerce. In this respect, it is anticipated that as the Internet population matures in each of the countries, there will be a mass call for a privacy or data protection legal framework to be established in each of the countries.

Telecommunications

Regulatory bodies in Asia-Pacific have taken to the trend of deregulating their respective telecommunications markets. The telecommunications policy has been driven by the recognition that development of an economic friendly environment depends on the quality of the telecommunications networks and services. In line with this approach, governments have sought to untie the apron strings to state-owned telecommunications companies, encourage fair competition in the marketplace and encourage the introduction of a wide range of telecommunications services with the liberalisation of the market.

China's telecommunications industry has been vaunted as one of the world's largest and indeed, China's GSM network is the largest network in the world. Telecommunications sector policy and regulation in China was driven up till the early 1990s by the interests of the state-owned carrier, China Telecom[36], with a monopoly over all telecommunications services made to the public. In light of China's accession to the World Trade Organisation[37] (WTO), China Telecom has since been carved into separate entities and the sector has been significantly liberalised; foreign ownership has been allowed in the provision of telecommunication services in line with the obligations under the WTO. The Chinese ministry of information industry[38] has a department specifically to address the issue of liberalising the telecommunications sector. As such, the Chinese government has laid down a litany of regulations governing the telecommunications industry, including management of licences, disputes as well as Internet networks.

The Philippines has also liberalised the telecommunications market with the enactment of the Telecommunications Act. The Act provides for the regulation of telecommunications operators as well as the rights of users and the ownership requirements, including foreign ownership, of telecommunications entities. The National Telecommunications Commission[39] is the designated regulatory authority in the country that administers the Act, including overseeing the interconnection between the public telecommunication carriers in the country.

Pakistan has also followed the trend of liberalising the telecommunications market by moving from a monopolistic market to a open regime. The move toward a deregulated telecommunications industry was propelled by the Pakistan Telecommunications (Reorganisation) Act, which provided the initial impetus toward a more liberalised market. Under the Act, the Pakistan Telecommunications Authority oversees the telecommunications carriers with the aim to promote the growth of the telecommunications sector.

Undoubtedly, in most Asia-Pacific countries the telecommunications sector is long overdue for liberalisation, and this is evident from the measures taken by most governments to provide for a gradual transition from a state monopoly on essential telecommunications services to a full competition telecommunications arena.

Whilst the countries have adopted a liberal stance, new challenges are constantly thrown up into the sphere: the growth of Wi-Fi[40] as a wireless communications standard has been met with varying responses by the regulators. While countries such as Australia have chosen to take a hands-off approach by allowing use of wireless equipment to be excluded from licensing, other countries have chosen to provide for guidelines and the like.

The rise of Voice over Internet Protocol (VoIP) as an alternative to Public Switch Telephone Network (PSTN) has also created a myriad of issues. Whilst VoIP in its infancy was plagued with a number of problems such as poor voice quality, recent developments have pushed VoIP as a contender to PSTN. Advocates have asserted that VoIP is ready for mainstream adoption and with the benefits of low cost and ease of set-up, telecommunications companies are facing a paradigm shift in coping with this new upstart. Regulators are also stymied by the convergence between voice and data services and most are still in a quandary as to how VoIP should be effectively regulated, or if it should be regulated in the first place.

The advent of technology has created a veritable mix of issues, which telecommunications regulatory authorities are still learning to grapple with. Wireless communications and VoIP are but two of the salient issues that have reared their head in recent times and it is inevitable that other issues should arise in the years to come.

Conclusion

The Asia-Pacific has experienced a paradigm shift in ICT, and the governments have reacted accordingly to provide for some semblance of legal structure and framework for regulating the issues thrown up by the advent of new technology.

Many governments and regulatory bodies in the Asia-Pacific have been slow on the uptake, though the momentum is picking up. While most legislatures have adopted pre-existing legislations—often reproduced from their Western counterparts—to address ICT issues, the governments have begun to experiment and customise ICT legislation in light of their native policies and economies unique to their respective cultures and societies.

The diverse range of opinions and the variety of methods of implementation have created a fascinating potpourri of statutory instruments with the common goal to unify the digital experience under a common legislative umbrella. The Indonesians have an apt national ideology that is applicable to the Asia-Pacific digital legislation experience—*Panca Sila*, or unity in diversity. Despite the cacophony of voices expressing national perspectives in creating a stable legal infrastructure, the countries in Asia-Pacific are generally united in their single aim in harnessing the digital age.

Resources

Asian Development Bank (http://www.adb.org)

Australian Broadcasting Authority (http://www.aba.gov.au)

Australian Competition and Consumer Commission (http://www.accc.gov.au)

Baker and McKenzie (2001) *China and the Internet: Essential Legislation* Publisher: Asia Information Associates Limited, Hong Kong.

Brunei Information Technology Council (http://www.bit.gov.bn)

China Internet Network Information Centre (http://www.cnnic.net.cn)
Cyberlaw India (http://www.cyberlawindia.com)
Department of Communications Information Technology and the Arts (Australia) (http://www.dcita.gov.au)
Department of Information Technology (Bhutan) (http://www.dit.gov.bt)
Chin Saik Yoon (2003) *Digital Review of Asia Pacific*, Published by UNDP-APDIP.
Hong Kong as an Information Society (2002)
Infocomm Development Authority of Singapore (http://www.ida.gov.sg)
Information Technology and E-Commerce Council (Philippines) (http://www.itecc.gov.ph)
International Telecommunication Union (http://www.itu.org)
InternetNZ (http://www.internetnz.net.nz)
Lawnet (Singapore) (http://www.lawnet.com.sg)
Malaysian Communications and Multimedia Commission (http://www.cmc.gov.my)
Malaysian Institute of Microelectronic Systems Berhard (http://www.mimos.com.my)
Media Development Authority of Singapore (http://www.mda.gov.sg)
Ministry of Economic Development (New Zealand) (http://www.med.govt.nz)
Ministry of Home Affairs, Public Management, Posts and Telecommunications (Japan) (http://www.soumu.go.jp)
Ministry of Industry and Trade (Indonesia) (http://www.dprin.go.id)
Ministry of Information and Communication (South Korea)
Ministry of Information and Industry (China) (http://www.mii.gov.cn)
New Zealand Trade & Enterprise (http://www.nzte.govt.nz)
National Electronics and Computer Technology Centre (Thailand)
National Information and Communication Initiative (http://www.nici.nat.gov.tw)
National IT Steering Committee (Vietnam) (http://www.itvietnam.gov.vn)
National Office for the Information Economy (Australia) (http://www.noie.gov.au)
National Telecommunications Commission (Philippines) (http://www.ntc.gov.ph)
New Zealand Government (http://www.govt.nz)
Office of the Telecommunications Authority (Hong Kong) (http://www.ofta.gov.hk)
Organisation for Economic Co-operation and Development (http://www.oecd.org)
Prime Minister's Office (Japan) (http://www.kantei.go.jp)
Public Access to Legislation Project (New Zealand) (http://www.legislation.govt.nz)
Singapore Statutes Online (http://statutes.agc.gov.sg)
Global Competition Review (2004), London. http://www.globalcompetitionreview.com Editors: Simon Bushell, Herbert Smith.

Thai ICT Laws (http://ictlaws.thaigov.net)
United Nations (http://www.un.org)
United Nations Commission on International Trade Law (http://www.uncitral.org)
Zhihua Online Information Network Law Research

Endnotes

[1] http://www.cid.harvard.edu/cr/profiles/Korea.pdf

[2] http://www.worldcybergames.org

[3] Tan Min-Liang. 'The Fifth Estate' *Infotech: Today* December 17, 2002

[4] Adopted at the 31st ordinary meeting of the State Council on September 20, 2000. Promulgated by the Premier with Order 292 on, and effective from, September 25, 2000 http://www.cnnic.net.cn/policy/18.shtml

[5] Adopted at the 4th Ministry Affairs Meeting of the Minstry of Information Industry on October 8, 2000, issued on, and effective from, November 6, 2000 http://www.cnnic.net.cn/policy/22.shtml

[6] Issued by the Press Office of the State Council and the Ministry of Information Industry on, and effective from, November 6, 2000 http://www.cnnic.net.cn/policy/21.shtml

[7] http://www.*businessweek*.com/magazine/content/02_38/b3800088.htm

[8] http://www.mda.gov.sg

[9] http://www.mda.gov.sg/medium/internet/i_policy.html

[10] http://www.mcmc.gov.my/mcmc/the_law/ViewAct.asp?cc=67919124&lg=e&arid=997371

[11] http://www.druknet.bt

[12] http://www.aba.gov.au/legislation/bsa/

[13] Tan Min-Liang 'If you blog, beware' *Infotech: Today* March 18, 2003.

[14] Tan Min-Liang 'Face the Music' *Infotech: Today* October 28, 2003

[15] http://www.riaa.org

[16] http://www.rias.org.sg

[17] http://asia.cnet.com/newstech/security/0,39001150,39156483,00.htm

[18] http://statutes.agc.gov.sg

[19] http://www.e-government.gov.bn/downloads/electronic.pdf

[20] http://www.info.gov.hk/citb/ctb/english/it/eto.htm

[21] http://www.austlii.edu.au/au/legis/cth/consol_act/tpa1974149/

[22] http://www.accc.gov.au

[23] http://www.uncitral.org/english/texts/electcom/ml-ecomm. htm

[24] http://www.uncitral.org

[25] http://www.uncitral.org/english/texts/electcom/ml-elecsig-e.pdf

[26] http://www.ictlaw.thaigov.net/ictlaws.html

[27] Rajesh Sreenivasan and Tan Min-Liang Cracking the Online Vault: Legal Issues and Liabilities for Internet Banking, *Law Gazette* October, 2002 http://www.lawgazette. com.sg/2002-10/Oct02-focus2.htm

[28] http://www.ktkm.gov.my/template01.asp?Content_ID=80&Cat_ID=1&CatType_ID=17&SubCat_ID=40&SubSubCat_ID=15

[29] http://statutes.agc.gov.sg

[30] http://www.e-government.gov.bn/downloads/computer.pdf

[31] http://www.sal.org.sg/Pdf/tldgComputer%20Output%20as%20Evidence.pdf

[32] http://www.pco.org.hk/english/ordinance/ordglance.html

[33] http://www.legislation.govt.nz

[34] http://statutes.agc.gov.sg

[35] http://www.trustsg.org.sg/pdf/pc_report_v3.2.pdf

[36] http://www.chinatelecom.com.cn/english/

[37] http://www.wto.org

[38] http://www.mii.gov.cn

[39] http://www.ntc.gov.ph

[40] Rajesh Sreenivasan and Tan Min-Liang (2002) 'Regulating Wi-Fi', *TeledotcomAsia*, December.

14. Convergence of Old and New Wireless Worlds

Challenges for Policymakers

Laina Raveendran Greene

Editors' Note: Successive waves of innovation in digital technologies and communication models have posed increasingly complex challenges for policymakers, national regulators and international regulatory bodies around the world. As if the convergence brought about by digitisation and broadband was not complex enough, the Internet and now wireless Internet are becoming even more tricky to handle in a unified framework. This chapter highlights regulatory approaches being adopted in Asia-Pacific, with case studies from Malaysia and Singapore, on how to handle the challenges of convergence, and identifies open questions surrounding traditional regulations of Wi-Fi and mesh networks. The next chapter highlights some of these issues with a detailed case study from the regulatory environment of India.

Convergence between the telecommunications, computer, audio-visual, and publishing technologies, services and industries will constitute the greatest challenge to policymakers in the years to come, a challenge more complex than the liberalisation of the telecommunications markets.

Robert Bangemman, European Commission; 1997

Introduction

To understand the complexities over mobile and wireless telecommunications regulations today, one really needs to understand how telecommunications policy and regulation evolved over the years. One has to look back to the early 1980s of the GATT Uruguay Rounds and the divestiture of AT&T in the United States. It could be said to some extent that the drive towards liberalisation in the 1980s was partially driven by new advanced wireless technologies defying the natural monopoly argument. Telecommunications up to that point was seen as a natural monopoly, which for national strategic reasons needed to be either run by a government enterprise or a private monopoly.

New cellular mobile technologies challenged this natural monopoly theory, leading to industry pressures on government to liberalise the provisioning of telecommunications. Many countries began licensing mobile operators either as value-added providers or as a separate licence, i.e., they were licensed as operators who were not offering a basic infrastructure for the public. However, it was only a matter of time before mobile became the 'basic' infrastructure in countries with poor wired communications. This led to confusion over how telecommunications should be regulated. Meanwhile, mobile networks were only supposed to provide voice services but, with the advent of the Internet, data services could also be offered, further challenging the way regulators tended to regulate. Regulations were usually based on different technologies or segments (voice versus data) used for telecommunications provisioning. Today, non-traditional mobile networks such as Wi-Fi (originally meant as a LAN network and thereby excluded from regulations since they were closed user groups), which are capable of being used as a public network, are further challenging the demarcations set by regulators.

This article will, therefore, look at the challenges of wireless world, within the larger context of challenges of convergence in the ICT field as a whole.

Early Convergence Challenges Faced by Policymakers

During the negotiations of the GATS Telecom Annex in the late 1980s, countries created artificial demarcations such as basic and value-added/enhanced as a way to limit the introduction of competition into telecommunications. These demarcations were economic and political creations, and had little to do with technology. For instance, while some countries considered packet switch as basic, it was considered as an enhanced service in the US. Only enhanced or value-added services back then were open to competition under the GATS Telecommunications Annex.

Technological demarcations were, however, attempted and date back to the World Administrative Telegraph and Telephone Conference (WATTC) in 1988 in Melbourne. This was the first global attempt to deal with convergence brought about by digitisation. In a simplified way, digitisation (modems and fibre optic technologies) allowed for the transmission of both analogue signals (voice) and digital language of computers to run over the same telecom network (which formerly was structured to carry only analogue signals, i.e., 'voice'). This was the advent of the digital revolution and the true birth of convergence.

Later, the bigger the pipes got (broadband), the more services could be offered through it—and soon convergence led to not just the voice and data revolution, but also video, audio convergence and the like. Nevertheless, being tied to digitisation and broadband, convergence early on was largely discussed as a very developed country issue. Developing countries were often still far removed from these issues and, therefore, not as involved in these debates.

In many developed countries, digitisation and broadband meant that a cable operator providing cable TV services over co-axial cables could also offer voice services over the cable network and a telecom network provider could offer video-on-demand services. This meant that previously separated industries now had the opportunity to provide the services of another industry and brought in new players as well. Often, however, these industry players were regulated by different regulatory bodies, e.g., broadcasters under the broadcasting authority and telecom under the telecom authority. This was not only confusing for the industry players, but the application of laws with a different basis to a different industry could also lead to discrepancies. There was clearly some need for coordination or even the merger of regulatory bodies.

Interestingly, developed countries—with their established regulatory bodies and with established regulations for different networks and services—faced this issue of convergence with greater complexity. Developed countries, on the other hand, were just beginning to face regulatory challenges brought on with liberalisation. For example, of all the Pacific Forum Island countries, only Fiji, Papua New Guinea and Vanuatu have detailed industry-specific regulatory measures for the telecommunications and broadcasting industries. These countries were focusing more on whether to have an independent regulatory body and how to introduce competition—although this also provides them the opportunity to begin with converged bodies from the start.

The WATTC 1988 conference was one of the earlier attempts to create a unified set of regulations to deal with convergence of all forms of telecommunications. New definitions were introduced such as regulating services as basic versus value added, or introducing new distinctions based on usage instead of technological demarcations, e.g., closed user groups versus services offered to the public.

Further Impacts on Convergence: Internet Protocol (IP)

Whilst digitisation and broadband forced the issue of convergence on many developed countries, developing countries were often not concerned as they operated with low-band and/or analogue networks only. However, IP has made convergence an issue for all countries, developed and developing alike. In the early days of IP/Internet, many however did not see it this way. The Internet was regulated as a value-added network or service (i.e., a network to offer services such as e-mail, bulletin board, Web pages) and therefore no convergence issues were involved. It was a separate data network.

Yet, today many countries are now migrating all their networks towards IP ('IP over everything, and everything over IP,' as the director general of the Singapore regulatory body put it in his keynote address at ITU's Telecom '99). This poses a new challenge to them in terms of how they regulate the Internet. The Internet was no longer seen as a separate data network, but rather a suite of protocols that created a more efficient telecom network to run convergent services.

The IP suite of protocols, which can run over any underlying or overlaying infrastructure (mobile, wireless, satellite, analogue or digital), unifies these different networks into one ubiquitous network or 'cloud' (a multi-point to multi-point network). This ubiquitous network can then offer numerous types of services such as voice, data, audio, video, which can be used for numerous applications such as e-commerce, e-government, and e-banking. This ubiquitous network also allows for a more efficient use of the underlying infrastructure. Unlike a traditional switched voice network, whereby the circuit is tied up during the entire one single call, the same circuit if TCP/IP enabled can handle multiple traffic flows simultaneously. It makes more sense for a developing country with limited bandwidth to IP-enable it rather than leave it as switched circuit voice only.

As a result, many countries in the Asia-Pacific region are migrating their legacy networks into IP-enabled networks. For example, some no longer have dedicated bandwidth for International Direct Dial (IDD). They migrate their IDD circuits over to IP, although some may still have dedicated IP bandwidth for VoIP to ensure quality of service. The point is TCP/IP creates a multi-point to multi-point 'cloud' network, where there can be point to point, point to multi-point and multi-point to multi-point communications happening all at the same time.

This creates new dilemmas for regulators. Telecommunications regulation was always about point to point communications and broadcast about point to multi-point communications. There were no regulations about multi-point to multi-point communications, and LAN/computer networks which operated on multi-point to multipoint basis were unregulated as a closed user group. How does one therefore regulate IP-enabled networks today?

IP-enabled networks have also created a situation where the 'network' can be managed by the customer premises equipment. In the telecom and broadcast world, the CPE (customer premise equipment) was a 'dumb' terminal that sat at the edge of the network and the 'intelligence' sat within the network. Take, for example, roaming cellular services. Sufficient base stations are required to ensure that a call gets transferred from base station to base station to ensure no drop in the call, and a continued service. Today, software and protocols are being developed to make the CPE/PDA part of the network. For instance, a US company called MeshNetworks is working on software and protocols that enable handheld devices to route calls to one another rather than on needing the network to do it. This turns the CPE or PDAs into temporary mesh networks rather than rolling out so many base stations to do the same thing. This may be a cheaper and quicker alternative to 3G, the company claims. This new P2P paradigm on how to define the network was also used by Napster. Combined with the development of the IP phone over Wi-Fi networks, this creates new possibilities for new players on the mobile operators field, further challenging regulators in the way they traditionally regulated.

Convergence Regulation and Regulatory Bodies

Whether convergence is an issue because of broadband and digitisation or IP-enabling legacy networks, it impacts regulations and regulatory bodies in several ways. To understand how it impacts traditional regulations, it would be useful to study the basis behind many regulations that exist today.

Traditional Regulations

Many countries have had different rules and different regulatory bodies for the various networks and services providers. Convergence has brought together three distinct areas of regulations, i.e., telecoms (network), broadcasting (content), and IT (applications).

i. *Telecommunications Industry Regulation (point to point premise-based regulations)*

Telecommunications regulation both domestically as well as internationally has primarily been about facilities regulation (regulating the infrastructure itself) and was based on the premise of point to point communications. Telecommunications regulations governed facilities—such as fibre optics, satellites, microwave, radio and copper wire cabling—basically, communication through electromagnetic means. Rules of the game were needed to ensure the smooth implementation of communications. Remember that national sovereignty was still an important consideration for most countries and, therefore, telecommunications between countries would meet at some halfway point to connect and complete a call. National and international rules of the game were, therefore, needed to ensure that calls were completed (this required settlement guidelines too).

In other words, these domestic and international rules related primarily to infrastructure carrying public services, which were regulated on the basis that they were point to point communications. Private voice networks, data and video were until recently not regulated as such since they were meant for a closed user group and not the public. This closed user group exception has become harder to keep with the expanding capabilities of closed user group technologies.

Since point to point communications meant that the users generated the content in the conversation, there was no need to regulate content, beyond regulating the provisioning of the infrastructure. Moreover, since the network provider had control over the point to point communication, there were strict standards applied to them, e.g., quality of service, such as 99.99999 per cent reliability. However, with convergence with IP for instance, some regulators have attempted to apply network standards over IP networks. But unlike the telecom network with point to point communications where the network provider had control of the network, public IP networks are different. The network is a network of networks and is as robust as its weakest point (under someone else's control).

In Singapore, for instance, initially all Internet Service Providers (ISPs) were regulated by the telecom regulator, and were expected to meet the 99.9999 per cent reliability standard. It was later when the regulator better understood the Internet network that these standards were relaxed. So we will find that in a converged world, as we bring the regulations together, we also have to understand and remember the basis for existing regulations before we attempt to apply them to new technologies.

Having said that, it is also interesting to see how countries have been creative in allowing the introduction of IP-based services. With monopolies existing in their countries, regulators sought to define the Internet as a value-added service or data service in order to allow for new players to offer these services. However, as IP networks were also capable of providing voice services, for instance, it became harder for the regulator to allow these services as the monopoly provider protested.

In the case of Singapore, before the government lifted the duopoly provision, ISPs were not allowed to offer voice over IP. However, the government recognised that they could not stop users from doing voice, and so allowed users to conduct voice over IP (VoIP) if the 'gateway' was at the user's site—but did not allow ISPs to offer the voice 'gateway'. This was a very clever way to get around the regulations although it did not make much technical sense. Later, the duopoly was lifted and the regulator now allows any provider to get a VoIP licence. Thus, how the regulator deals with new network services largely will depend on its current state of liberalisation. There are many interesting examples in Asia-Pacific of creative ways to get around these limitations.

ii. *The Public Broadcasting and Media Industry Regulation (point to multi-point premise based regulations)*

Broadcasting or media content regulation was based on the premise of point to multi-point communications. The public here were considered as merely passive receivers of the information. They had no control over the information and often this information may have been thrust on them unwillingly (e.g., through billboards or on television). Therefore, the laws were enacted to protect public interest. Broadcasters who used these broadcast facilities for point to multi-point transmissions (to reach out to the public), whether or not they were the creators of the information, were regulated by content laws.

Laws such as privacy, libel, copyright, pornography, decency and defamation existed to ensure a certain level of community standards for society. They were enforced even if the broadcasters were not the generators of the information, but became liable as 'republishers'. These rules similarly apply to other non-telecommunications media such as print publications. Often there were also separate regulatory bodies regulating this content industry, usually falling within the purview of the ministry of information and the arts as in the case of Singapore.

With IP convergence, however, the application of these laws to the Internet can lead to discrepancies. Web pages are not thrust on people—the Web is largely a pull technology not a push technology. Applying a point to multipoint law to a multipoint to multipoint network may not be realistic. Nevertheless, since they had not created a multipoint to point or multipoint to multipoint based regulations, many regulators still chose to apply broadcasting laws to the Internet.

Singapore is a good example of this, leading sometimes to discrepancies. Malaysia, on the other hand, with its new convergent body, has the opportunity to look at the Internet in a new way and avoid discrepancies. On the other hand, there are many people who feel that the Internet falls within the computing world, and should not be regulated at all.

iii. *The Information Technology or Computer Industry Regulations (closed user group)*

This has been a relatively unregulated area. Computers were previously stand-alone devices and, therefore, only laws relating to industrial and property rights for hardware and software applied here. Later, distributed computing and smaller PCs were developed, configured as Local Area Networks (LANs), Wide Area Networks (WANs) and Metropolitan Area Networks (MANs). Even then, however, communications were carried out within a company either located in a building or several buildings connected virtually as one. Such communications were considered internal and no broadcasting/content regulations applied. The network itself, usually located within a building, was considered outside the ambit of public telecommunications regulations.

Today, digitisation, broadband, and IP-enabling of networks has allowed these internal networks to communicate virtually with other networks outside the 'closed user group.' This has changed the concept of 'closed user groups' and forces regulators to re-examine these networks. In fact, many closed user group network operators such as SITA became service providers to users outside of their core user group. Furthermore, wireless LAN technology which started out as a LAN technology is now being used by some countries to provide basic network services to the public—and Napster-type technologies (P2P) are providing 'networking' power back to the user. Fine lines are becoming harder to maintain in this age of rapidly changing technologies.

Approaches to Convergent Regulations and Regulatory Bodies

Merging of all ICT-related regulatory bodies Some countries look at convergence from an industry angle and see how industries either merge together (e.g., AOL and Time Warner) or how each industry player can now encroach into another player's field. Since these players were often regulated by different regulatory bodies, some countries look at merging regulatory bodies. This poses questions: who will regulate a cable company and who can now provide a telephone service and how; who should regulate a telephone company that can now provide video on demand/broadcasting services, and how; and so on. Countries who take an industry approach to convergence are more likely, therefore, to merge regulations and regulatory bodies.

For example, in 1998, Malaysia passed the Communications and Multimedia Act and the Communications and Multimedia Act Commission which deals with convergence and converging regulatory functions. Jabatan Telekom Malaysia (the former telecom regulatory body) has now been absorbed under the new Malaysian Communications and Multimedia Commission (MCMC), which has become operational as of April 1, 1999. The Commission is responsible for all aspects of the info-communications sector regulation and licensing is done on the basis of the distinctions such as network provider, services provider, applications provider, and so on.

Regulation-wise, many countries are moving away from technology specific regulations but still maintain divisions between network/facilities-based providers and service-based providers. Whilst this distinction may seem clear to some, there will be overlaps—but since they are now under one ministry, this should minimise the risk of duplication or the framing of regulations that make no sense. Having said that, merging all regulatory functions under one ministry could also be more than any one regulator can handle. Suddenly, content-related laws, cyberlaws, e-commerce laws and other issues that a ministry was not used to handling now come within its purview.

Partial merger only Singapore, on the other hand, decided not to go all the way in merging regulations and regulatory bodies. It did recognise some level of industry convergence, thereby merging the National Computer Board (responsible for promotion of the IT sector) and the Telecommunications Authority of Singapore (which was the telecom regulator), into the Infocomm Development Authority (IDA) in 1999. However, unlike Malaysia, the government decided not to merge them with the Singapore Broadcasting Authority (the broadcasting regulatory body) recognising that content regulation is this body's core competence.

They have since, however, merged the Singapore Broadcasting Authority with the Films and Publications Department and the Singapore Films Commission on January 1, 2003 to create the Media Development Authority (MDA). Radio and TV licensing still fall under the MDA.

Thus, Internet infrastructure is regulated by the IDA, while Internet content is regulated by the MDA. ISPs literally, therefore, have two 'licences' from two different agencies for the same activity. In other words, it seems to take the traditional approach of the 'point to point' telecom regulations where the telecom regulator does not regulate content. Instead, content is treated as a point to multipoint activity and, therefore, the broadcasting regulator regulates content over networks licensed by the telecom regulator, especially for the Internet. There are still controversies over this approach mainly because of the differing approaches to regulation. To avoid inconsistencies, rather than merging the bodies, IDA and MDA now report to the same ministry and minister (they used to be under different ministries).

As legacy networks go the way of IP, this will offer more challenges. The Singapore government is constantly reviewing its approach to convergence to ensure there are no discrepancies. They are moving away from technology-based distinctions for licensing purposes and have lifted many restrictions on network and services provisioning.

Convergence in the Wireless World

Policymakers and regulators who dealt with the turbulent years of convergence brought about by digitisation and convergence may be even more bewildered with today's alphabet soup of acronyms in wireless and mobile communications: 2.5G, 3G, 4G, Wi-Fi, GSM, GPRS, and so on. This is especially so given the fast pace of change unlike the status of affairs from the early days of the International Telecommunications Union, which was established in 1865.

With the Internet taking off on the wired network, and broadband wired services offering voice, data, and video, there was pressure on the wireless networks to do the same. Wireless technologies were originally built specifically to handle voice communications, i.e., primarily up to 9.6 Kbps. Even the ambitious Iridium project with a projected 66 low earth orbit satellites to provide global wireless services was built to carry 9.6 Kbps only. Today, the quest for broadband wireless has begun.

Second generation wireless or 2G includes GSM (Global System for Mobile) networks. The 2.5G networks usually refer to systems like GPRS (General Packet Radio Services), or anything between the second and third generation networks. Third generation or 3G networks are usually identified as UMTS (Universal Mobile Telecommunication Service) or other higher bandwidth data networks like the NTT DoCoMo W-CDMA service. China has tried to launch its own 3G standards by adopting TD-SCDMA (time-division synchronous code-division multiple access). Basically, each generation provides a higher data rate and additional capabilities. Meanwhile, the promise of 3G has been around for a long time, but the costs that many industry players had to pay for spectrum and the costs of rolling out the 3G network made it hard to build a business case to roll out the services quickly. Several operators have become disillusioned about near-term rollout of 3G.

Then along comes the promise of the IT/Internet world with newer technologies such as Wi-Fi (IEEE 802.11). What is new and exciting about Wi-Fi, in addition to its ability to provide throughputs of 10 Mbps and greater, is that it is also relatively cheap and easy to roll out. There is a huge opportunity here to bridge the digital divide (given the lower costs), especially when rolled out as a public network. Despite the promise of Wi-Fi, some incumbent players, especially those who have invested in 3G, are reluctant to roll out Wi-Fi. Strangely, it is the incumbent player that has an advantage in rolling out Wi-Fi as it has billing systems in place (as proven in the US).

In addition to the potential opportunity to bridge the digital divide, what Wi-Fi has also done to the telecommunications industry is to introduce new players who are rolling out Wi-Fi for wireless LANs and hotspots. Some are already experimenting on seeing how to provide it as the public telecommunications infrastructure. Meanwhile, developments on IP Wi-Fi phones make these developments even more interesting as the network opens up as a substitute to cellular networks. There are still issues to be resolved such as roaming between base stations, billing systems and security, but these are issues that are rapidly being dealt with. Wi-Fi is bringing in many new vendors; even application providers who previously only focused on traditional wireless services are focusing on Wi-Fi applications.

New players, however, are not as savvy about the telecommunications industry, especially when it comes to policy and regulations. The traditional players are very used to the world of national and international telecommunications policy and regulation. The newer players are more used to an unregulated arena. They often underestimate the importance of national and international policy and regulation to global communications.

The wireless world does not have the same tradition of common global standards as the wired telecom world. We have ended up with GSM, GPRS, CDMA, PHS, and the like. Users cannot take their CDMA cellphones when they travel and expect to use them in a country with a GSM network. This is so unlike the wired telephone world where a telephone set from almost anywhere in the world can be plugged in and used almost anywhere else in the world. There are attempts to overcome this problem using multiple band phones, but they tend to work within the same network standard, e.g., GSM 1900 and GSM 800. Nevertheless, in the wireless world, one of the main themes in policy and regulation relate to the allocation of frequencies within which these networks can operate and to avoid harmful interference. However, whether allocating, auctioning, coordinating or some other means, both the ITU and many national regulators tend to only regulate or coordinate public telecommunications services. In the case of closed user groups, (e.g., LAN networks), it is left to the users to regulate their own services.

This is where Wi-Fi may fall outside this spectrum, which makes it an interesting issue. Wi-Fi started off as a wireless LAN technology that is now being used in hotspots or closed user groups. What happens when it is used for public access? What happens when it is used for public IP phone services? In countries where the frequency used for Wi-Fi is clearly stipulated as unregulated, this may or may not be a question. But in many others, regulators are deciding whether it is or not since all public telecommunications have to be regulated. This is where we will find a very interesting coming together of the old wireless world and the new.

Conclusion

The bottom line seems to be that whether in terms of digitisation and broadband or mobile and IP-enabling legacy and new networks, regulators are having a hard time to keep up. There is often a need to exchange information among different regulatory bodies to ensure consistency in national planning. Converging regulatory bodies may open the Pandora's box for issues that the regulator may not be equipped to handle and, yet, it may seem the more logical way to avoid discrepancies.

Whatever approach the countries decide to adopt, industry needs clarity so that it can develop cost-effective services to benefit the public. Some clarity is important otherwise the industry is never clear which regulatory body will hand them up and hamper growth in the long run. The issue of convergence is closely tied to liberalisation and the introduction of new players. While countries such as Malaysia have taken the ambitious leap to look at convergence in a comprehensive manner, other countries such as China and Hong Kong have merged only their different ministries and left separate regulatory bodies to reflect some level of convergence. Other countries such as Mongolia have set up inter-governmental agencies to look into overlapping issues, while Singapore stopped short of merging content regulation with telecom and IT industry regulation.

The clash of cultures over how to regulate and what type of regulatory frameworks to have is heating up, both on the national and international fronts. IP has brought the developing countries into the debate that has been taking place since the 1980s. The difficulty in coming up with 'best practices' or 'models' is another reason why people have different definitions of convergence of the Internet, and even of the network versus services. For example, while the telecom world calls Internet working 'interconnection,' the Internet world calls it 'peering agreements'. 'Mesh networks' mean different things to a telecom provider and a software developer. The different approaches, terminologies and approaches to regulation make it hard to create universal models to emulate for new countries adopting convergence.

Many developing countries can take hope in the fact that their regulations are often carried out by the same ministries and so their current state of underdevelopment should be an ideal setting for convergence of the telecommunications, broadcasting and information industries. Moreover, they are being forced by the demands for liberalisation to adopt immediate adequate regulatory measures. Yet, convergence will also force them to look carefully at the implications of convergence and possible discrepancies that may occur within any structure that they put in place. Since wireless and mobile hold the promise for developing countries to leapfrog the digital divide, they will also need to be cognisant of how convergence affects this sector and how to encourage the growth of these industries.

The worlds of mobile telephony and wireless Internet throw up a battery of new issues for regulators in terms of licence allocation, service delivery and interconnect agreements. Ultimately, each country will have to access their current situations as they decide how to change ICT and telecommunications regulations to meet their changing needs.

Resources

International Telecommunications Union: www.itu.int, www.itu.int/itu-d/ict www.regulateonline.org/2002/resources/convergence
InfoComm Development Authority, Singapore: www.ida.gov.sg
Media Development Authority, Singapore: www.mda.gov.sg
Multimedia Development Commission, Malaysia: www.mdc. gov.my
Bridges.org, South Africa: www.bridges.org/e-policy/convergence
Asian Development Bank: www.adb.org/documents/events/2001/ict/ict_links.asp

15. Radio Communications, Mobile Telephony and Regulation in India—Case Study

T.H. Chowdary

Editors' Note: Governments around the world have often bungled the regulation of telecommunications services, radio spectrum and media content—and India is no exception. To be fair, the rapid pace of technology change and the convergence between previously separate realms of technology have thrown up new opportunities (e.g., increasing bandwidth and switching efficiency) as well as challenges (e.g., maximising benefits to consumers while also promoting domestic/global industry players and investors). This chapter highlights some of the regulatory challenges faced in India in terms of mobile operator licensing, interconnect agreements, tariff levels, managing the concerns of incumbents and new entrants, allocation of spectrum for Wi-Fi/Wi-Max, devising progressive regulatory schemes, and creating impartial regulatory bodies. (See the Singapore country profile for related issues and discussion.)

Introduction

This chapter traces key regulatory developments and challenges in the growth of radio communications and mobile telephony in India; many of these issues would have parallels in other developing countries as well. The rapid proliferation of a wide range of media including radio paging, TV, FM and Internet have posed special challenges to the creation and operation of an impartial regulatory authority in areas like spectrum allocation, awarding of licences and determining tariffs. With a backdrop of increasing privatisation of incumbent government telcos, key regulatory objectives include guaranteeing a level playing field between various operators, dealing with media convergence, ensuring financial viability of the market, and meeting public interest concerns.

Evolution of Radio Communications in India

Not long after the invention of radio transmissions by Marconi, India—started using radio under the imperial rule of Britain. In 1902, wireless telegraphy was used between Saugar islands (Bay of Bengal) and Sandheads (near Calcutta). Radio telegraph services for the public were introduced by the Overseas Commission Service (OCS), later renamed as the Videsh Sanchar Nigam Limited (VSNL), in 1927, and radio telephony in 1933. Both services were first to the United Kingdom and progressively to many other countries. OCS/VSNL was also the first to use satellite communications for telegraphy, telex and voice with a satellite earth station in 1971.

During the Second World War, radio communications expanded under OCS. A private company in Mumbai for the first time used radio broadcasting in 1926. It was taken over a few years later as a monopoly of the then imperial government which established All India Radio (AIR), now a part of Prasar Bharati (Broadcasting India).

Radio TV transmissions began in India in mid-1960s. Almost the entire territory of India received TV transmission by the mid-1980s using both satellite and terrestrial radio links to feed the transmitting stations. Departments like civil aviation, coast guard, forest, police, and, of course, the armed forces had been using radio on a very extensive scale and, for understandable reasons, had been given an extraordinarily large spectrum.

For public telecommunications like telephony, radio spectrum was required for long distance, very high frequency (VHF), ultra high frequency (UHF), terrestrial and satellite microwave transmission highways from the mid-1960s onwards. The use of radio spectrum for telecommunications for the public kept increasing. The first microwave radio link was established between Calcutta and Asansol in 1965; the first satellite earth station for international telecommunications was commissioned in 1971 (Arvi, near Pune) and for domestic purposes in November 1980. UHF radio links began to be established from 1974 (Mangalore–Udipi). Bharat Sanchar has now 110,000 route kilometer (RKM) of terrestrial digital microwave, 45,000 RKM of

UHF, and 28,000 RKM of analogue radio systems—linking over 4,000 towns and cities across the country.

Regulation of Radio Communications: The Impact of Mobile Telephony

In order to minimise the interference between various radio transmissions by different users and in order to plan and allot the spectrum and monitor its proper use (both in regard to the frequency band allotted and the transmitted power), the Indian government created the office of Central Wireless Planning and Co-ordination (CWPC) as well as a Wireless Advisor.

The explosive requirement of spectrum had arisen with the introduction of cellular mobile radio telephony in the early 1990s (mobile telephony was introduced in Delhi in December 1985) and the use of radio or wireless in the access segment (from the customer premises to the network point, e.g., a telephone exchange or an Internet point of presence). India is a vast country and the spectrum that is given to it for different uses is just as much as is given for a very small country. Therefore, the complexity of planning, allotment and monitoring should be kept in perspective.

Since urban users have a great requirement for many wireless services, coordination for limiting interference between various transmissions and the availability and site location of towers for mounting antennas is extending the problem from one of radio spectrum management to one of city planning, architecture and environment. Aggravating this problem is the presence of competing multiple operators in mass markets like fixed telephony and mobile telephony. In any given area, if the radiation from several sources exceeds a certain level, there is fear of harmful effects for citizens. Italy has adopted drastic regulation under which, irrespective of the number of radio transmission enterprises, the radiation strength at any site at any time should not exceed a certain number of millivolts. This is a portent for what may have to be enforced in India as well.

As in other countries, India has a problem arising from the fact that the armed and security forces have been allocated large chunks of radio spectrum, only a part of which is utilised for very critical purposes. If they vacate these bands for civilian use and move to other bands, the question of compensation and the time to be allowed for migration constitutes another critical problem.

Radio spectrum is a limited resource and, by tradition, has come to be owned by the state and can be given for private users under strict conditions both with regard to the purpose of use and place as well as the transmitted power and coverage including directivity. The government has been issuing licences for use of radio frequency carriers and bandwidth to private and semi-government users for several purposes: communications between mines and company headquarters and seaports; flood control along the river courses; forestry communications for forest rangers; and for connecting project sites to the public telephone network (because there is no coverage of the public switched telephone network in such greenfield sites).

As long as economic development was slow and confined to some areas only, this was not a big problem. But as development spreads all over the territory, radio spectrum has to be used not only for telegraphy and telephony but for every form of communication. Satellite communications through V-SATs (very small aperture terminals) is the best answer in the last seven to eight years, there has been a spurt in V-SAT usage. Currently, there are over 15,000 V-SAT earth stations in India growing at the rate of 20–25 per cent a year.

Bandwidth and Licensing

All telecommunications was the monopoly of the Government of India, and the prime consideration of the Department of Telecommunication (DoT), now BSNL, was the preservation of its revenue base. At the same time, DoT itself was the licensor of V-SAT facilities, thus imposing certain distorting conditions for V-SAT use. For example, voice transmission was prohibited and therefore the speed of data transmission was limited to 64 Kbps from the V-SATs. This was unrealistic because with modern techniques of digitisation and compression, voice could be transmitted even at 9.6 Kbps! So, technology could undermine the administratively imposed licence conditions, making the users likely criminals.

As video-conferencing and high speed Internet access are also required by enterprises which had to go in for V-SATs to interlink their India-wide offices, constant friction arose—with the users demanding more and more bandwidth, the incumbent monopoly for PSTN resisting these demands, and the minister having to adjudicate on these rival claims.

The real and extensive problem with regard to radio bandwidth started intensifying with the ending of the monopoly in public telecom services. Radio paging, cellular mobile radio telephony and electronic mail (all since 1993) and Internet (since 1998) were services that private companies were first permitted to provide under a competitive duopoly in some cases and multiplicity in some other services.

This threw up certain questions: How much bandwidth would be required for citywide-only cellular mobile telephony? What would happen if the bandwidth allotted got exhausted due to rapid growth of customers? What is the price for the bandwidth allocated? These questions became crucial as the cellular mobile telephony licences—which were first issued only in the metro cities (Mumbai, Delhi, Kolkata and Chennai) without any roaming in the first half of the 1990s—were within a few years given to cover the entire country (but initially with two operators per state and now four operators, with the prospect of a fifth and more being added).

The licences were initially for intra-state roaming. The number of licences a company could hold was also limited. Roaming between two states initially required one more licence. Even if a company had a licence to provide mobile services in two adjacent states which could be easily interconnected by the telco, the licensor, who was also the incumbent government operator, imposed the condition that the calls should go through a transit exchange of the incumbent.

Where the interconnection would take place, at what charges, with what quality of interconnection, and in what time the interconnection could be completed have all become matters of contention between the incumbent (who was for quite some time restrained from providing mobile telephony services) and the private telcos. The deprived incumbent operator in his capacity as the licensor had, thus, all the power to restrict the growth of the upstart private telephone companies.

An independent regulatory body, the Telecom Regulatory Authority of India (TRAI), was created in 1997 by which time almost all the licences had been given away with conditions that were imposed by the incumbent operator in his capacity as licensor and regulator. Even after TRAI was created, the incumbent operator continued to be a government department operating all over India and providing the interconnection between cellular networks within a state and PSTN as well as between two or more cellular networks in different states and the PSTN.

The incumbent was corporatised as Bharat Sanchar Nigam Limited (BSNL) in 2000, but is still owned by the Government of India. Thus, the telecom minister has an unenviable task, as the owner of the incumbent company, on the one hand, and as the minister who must ensure the health of all the private telcos, on the other as well. It is this duality of roles that creates a dilemma for the minister who should really be a saint to reconcile the considerations of equity as a public policymaker and as the de facto head of the incumbent state-owned company!

Regulatory Disputes

Even as the problems of interconnection and adequacy of bandwidth for the rapidly growing cellular mobile subscriber have been plaguing the telecoms, the issuance of licences for competitive private telephone companies for basic (i.e., non-mobile) telephony is further complicating the inter-telco struggles. Technological developments have been enabling the use of WLL (Wireless in the Local Loop) to connect customer premises with fixed telephones to the telephone exchanges without the use of underground copper conductor cables.

Frequency spectrum was allotted to the private telephony licensees with nominal charge and the bandwidth allotted was as much as that for cellular mobile telephony companies. When the basic telephone companies scouted for WLL technology, they found that it was possible to have a dual-use technology which allows fixed point-to-point WLL as well as mobility just like in cellular telephony (with a different type of handset).

The licensor, who is the owner of the state-owned telecom company, found a great opportunity to overcome the deprivation (i.e., not having allowed BSNL to provide state-wide mobile services earlier). The licensor termed WLL use as a limited mobility service. This is just administrative jargon, which has no relation to the technology which is dual use—that is, WLL as well as full mobility in the specified area—just as in 1992 for mobile telephony licences which limited service to the metro-area only.

This has sparked an interminable dispute between limited mobility-providing basic telephone companies and the full mobility cellular-only telephone companies, licensed statewide. A further complication arose when the licensee in a given state won both basic telephony and cellular mobile telephony licenses. About 90 per cent of users of radio telephone users are satisfied with mobility within a city. That means the cellular mobile telephony companies can lose all this segment of the market. That has led to a ferocious dispute between the two sets of telcos. The financial obligations of the two sets of companies by way of entry and licence fees are different. This problem and related ones will eventually be resolved in judicial courts. One complicating factor is that the politician minister has been able to influence the outcome in his capacity as licensor and regulator of last resort (the state's sovereign power).

It was not radio frequency that was sold, auctioned or licensed—it was the service provision that was licensed on the basis of who paid most to the government to acquire the licence. The radio spectrum needed by the service provider was priced in addition to the licence. Till recently the technology by which the spectrum could be used was also specified by the licensor; for example, cellular mobile services had to use GSM; WLL access had to be by CDMA or corDECT (an indigenously developed technology).

Radio spectrum allotments in India are thus in contrast to those which were given free for AMPS and put to multi-stage bidding for PCS in the US, and auctioned for 3G licences in some countries of Europe. Radio frequency spectrum would be needed for certain broadband Internet access applications (point to multi-point distribution). The high costs of spectrum detract from mass use of broadcast Internet.

The unwise management of radio spectrum in India has not been confined to telecommunications but also extends to FM (frequency modulation) broadcasting. Just as the greed to garner money made the DoT auction telecom licences (more or less), FM broadcast licences were also auctioned. Just as the gambler private telephone companies bid fantabulous amounts and then started complaining and eventually migrated to a revenue-sharing arrangement with the government, the private companies which got the FM broadcasting licences are now facing financial death and are petitioning the government to migrate to a revenue-sharing regime as well.

Outlook and Recommendations

This greed in government is not unique to India. It was in evidence in Europe and the US as well with the same harmful consequences and the same remedies, with some politicians benefiting in the process. The few private FM broadcasters in Mumbai city alone are required to pay an annual licence fee which is much more than the licence fee that all the hundreds of FM stations in the US together pay their government! Just as the basic telephone operators withheld implementing their licences until they were migrated to a revenue-sharing, arrangement, those who won licences for FM broadcasting in India are also not implementing their licences. Instead of 108 licences for 40 cities, only 13 have been implemented in nine cities.

Co-ordination and clearance for site locations of radio-base stations and V-SATs is excruciatingly slow. Over a score of government agencies have to 'clear' the site; if any one objects, the search for another site starts. Geographical information systems, computerised procedures and management for spectrum should have been undertaken. If all the 75 per cent of the 3.3 million square kilometres of territory in India are to be covered by five competing cellular operators (not to talk of limited mobility basic telephone operators), the number of radio base stations could be not less than 50,000! To co-ordinate their site location to avoid mutual interference will be a stupendous task.

Another development with huge potential for India is the emergence of Wi-Fi and Wi-Max. The deregulation (that is, no licence is required) of frequency bands in the 2.4 and 5.4 Ghz bands will greatly help the installation of Wi-Fi hot spots for high speed interconnection to the Internet. The government is getting the defence services to vacate certain portions of the spectrum to make it available for telecommunications use. The cost of shifting to another spectrum by the defence services will have to be reimbursed to them from a charge on the beneficiaries. The Indian government should expedite the computerisation of radio spectrum management operations and decentralise the allocation. It should franchise selected, competent companies in different states to advise the applicants as to how much bandwidth is available, and in what RF range, for the specific application.

The allocation, assignment and management of radio spectrum—which is not an unlimited resource—as well as its pricing for good and efficient use are very important public policy issues. Fortunately, technological discoveries are extending the limits of the carrier frequencies, and new modulation and compression techniques are packing more traffic capacity into a given bandwidth.

If governments can give up greed (to collect huge revenues by auctioning bandwidth); if companies search for technologies to reduce interference and maximise traffic-carrying capacity; and if regulators/licensors care for public good and not merely company good, then there could be less strife and delays in bringing the fruits of communications to more and more people.

16. The Cellphone and the Crowd

Messianic Politics in the Contemporary Philippines

Vicente Rafael

Editors' Note: This chapter portrays a fascinating development in the mobile ecosystem: the use of SMS to coordinate popular movements and protests which eventually unseated the president of the Philippines in 2001. In the larger context, cellphones and low cost messaging services helped meet the pent-up communication demand in the Philippines. For 'Generation Txt' users in crowded cities like Manila, cellphones are almost treated like an additional limb. Like many other developing countries, the Philippines on the one hand has access to some of the latest wireless devices while also being, paradoxically, mired in deteriorating terrestrial infrastructure. SMS helped spread rumours, jokes and information that steadily eroded whatever legitimacy President Estrada still had during impeachment proceedings, and cellphone users themselves became editors and broadcasters. This superbly detailed chapter captures the complexity as well as power of humble SMS as a social tool, and provides informative insights into the connections between new media and perceived political notions of empowerment.

Introduction

This essay explores a set of telecommunicative fantasies among the middle classes in the contemporary Philippines within the context of a recent historical occurrence: the civilian-backed coup that overthrew President Joseph Estrada in January 2001. It does so with reference to two distinct media: the cellphone and the crowd.

Various accounts of what has come to be known as 'People Power II' (as distinguished from the populist coup that unseated Ferdinand and Imelda Marcos in 1986) reveal certain pervasive beliefs on the part of the middle classes. They believed, for example, in the power of communication technologies to transmit messages at a distance and in their ability to possess that power. In the same vein, they had faith in their ability to establish their relationship with the masses with whom they regularly shared Manila's crowded streets, utilising the power of crowds to speak to the state. Thus, they saw themselves capable of communicating with the crowd, and also beyond it, transcending the sheer physical density of the latter by technological means while at the same time ordering its movement and using its energy to transmit middle-class demands. At its most utopian, the fetish of communication suggested the possibility of dissolving, however provisionally, existing class divisions. Communication from this perspective held the messianic promise of refashioning the heterogenous crowd into people addressing and addressed by the promise of justice. But as we shall see, such telecommunicative notions were predicated on the putative 'voicelessness' of the masses. For once heard, the masses called attention to the fragility of bourgeois claims to shape the sending and reception of messages about the proper practice of politics in the nation-state. Media politics (understood in both senses of that phrase as the politics of media systems but also politics as the inescapable event of mediation) in this context reveals the unstable workings of Filipino middle-class sentiments. Unsettled in its relationship to social hierarchy, such sentiments at times re-drew class divisions, at other moments anticipated their abolition, and still at others called for their reinstatement and consolidation.[1]

Calling

Telephones were introduced in the Philippines as early as 1885, during the last decade and a half of Spanish colonial rule[2] Like telegraphy before it, telephony provoked fantasies of direct communication among the colonial bourgeoisie. They imagined that these new technologies would afford them access to those on top, enabling them to hear and be heard directly by the colonial state. We can see this telecommunicative notion, for example, in a satirical piece written by the Filipino national hero, Jose Rizal in 1889. Entitled 'Por Telefono', it situates the narrator as an eavesdropper. He listens intently to the sounds and voices that travel between the Spanish friars in Manila—regarded as the real power in the colony—and their superiors in Madrid.[3] The nationalist writer wire-taps his way, as it were, into the walls of the clerical residences, exposing their hypocrisy and excesses. In this sense, the telephone shares in the capacity of that other telecommunicative technology, print, to reveal what was once hidden, to repeat what was meant to be secret, and to pass on messages that were not meant for those outside of a particular circle.[4] It is this history of tapping into and forwarding messages, often in the form of ironic commentaries, jokes, and rumours that figured recently in the civilian-led coup in the Philippines known as 'People Power II'. From the evening of January 16 to January 20, 2001, over a million people massed at one of Manila's major highways, Epifanio de los Santos Avenue, commonly called Edsa, site of the first People Power revolt in 1986 that overthrew the Marcos regime. A large cross-section of Philippine society gathered to demand the resignation of President Joseph 'Erap' Estrada after his impeachment trial was suddenly aborted by the 11 senators widely believed to be under his influence. These senators had refused to include key evidence that would have shown the wealth Estrada had amassed from illegal numbers game while in office. The impeachment proceedings had been watched avidly on national TV and heard to on radio. Most viewers and listeners were keenly aware of the evidence of theft and corruption on the part of Estrada and his family.[5] Once the pro-Estrada senators put an abrupt end to the hearing, however, hundreds of thousands of viewers and listeners were moved to protest in the streets. Television and radio had fixed them in their homes and offices attending to the court proceedings. But at a critical moment, these media also drew them away from their seats. Giving up their position as spectators, they now became part of a crowd that had formed around a common wish: the resignation of the president.

Aside from TV and radio, another communication medium was given credit for spurring the coup: the cellphone. Nearly all accounts of 'People Power II' available to us come from middle class writers or by way of middle class-controlled media with strong nationalist sentiments. And nearly all point to the crucial importance of the cellphone in the rapid mobilisation of people. 'The phone is our weapon now,' one unemployed construction worker is quoted in a newspaper article. 'The power of our cellphones and computers were among the things that lit the fuse which set off the second uprising, or People Power Revolution II, (*Ang lakas ng aming mga celfon (sic) at computer ay isa sa mga nagsilbing mitsa upang pumutok ang ikalawang pag-aalsa o people power revolution II*), according to a college student in Manila. And a newspaper columnist relayed this advice to 'would-be foot-soldiers in any future revolution: As long as your cellphone is not low on battery, you are in the groove, in a fighting *mood*.'[6] 'A technological thing was thus idealised as an agent of change, invested with the power to bring forth new forms of sociality.

Introduced in the latter half of the 1990s, cellphones in the Philippines had become remarkably popular by around 1999.[7] There are a number of reasons for their ubiquity. To begin with, there is the perennial difficulty and expense of acquiring land-line phones in the Philippines along with the erratic service provided by the Philippine Long Distance Company (PLDT) and the more recent, smaller Bayan Tel. Cellphones seemed to promise to fill this pent-up need for connectivity. Additionally, cellphones cost far less than personal computers, of which less than one per cent of the population own, though a larger proportion has access through Internet cafes. By contrast, there are over eight million cellphone users in a population of about 77 million. The great majority of them buy pre-paid phone cards which, combined with the relatively low cost of the phone (as low as $50 in the open market and half this amount in secondary markets) makes this form of wireless communication more accessible and affordable than regular telephones or computers.

Even more significant, cellphones allow users to reach beyond traffic-clogged streets and serve as a quicker alternative to slow, unreliable and expensive postal services. Like many third world countries recently opened to more liberal trade policies, the Philippines shares in the paradox of being awash in the latest technologies of communication such as the cellphone while mired in deteriorating infrastructures such as roads, postal services, railroads, power generators and land lines. With the cellphone, one seems able to pass beyond these obstacles. Since such infrastructures are state run so that their break down and inefficiencies are a direct function of governmental ineptitude, passing beyond them also feels like overcoming the state which, to begin with has long been overcome by corruption.[8] It is small wonder then that cellphones could prove literally handy in spreading rumours, jokes and information that steadily eroded whatever legitimacy President Estrada still had amid his impeachment hearings, along with those of his congressional supporters. Bypassing the broadcasting media, cellphone users themselves became broadcasters, receiving and transmitting both news and gossip and often confounding the two. Indeed, one could imagine each user becoming a broadcasting station unto him or herself, a node in a wider network of communication that the state could not possibly even begin to monitor much less control.[9] Hence, once the call was made for people to mass at Edsa, cellphone users readily forwarded messages they received even as they followed what was asked of them.

Cellphones then were invested not only with the power to surpass crowded conditions and congested surroundings brought about by the state's inability to order everyday life. They were also seen to bring a new kind of crowd about, one that was thoroughly conscious of itself as a movement headed towards a common goal. While telecommunication allows one to escape the crowd, it also opens up the possibility of finding oneself moving in concert with it, filled with its desire and consumed by its energy. In the first case, cellphone users define themselves against a mass of anonymous others. In the second, they become those others, assuming anonymity as a condition of possibility for sociality. To understand how the first is transformed into the second, it helps to note the specific form in which the vast majority of cellphone messages are transmitted in the Philippines: as text messages.

Texting

Text messages are e-mails sent over mobile phones and transferrable to the Internet. Recently, a verb, 'texting' has emerged to designate the act of sending such messages indicating its popularity in such places as England, Japan and Finland (where Nokia first began offering it and where it is referred to as 'SMS' or short message service). In the Philippines, texting became the preferred mode of cellphone use once the two major networks, Globe and Smart introduced free, and then later on, low-cost text messaging as part of their regular service in 1999. Unlike voice messages, text messages take up less bandwidth and require far less time to convert into digitised packets available for transmission. It thus makes economic sense for service providers to encourage the use of text messaging in order to reserve greater bandwidth space for the more expensive and profitable voice messages. From an economic stand point then, texting offers one of those rare points of convergence between the interests of users and providers.[10] But there is obviously more than costs that makes text messaging popular among Filipino users. In an essay sent over the Internet signed 'an anonymous Filipino,' the use of cellphones in Manila is described as a form of 'mania'. Using Taglish, the urban lingua franca that combines Tagalog, English and Spanish, this writer, a Filipino 'balikbayan' (that is, one who resides or works elsewhere and periodically returns to visit the motherland) remarks:

HI! WNA B MY TXT PAL? (Sic) They're everywhere! In the malls, the office, school, the MRT (Manila Railroad Transit), what-have-you, the cellphone (sic) mania's on the loose! Why, even Manang Fishball (i.e., Mrs Fishball, a reference to older working class women vendors who sell fishballs, a popular roadside snack), is texting! I even asked my sisters how important they think they are that they should have cells? Even my nephew in highschool has a cellphone. My mom in fact told me that even in his sleep, my brother's got his cell, and even when they have a PLDT (i.e., land line) phone in the house, they still use the cellphone.[11]

'Mania' according to the *Oxford English Dictionary* is a kind of madness characterised 'by great excitement, extravagant delusions and hallucinations and its acute stage, great violence.' The insistence of having cellphones nearby, the fact that they always seem to be on hand, indicates an attachment to them that surpasses the rational and the utilitarian, as the remarks above indicate. It lends to its holder a sense of being someone, even if he or she is only a street vendor or a high school student. Someone, in this case, who can reach and be reached and is thus always in touch. The 'manic' relationship to cellphones is thus this ready willingness to identify with it, or more precisely with what the machine is thought capable of doing. One not only has access to it; by virtue of its omnipresence and proximity, one becomes like it. That is to say, one becomes an apparatus for sending and receiving messages at all times of the day and night. An American journalist writing in the *New York Times* observes as much in an article on Manila society:

'Texting?' Yes, texting—as in exchanging short typed messages over a cellphone. All over the Philippines, a verb has been born, and Filipinos use it whether they are speaking English or Tagalog.... The difference [between sending e-mail by computers and texting] is that while chat-room denizens sit in contemplative isolation, glued to computer screens, in the Philippines, 'texters' are right out in the throng. Malls are infested with shoppers who appear to be navigating by cellular compass. Groups of diners sit ignoring one another, staring down at their phones as if fumbling with rosaries. Commuters, jaywalkers, even mourners—everyone in the Philippines seems to be texting over the phone.. .. Faye Siytangco, a 23-year-old airline sales representative, was not surprised when at the wake for a friend's father she saw people bowing their heads and gazing toward folded hands. But when their hands started beeping and their thumbs began to move, she realised to her astonishment that they were not in fact praying. 'People were actually sitting there and texting,' Siytangco said. 'Filipinos don't see it as rude anymore.'[12]

Unlike those on computers, cellphone users are mobile, immersed in the crowd, yet communicating beyond it. Texting provides them a way out of the very surroundings they find themselves in. Thanks to the cellphone, they need not be present to others around them. Even when they are part of a socially defined group—say, commuters or mourners—they are always someone and somewhere else, receiving and transmitting messages from beyond their physical location. It is in this sense that they become other than their socially delineated identity: not only cellphone users but cellphone 'maniacs'. Since it rarely leaves them, the phone becomes part of the hand, the digits an extension of their fingers. In certain cases, the hand takes the place of the mouth, the fingers that of the tongue. Writing about his Filipino relative, one Filipino-American contributor to Plaridel, an on-line discussion group dealing with Philippine politics, referred to the former's cellphone as 'almost a new limb.'[13] It is not surprising then that the consciousness of users assumes the mobility and alertness of their gadgets. We can see how this process of taking on of the qualities of the cellphone comes across in the practice of sending and receiving messages.

The craze for sending text message by phone started [in 1999] when Globe introduced pre-paid cards that enabled students, soldiers [and others] too poor for a long-term subscription to start using cellular phones.... People quickly figured out how to express themselves on the phone's alphanumeric keypad.... 'Generation Txt,' as the media dubbed it, was born. When sending text messages one does not need to make a call. People merely type in a message and the recipient's phone number, hit the phone's send key and off it goes to the operator's message centre, which forwards it to the recipient. Since messages are exchanged over the frequency the network uses to identify phones rather than the frequencies their owners talk on, messages can be sent and received the instant a phone is turned on—and can even be received when a phone call is in progress.

Sending text messages by phone is an irritating skill to master, largely because 26 letters plus punctuation have to be created with only 10 buttons. To type the letter C, for example, you need to press the No.2 button three times; an E is the No. 3 button pressed twice; and so on. After the message is composed it can be sent immediately to the phone number of the recipient, who can respond immediately by the same process. People using phones for text messages have developed a shorthand.. 'Where are you?' becomes 'WRU'. And 'See you tonight' becomes 'CU 2NYT.' People have different styles of keying in their messages. Some use their index fingers, some thumb, others both... [Others] tap away with one hand without even looking at [their] phone.[14]

As is frequently the case with e-mail, conventions of grammar, spelling and punctuation are evaded and rearticulated with texting. The constraints of an alphanumeric key pad means that one goes through numbers to get to letters. As a result, counting and writing become closely associated. Digital communication require the use of digits, both one's own and that of the machine, as one taps away. But it is a tapping that is done not to the rhythm of one's speech and in tempo with one's thoughts, but in coordination with the numbers through which one reaches a letter: three taps on 2 to get C, for example, or two taps on 3 to get to E. It is almost as if texting reduces all speech to writing, and all writing to a kind of mechanical percussiveness, a drumming that responds to an external constraint rather than one that emerges from and expresses an internal source. In addition, as it were, there are no prescribed styles for texting: one or two fingers will do, or one can use a thumb and look at the screen, while those adept enough can text while looking elsewhere as in the case of skilled typists. Neither standardised body postures are required with texting: one can sit or walk or drive while sending messages. Where hand writing in the conventional sense requires learning proper penmanship and body postures under the supervision of teachers within the confines of desk and classroom, texting frees the body, or so it seems, from these old constraints.

Mimicking the mobility of their phones, texters move about, unmoored to anything except the technological forms and limits of the medium. The messages they receive and send are condensed versions of whatever language—English or Tagalog and more frequently, Taglish—they are using and so belong to neither. The hybrid form of this language comes from the demands of the medium itself rather than reflecting the idiosyncrasies of their users. The introduction of a limit on the number of free text messages one can send and the assessment of a fee per character of text has meant the further shortening of words and messages. Instant messaging along with mechanical storage and recall of prior messages require only the most drastically abbreviated narrative constructions with little semantic deferral or delay. Using the cellphone, one begins to incorporate its logic and technics to the extent of becoming identified with what appears to be a novel social category: 'Generation Txt' (sic).

An obvious pun on Generation X, 'Generation Txt' first began as an advertising gimmick among cellphone providers in order to attract young users to their products. Defined by their attachment to and skill with the cellphone, Generation Txt also troubled the older generation which was uneasy about the rise of texting. An anthropologist from the University of the Philippines , for example, writes about the dangers of texting in terms that have appeared in other countries where the practice has become popular, especially among youth: its propensity to stifle literacy by '[wreaking] havoc' on spelling and grammar, and, 'working in tandem with mindless computer games and Internet chat rooms, are eroding young people's ability to communicate in the real world in real time.'[15] Rather than promote communication, texting in this view actually obstructs it, cultivating instead a kind of stupidity. This can be seen from the young people's gullibility and willingness to surrender to the marketing ploys of cellphone providers so that they end up spending more, not less, in sending messages of little or no consequence. Furthermore, cellphones actually lead to 'anti-social' behaviour, as users 'retreat to their own cocoons,' while parents who give their children cellphones in effect evade the responsibility of 'interacting' with them in any meaningful way[16]. Other writers report the occasional use of texting by students to cheat on exams, or the use of cellphones to spread rumours and gossip that may ruin someone's reputation.[17] As one Filipino on-line writer put it, cellphones are like 'loaded weapons' and its avid use needs to be tempered with some caution. Another writes that 'if the text [I received] felt like a rumour masquerading as news, I didn't forward it.' An office worker from Manila writes, 'Sometimes whenever you receive serious msgs (sic), sometimes you have to think twice if it is true or if perhaps someone is fooling you since there is so much joking [that goes on] in txt (sic).'[18]

Part of the anxiety surrounding texting arises from its perceived tendency to disrupt protocols of recognition and accountability. Parents are disconnected from their children while children are able to defy parental authority. Cheating is symptomatic of the inability of teachers to monitor the communication of students via cellphone. And the spread of rumours and gossip, along with irreverent jokes, means that the senders of messages readily give in to the compulsion to forward messages without, as the writers above advice, weighing their consequences or veracity. Indeed, it is this capacity to forward messages almost instantaneously that proves to be the most dangerous feature of this 'weapon'. The urge to forward messages one receives seems difficult to resist. Under certain conditions, this urge becomes irrepressible as the events leading to People Power II proved. We can see this happening, for example, in a posting by theatre actor and writer, Bart Guingona to the Plaridel Listserv. As part of a group that planned to stage demonstrations at Edsa on January 18, he initially expressed doubts about the effectiveness of texting for popular mobilisation. 'I was certain it would not be taken seriously unless it was backed up by some kind of authority figure to give it some sort of legitimacy. A priest who was with us suggested that [the Church-owned broadcasting station] Radio Veritas should get involved in disseminating the particulars.... We [then] formulated a test message...and sent it out that night and I turned off my phone.... By the time I turned it on in the morning, the message had come back to me three times.... I am now a firm believer in the power of the text!'[19]

The writer is initially hesitant to resort to texting, thinking that messages sent in this way would be no different from rumours. They would lack authority by themselves. Anonymously passed on from phone to phone, texts seemed unanchored to any particular author that could be held accountable for their contents. Only when the Church-owned radio station offered to disseminate the same information did he agree to sending a text. Waking up the next day, he sees the effect of this transmission. Not only does his message reach others at a distance; they return to him three-fold. From a doubter, he is converted into a 'believer' in the 'power of the text.' Such a power has to do with the capacity to elicit numerous replies.

There are two things worth noting, however, in this notion of the power of texting: first, that it requires, at least in the eyes of this writer and those he sends messages to, another power to legitimise the text's meaning; and second, that such a power is felt precisely in the multiple transmissions of the same text. The power of texting here has less to do with the capacity to open interpretation and stir public debate as it does with compelling others to keep the message in circulation. Receiving a message, one responds by repeating it. One forwards it to others who, it is expected, will do the same. Repeatedly forwarding messages, one gets back one's exact message, mechanically augmented but semantically unaltered. They crowd one's phone mailbox just as those who read and believe in the truth of the call they've received end up crowding the streets of Metro Manila. In this view, the formation of crowds is a direct response to the repeated call of texts now deemed to have legitimacy by virtue of being grounded in an authority outside the text messages themselves: the electronic voice of the Catholic church. Such a voice in effect domesticates the dangers associated with texting. Users can then forward texts and feel themselves similarly forwarded by the expectations they give rise to. Finding themselves called by the message and its constant repetition, they become 'believers,' part of Generation Txt.

Generation Txt thus does not so much name a new social identity as it designates a desire for seeing in messages a meaning guaranteed by an unimpeachable source residing outside the text. Most of those who gathered at Edsa and marched towards Mendiola, the road leading the Presidential Palace, were united in their anger at the corrupt regime of President Estrada and their wish to replace him with a more honest leader. Doing so, however, did not mean changing the nature of the state or doing away with class divisions. Indeed, everything I have read about these events is at pains to stress the legality and constitutionality of these transitions, looking towards the Supreme Court and the Catholic church (rather than either the army or left-wing groups) for institutional legitimacy. In the end, Estrada was replaced not by a new leader, but by one who was part of the same old leadership: his vice-president, and daughter of a former Philippine president, Gloria Macapagal-Arroyo. It would appear then that Generation Txt comes out of what its 'believers'claim to be a 'technological revolution' that sets the question of social revolution aside.

Texting is thus 'revolutionary' in a reformist sense. If it can be said to have a politics, it includes seeking the cleaning up and consolidation of authority, both that of the state and the source of messages. We can see an instructive instance of this politics in a manifesto that appeared in what was till recently one of Manila's more widely read tabloids, *Pinoy Times*. 'Voice of Generation Txt' (*Tinig ng Generation Txt*) by a twenty-something University of the Philippines graduate, Ederic Penaflor Eder, credits the 'power' (*lakas*) of 'our cellphones (sic) and computers' for contributing to the 'explosion' of People Power II. Texting became the medium with which 'we' responded quickly to the 'betrayal' (*kataksilan*) of the pro-Estrada senators who had sought to block the impeachment hearings. Elaborating on this 'we' that is Generation Txt, Eder writes in Taglish:

We are Generation Txt (sic). Free, fun-loving, restless, insistent, hard-working, strong and patriotic.

We warmly receive and embrace with enthusiasm the revolution in new technology. Isn't it said that the Philippines rules Cyberspace and that the Philippines is the text messaging capital of the world? Our response was rapid to the betrayal of the eleven running dogs (*tuta*) of Jose Velarde (a.k.a. Joseph Estrada). The information and calls that reached us by way of text and e-mail was what brought together the organised as well as unorganised protests. From our homes, schools, dormitories, factories, churches, we poured into the streets there to continue the trial—the impeachment trial that had lost its meaning.

...Our wish is for an honest government, and a step towards this is the resignation of Estrada. We are patriotic and strong and with principles, since our coming together is not merely because we want to hang out with our friends, but rather to attain a truly free and clean society brought by our love for the Philippine nation....

There were those from our generation that have long since before the second uprising chosen to struggle and fight in the hills and take up arms, trekking on the harsh road towards real change. Most of us, before and after the second uprising, can be found in schools, offices, or factories, going about our everyday lives. Dreaming, working hard for a future. Texting, internetting, entertaining ourselves in the present.

But when the times call, we are ready to respond. Again and again, we will use our youth and our gadgets (*gadyet*) to insure the freedom of our Motherland.... After the second uprising, we promise to militantly watch over the administration of Gloria Macapagal Arroyo while we happily push Asiong Salonga (a.k.a. Joseph Estrada) into the doors of prison.

We are Generation Txt (sic).[20]

This statement of identity curiously enough does not identify who this 'we' is except as those who 'warmly accept and embrace'the 'revolution' in new technology. The 'we' that is invoked here comes about through its identification with technology and its purported new-ness that situates the country globally as the 'text messaging' capital of the world. It is perhaps for this reason that the message reads as if it was meant to be received then forwarded: it begins and ends with exactly the same lines: *Kami ang Generation Txt* (We are Generation Txt). Rather than develop ideas or put forth an analysis of social relations, Generation Txt has attitudes and affects: *malaya* (free), *masayahin* (fun-loving), *malikot* (restless), *makulit* (insistent), *masipag* (hardworking) and so forth. They pride themselves in having principles and courage, and unlike the rudderless and westernised Generation X, they have direction. They stand for 'transparent' government, and a 'free' and 'clean' society. In this sense, they are really not that different from their elders for they are patriots (*makabayan*) dedicated to using their 'gadgets' for the sake of the motherland (*Inang Bayan*). Such commitment comes in the form of a 'militant' readiness to watch over the workings of the new government in order to insure 'justice'(*katarungan*). Unlike those who have chosen to take up arms and go to the mountains, Generation Txt can be found in schools, offices, and factories, ready to respond to the call of the times. They watch, they wait, and they are always ready to receive and forward messages.

The interest of Generation Txt lies not in challenging the structures of authority but in making sure they function to serve the country's needs. This reformist impetus is spelled out in terms of their demand for accountability and their intention of holding leaders under scrutiny. Through their gadgets, they hold on to this holding, keeping watch over leaders rather than taking their place or putting forth other notions of leadership. Thus does Generation Txt conceptualise its historical agency: as speedy (*mabilis*) transmitters of calls (*panawagan*) that come from elsewhere and which have the effect of calling out those in their 'homes, schools, dormitories, factories, churches' to flood the streets in protest. Rather than originate such calls, they are able to trace them to their destination which, in this case, is the nation of middle class citizens as it seeks to renovate and keep watch over the state. Like the first generation of bourgeois nationalists in the 19th century I cited earlier, Generation Txt discovers yet again the fetish of technology as that which endows one with the capacity to seek access to and recognition from authority.[21]

Crowding

From the perspective of Generation Txt, a certain kind of crowd comes about in response to texting. It is one that bears, in both senses of that word, the hegemony of middle class intentions. Texting in its apolitical mode, sought to evade the crowd. But in its reformist mode, it is credited with converting the crowd into the concerted movement of an aggrieved people. In the latter case, the middle class invests the crowd with a power analogous to their cellphones: that of transmitting their wish for a moral community, whereby the act of transmission itself amounts to the realisation of such a community. Such a notion assumes the possibility of endowing the crowd with an identity continuous with that of middle class texters. However, this assumption had another aspect. Not only did it lead to the fantasy of ordering of the masses under bourgeois direction. As I demonstrate below, the middle class interest in ordering the crowd also tended to give way to a different development. At certain moments, we also see the materialisation of another kind of desire this time for the dissolution of class hierarchy altogether. How so?

To understand the contradictory nature of middle class ideas about crowds, it helps to look at the streets of Manila at the turn of the 21st century. The city has a population of over 10 million, a large number of whom are rural migrants in search of jobs, education, and other opportunities unavailable in the provinces. Congested conditions—packed commuter trains, traffic-clogged roads, crowded sidewalks, teeming shopping malls-characterise everyday life in the city making travel from one place to another slow and tedious throughout the day and late into the night. Such conditions affect all social classes. And because there is no way of definitively escaping them, they constitute the most common and widely shared experience of city life.

Just as the roads are clogged with vehicles, so the sidewalks seem unable to contain the unending tide of pedestrians who spill out onto the highways, weaving in and out of vehicular traffic. Indeed, one of the most anomalous sights on Manila sidewalks are signs for wheelchair access. Given the uneven surfaces and packed conditions of sidewalks, such signs can only be the traces of a possibility that has never been realised, a future overlooked and forgotten. It is as if at one point, someone had the thought of organising urban space along the lines of a liberal notion of accommodation. Instead, that thought itself quickly gave way to what everywhere seems like the inexorable surrender of space to people who use it and use it up.

Urban space in Manila thus seems haphazardly planned. It is as if no central design had been put in place and no rationalising authority at work to organise and coordinate the movement of people and things.[22] Instead, such movement occurs seemingly on its own accord. Pedestrians habitually jaywalk and jump over street barriers. Cars and busses belch smoke, criss-crossing dividing medians, if these exist at all, inching their way to their destinations. Drivers and passengers find it difficult to see beyond a few feet of their vehicles. The windshields and windows of jeepneys, tricycles and cabs are usually filled with decals, curtains, detachable sun shades, and other ornaments that make it difficult to get a view of the road, in effect obstructing one's vision and further heightening the sensation of congestion. Indeed, given Manila's topographical flatness, it is impossible to get a panoramic view of the city except on the commuter trains and from on top of tall buildings. In the West, the 'view' is understood as the site for evacuating a sense of internal unease and a resource for relieving oneself of pressure, both social and psychic. Such a notion of a view is not possible in Manila's streets. Caught in traffic, one looks out to see the view of more stalled traffic so that the inside and the outside of vehicles seem to mirror one another.

Adding to the sense of congestion is the presence of garbage. The disposal of garbage has long been a problem in Manila owing to, among other reasons, the difficulty in finding adequate landfills . As a result, trash seems to be everywhere, as if it were dumped indiscriminately on street corners or around telephone poles, some of which have signs that explicitly ask people not to urinate or dump garbage there. What appears are thus scenes of near ruin and rubble. While certainly not exclusive to Manila, such scenes bespeak of a city giving in to the pressures of a swelling population. Rather than regulate contact and channel the efficient movement of people and things, the city's design, such as it is, seems to be under constant construction from the ground up and from so many different directions. No singular and overarching authority seems to be in charge. To walk or ride around in Manila then is to be impressed by the power of crowds. Their hold on urban space appear to elude any attempt at centralising control. It is perhaps for this reason that the largest private spaces open to the public in Manila, shopping malls, play what to an outsider might seem to be extremely loud background music. A shopping mall manager once told me that turning the volume up was a way of reminding the crowd in malls that unlike the streets, someone was in charge and therefore watching their actions.[23]

The anonymity characteristic of crowds makes it difficult, if not impossible, to differentiate individuals into precise social categories. Clothes are at times clues to the social origins of people, but with the exception of beggars, it is difficult to tell on the basis of looks alone. The sense that one gets from moving in and through crowds is of a relentless and indeterminable mixing of social groups. This pervasive sense of social mixing contrasts sharply with the class and linguistic hierarchies that govern political structures and social relations in middle class homes, schools, churches and other urban spaces.[24] One becomes part of the crowd by becoming other than one's social self. Estranged, one becomes like everyone else. Social hierarchy certainly does not disappear on the streets. But like the police who are barely visible, appearing mostly to collect pay-offs (*tong* or *lagay*) from jeepney drivers and sidewalk vendors, hierarchy feels more arbitrary, its hold loosened by the anonymous sway of the crowd.

The power of the crowd thus comes across in its capacity to overwhelm the physical constraints of urban planning and to blur social distinctions by provoking a sense of estrangement. Its authority rests on its ability to promote restlessness and movement, thereby undermining the pressure from state technocrats, church authorities and corporate interests to regulate and contain such movements. In this sense, the crowd is a sort of medium; if by that word one the means for gathering and transforming elements, objects, people and things. As a medium, the crowd is also the site for the generation expectations and the circulation of messages. It is in this sense that we might also think of the crowd not merely as an effect of technological devices, but as a kind of technology itself. It calls incessantly and we find ourselves compelled to respond to it. The crowd as a kind of technology refers then not merely to its potential as instruments of production or as an exploitable surplus for the formation of social order. It also constitutes the context of and the content for a technic of engaging the world. The insistent and recurring proximity of anonymous others creates a current of expectation, of something that might arrive, of events that might happen. As a site of potential happenings, it is a kind of place for the generation of the unknown and the unexpected. Centralised urban planning and technologies of policing seek to routinise the sense of contingency generated in crowding. But at moments and in areas where such planning chronically fails, routine can at times give way to the epochal. At such moments, the crowd as I hope to show below takes on a kind of telecommunicative power, serving up channels for sending messages at a distance while bringing distances up close. Enmeshed in a crowd, one feels the potential for reaching out across social space and temporal divides.[25]

As we saw, middle class discourses on the cellphone tend to set texting in opposition to the crowd precisely as that which overcomes the latter during normal times. But in more politically charged moments such as People Power II, cellphones were credited along with radio, TV and the Internet for calling forth the crowd and organising the flow of its desire, turning it into a resource for the reformation of social order. Other accounts, however, indicate the crowd's potential for bringing about something else, transmitting messages which at times converged with, but at other times submerged those emanating from cellphones. For at times, the crowd made possible a different kind of experience for the middle class. Such had to do less with representing the masses as becoming one with them. In so doing, the crowd becomes a media for the recurrence of another fantasy which emanates from the utopic side of bourgeois nationalist wishfulness: the abolition of social hierarchy.[26] We can see the recurrence of this fantasy and the desire to do away with hierarchy in one of the more lucid accounts of the crowd's power from a posting by 'Flor C' on the Internet discussion group, Plaridel.[27] The text originally in Taglish is worth following at some length.

'I just want to share my own way of rallying at the Edsa Shrine,' (*Gusto ko lang ibahagi ang sarili kong siste sa pagrali sa Edsa Shrine*), Flor C begins. She invites others do the same, adding, 'I am also eager to see the personal stories of the 'veterans' of Mendiola.' (*Sabik din akong makita ang mga personal na kuwento ng mga beteranong Mendiola*). Here, the urge to relate her experiences at the protests comes with the desire to hear others tell their own stories. What she transmits is a text specific to her life, not one that comes from somewhere else and which merely passes through her. Yet, by signing herself as 'Flor C', it is difficult to tell who this story pertains to outside of that signature. Neither is it possible to tell who authorises its telling. In this way, she remains anonymous to her readers, the vast majority of whom similarly remain unknown to her.[28] What is the relationship between anonymity and the eagerness to tell and hear about experiences, one's own as well as that of others'?

Flor C recalls the practice of protest marchers from the 1970s and 1980s of having what is called a 'buddy-system' (sic) for guarding against infiltration from fifth columnists and harassment by the military and police. But because 'my feet were too itchy so that I could not stay in the place that we agreed to meet,' (*masyadong makati ang talampakan ko imbes na tumigil sa puwesto namin*), she ends up without a 'buddy' at Edsa. Instead, she finds herself swimming in the 'undulating river, without let-up from Edsa and Ortigas Avenue that formed the sea at the Shrine,' (*ilog na dumadaloy, walang patid, mula sa Edsa sa Ortigas Avenue at bumubuo ng dagat sa Shrine*). She can't keep still. She feels compelled to keep moving, allowing herself to be carried away from those who recognize her. At Edsa, she knows no one and no one knows her. Yet the absence of recognition is cause neither for dismay nor longing for some sort of identity. Instead, she relishes the loss of place brought about by her absorption into the movement of the crowd. She finds herself in a community outside of any community. It fills her with excitement (*sabik*). But rather than reach for a cellphone, she does something else: she takes out her camera.

And so I was eager to witness (*kaya nga sabik akong masaksihan*) everything that was happening and took photographs. Walking, aiming the camera here and there, inserted into the thick waves of people who also kept moving and changing places, walked all day until midnight the interiors of the Galleria [shopping mall], around the stage and the whole length of the Edsa-Ortigas flyover. Sometimes stopping to listen for a while to the programme on stage, shouting 'Erap resign!,' and taking close-ups of the angry, cussing placards, T-shirts, and posters and other scenes; 'Good Samaritans' giving away mineral water and candy bars, a poor family where the mother and child were laying on a mat while the father watched over, a group of rich folks on their Harley Davidsons, Honda 500's, and Sym scooters that sparkled.... And many other different scenes that were vibrant in their similarities but also in their differences.

Immersed in the crowd, Flor C begins to take photographs. The camera replaces the cellphone as the medium for registering experience. In the passage above, she initially refers to herself as *ako*, or 'I,' the first person pronoun singular in Tagalog. But once she starts to take photographs, the 'I' disappears. The sentences that follow do not contain any pronouns at all. It is as if there is no person performing the acts of walking, moving, listening and looking. While we can certainly read these sentences to imply a person carrying out these activities, we could just as easily infer the agency of some other thing at work: an 'it' rather than an 'I.' That 'it' of course is the camera that Flor C takes out and begins to aim (*tinutok*). Led by her desire to be among the crowd, she begins to act and see like her camera. She stops to listen, then moves on, taking close-ups of 'scenes' (*eksenas*) made up of the juxtaposition of various social classes. She is thus drawn to the appearance of sharp 'contrasts' (*pagkaiba*) that are thrown together, existing side by side without one seeming to dominate the other. The juxtaposition of contrasts, the proximity of social distances, the desire to come up close to all manner of expressions and signs, to bring these within a common visual field, but one whose boundaries and focus keep shifting: such becomes the vocation of Flor C's camera. These are also the very features associated with the crowd. The crowd drives Flor C to take out her camera; and the camera in registering the mixing of differences reiterates the workings of the crowd. Becoming the camera that brings distances up close and holds differences in sharp juxtaposition, Flor C begins to take on the telecommunicative power of the crowd. Yet, unlike the cellphone whose political usefulness requires the legitimation of messages by an outside authority, the crowd in Flor C's account seems to draw its power from itself. It does not look outside of itself, at least in this instance, precisely insofar as the crowd tends to erode the border between inside and outside. We can see further this blurring of boundaries in Flor C's account of entering the Galleria shopping mall next to the center stage of the Edsa protest:

Many times I entered the Galleria to line up for the restroom and at the juice store. During one of my trips there, I was shocked and thrilled (*kinilabutan ako*) when I heard 'Erap resign!' resonating from the food center, cresting upwards the escalator, aisles and stores. The mall became black from the 'advance' of middle-class rallyists wearing the uniform symbolic of the death of justice. But the whole place was happy (*masaya*). Even the security guards at the entrance simply smiled since they could not individually inspect the bags that came before them....

She is thrilled and shocked (*kinilabutan ako*) by a sonic wave making its way from the bottom of the shopping mall. Middle class 'rallyists' dressed in black surged through the aisles, protesting rather than shopping. Like all modern retail spaces, the shopping mall is designed to manufacture novelty and surprise only to contain them within the limits of surveillance and commodity consumption. But during these days, it is converted into a site for the wholly unexpected and unforseen. Ordinarily, the mall is meant to keep the streets at bay. Now it suddenly merges with them, creating a kind of uncanny enjoyment that even the security guards cannot resist. Formerly anonymous shoppers, middle class protestors now come across en masse. As shoppers, they had consumed the products of others' labor. But as demonstrators, they now shed what made them distinct. They set aside their identity as consumers. They are instead consumed and transformed by the crowd. While still recognisably middle class, they nonetheless appear otherwise, advancing in their black shirts and chanting their slogans. To Flor C, their unfamiliar familiarity produces powerful effects. In the mall, Flor C finds herself to be somewhere else. And as with the scenes in the streets, the intensification of this sense of displacement becomes the basis for the sensation of a fleeting and pleasurable connection with the crowd.

It is worth noting, however, that displacement as the source of pleasure can also, at certain moments, become the occasion for anxiety and fear. What is remarkable about Flor C's narrative is the way in which she takes on rather than evades this fear. The result, as we will see in the concluding section of her story, is neither the mastery nor overcoming of the crowd's disorienting pull. Rather, it is the realisation of what she conceives to be the saving power of the crowd. Back on the streets, she wanders onto a flyover, or an on-ramp at the Edsa highway.

When I first went to the flyover, I was caught in the thick waves of people far from the center of the rally. I could barely breath from the weight of the bodies pressing on my back and sides. I started to regret going to this place that was so packed that not even a needle could have gone through the spaces between the bodies. After what seemed like an eternity of extremely small movements, slowly, slowly, there appeared a clearing before me (*lumuwag bigla sa harap ko*). I was grateful not because I survived but because I experienced the discipline and respect of one for the other of the people—there was no pushing, no insulting, everyone even helped each other, and a collective patience and giving way ruled (*kolektibong pasensiya at pagbibigayan ang umiral*).

The night deepened. Hungry again. Legs and feet hurting. I bought squid balls and sat on the edge of the sidewalk.... While resting on the sidewalk, I felt such immense pleasure, safe from danger, free, happy in the middle of thousands and thousands of anonymous buddies.

Finding herself in the midst of a particularly dense gathering of bodies, Flor C momentarily fears for her life. She can barely breathe, overwhelmed by the weight of bodies pressed up against her. Rather than a medium for movement, the crowd in this instance becomes a kind of trap, fixing her in place. But ever so slowly, the crowd moves as if on its own accord. No one says anything, no directives are issued, no leader appears to reposition bodies. Instead a kind of 'collective patience and giving way ruled' (*kolektibong pasyensya at pagbibgayan ang umiral*). The crowd gives and takes, taking while giving, giving while taking and so suffers the presence of all those that comprise it. It is for this reason 'patient,' which is to say, forbearing and forgiving while forgetting the identities of those it holds and who hold on to it. Forbearance, forgiveness and forgetting are always slow, so slow in coming. They thus share in, if not constitute, the rhythm of the work of mourning which in turn always entails the sharing of work.

After what seemed like an eternity of waiting and very little moving, Flor C suddenly arrives at a clearing. *Lumuwag bigla sa harap ko* 'it suddenly cleared in front of me' she says, which can also be glossed as 'the clearing came before me.' Who or what came before whom or what remains tantalisingly uncertain. Earlier, she had started to regret being trapped in the crowd. But thrown into a sudden clearing by a force which came from within that which was radically outside of her yet which she had become an ineluctable part of, Flor C is grateful. She survives, but that is not the most important thing for her. Rather, what matters is that she was given the chance to experience the 'discipline and respect' of the crowd where no one was pushed or pushing, no one was insulted or insulting, and everyone seemed to help one another, a condition that in Tagalog is referred to as *damayan*, or cooperation, the very same word used to connote the work of mourning.[29] Flor C's account also brings to mind the experience of crowding in certain religious gatherings, notably the all male procession of the image of Black Nazarene that marks the high point of the fiesta of Quiapo, a district of Manila on 9 January. For a description of the 1995 procession that conveys some sense of the dangers and pleasures experienced by both on-lookers and practitioners alike in the experience of crowding, see Jaime C. Laya, 'The Black Nazarene of Quiapo,' in *Letras y Figuras: Business in Culture, Culture in Business,* Manila: Anvil Publishing, Inc., 2001, 86–90. It is a strange sort of discipline that she undergoes. It is one that does not form subjects through systematic subjugation en route to establishing hierarchies of recognition. Instead, it is a kind of discipline borne of mutual restraint and deference which, inasmuch as it does not consolidate identity, sets aside social distinctions.

Crowding gives rise to an experience of forbearance and a general economy of deference. At the same time it does not result in the conservation of social identity. Rather, it gives way to a kind of saving which Flor C refers to as the experience of 'freedom' (*kalayaan*). Far from being a mob, the crowd here is a principle of freedom and incalculable pleasure. It is where a different sense of collectivity resides, one that does away momentarily with hierarchy and the need for recognition. Constraint gives way to an unexpected clearing, to a giving way that opens the way for the other to be free, the other which now includes the self caught in the crowd. And because it is unexpected, this freeing cannot last just as it cannot be the last experience of freedom. Emancipation, however brief—and perhaps because it is felt to be so—depends here not on submission to a higher authority that guarantees the veracity of messages. Rather, it relies on the dense gathering of bodies held in patient anticipation of a clearing and release.

Accounts of People Power II indicate that over a million people gathered in the course of four days at Edsa. These included not only the middle classes. As Flor C's earlier remarks show, many from the ranks of the working classes as well as the urban and rural poor who opposed Estrada were also there. A heterogenous crowd formed not simply in response to texting, for obviously not everyone had cellphones. It emerged primarily, we might imagine, in response to a call for and the call of justice. Put another way, the crowd at Edsa was held together by the promise of justice's arrival. Here, justice is understood not simply in terms of a re-distributive force acting to avenge past wrongs, one that in its use of violence is productive of more injustice. The non-violent nature of People Power II suggests instead that the crowd formed not to exact revenge but to await justice. To do so is to dwell in a promise which, qua promise, is always yet to be realised. Like freedom and no doubt inseparable from it, justice is thus always poised to arrive from the future. And it is the unceasing uncertainty of its arrival that constitutes the present waiting of the crowd. The crowd in this case is a gathering which greets that whose arrival is never fully completed, forbearing this coming that is always deferred. Yet, it is precisely because justice comes by not fully coming, and coming in ways unexpected that it comes across as that which is free from any particular socio-technical determination. It is this promise of justice that is conveyed by Flor C's experience of the crowd. The promissary nature of justice means that it is an event whose eventfulness occurs in advance of and beyond any given political and social order. Evading reification and exceeding institutional consolidation, such an event entails a telecommunication of sorts. It is what Jacques Derrida might call the messianic without a messiah. It would be 'the opening up to the future or to the coming of the other as the advent of justice.... It follows no determinable revelation.... This messianicity stripped of everything, this faith without dogma....'[30] In the midst of messianic transmissions, Flor C along with others around her imagine the dissolution of class differences and feel, at least momentarily, as if it were possible to overcome social inequities. She sees in crowding therefore a power that levels the power of the social as such. Past midnight, Flor C finds herself no longer simply herself. Her body hurting, bearing the traces of the crowd's saving power, she sits on the sidewalk, eating squid balls, happy and safe, free in the midst of countless and anonymous 'buddies.'

Postscript

Utopias of course do not last even if their occasional and unexpected happenings are never the last.

Some three months after People Power II, the recently installed government of President Gloria Macapagal-Arroyo made good on its promise to arrest former President Estrada on charges of graft and corruption. On April 25, 2001, he was taken from his residence, fingerprinted and photographed, his mug shot displayed for all to see in the media. The sight of Estrada treated as a common criminal infuriated his numerous supporters, many of whom came from the ranks of the urban poor and who had given him the largest majority in any presidential election in the country. Spurred by the middle class leaders of Estrada's party, Puwersa ng Masa (Force of the Masses) and swelled by the ranks of the pro-Estrada Protestant sect, Iglesia ni Cristo and the populist Catholic group, El Shaddai, a crowd of up to one hundred thousand formed at Edsa raucously demanding Estrada's release and reinstatement. Unlike those who gathered there during People Power II, the crowd in what came to be billed as the 'Poor People Power' were brought in by the truck loads by Estrada's political operatives from the slums and near by provinces, provided with money, food and on at least certain occasions, alcohol. Rather than cellphones, many were reportedly armed with sling shots, home made guns, knives and steel pipes. English-language news reports described this crowd as unruly and uncivilised and castigated them for strewing garbage on the Edsa Shrine, cussing at reporters and urinating on the walls by the giant statue of the Virgin Mary of Edsa.[31]

Other accounts qualified these depictions by pointing out that many of those in the crowd were not merely hired thugs or demented loyalists but poor people who had legitimate complaints. They had been largely ignored by the elite politicians, the Catholic Church hierarchy, the middle-class dominated left-wing groups and the NGOs. They saw in Estrada a kind of patron who had given them hope by way of occasional hand-outs and who spoke to them in their vernacular even as he manipulated them. And unenlightened as they presently were, they deserved 'compassion' from 'us,' the nationalist middle class whose duty it was to uplift them to the latter's level of political and moral consciousness. The great majority of middle class opinion thus shared in the view that the pro-Estrada crowd was profoundly different from that which gathered in January during People Power II. Where the latter was technologically savvy and politically sophisticated, the former was retrograde and reactionary. In the earlier case, Generation Txt spoke of democratisation, accountability and civil society; in the other, the 'tsinelas crowd', so-called because of the cheap rubber slippers many of them wore, were fixated on their 'idol.' In their mystified state, they seemed to the middle class barely articulate, incapable of formulating their sentiments except in terms of seeking vengeance on those they deemed responsible for victimising their leader. If those in People Power II responded to the circulation of messages sanctioned by a higher authority, as well as to the prospect of justice as the promise of freedom, the 'masa' in People Power III were merely playing out its tragically mistaken identification with Estrada. They sought, or so it was assumed, a kind of crude payback characteristic of many of the former president's movie plots.[32]

Middle class accounts of this other crowd regularly made mention of the 'voicelessness' of the urban poor. At the same time, such accounts showed a relative lack of interest in actually hearing much less recording any distinctive voices. By remarking on the masses' lack of voice, the middle class in effect redoubled the former's seeming inarticulateness. It is almost as if the masses, without anything intelligible to say, could only say the same thing. 'Voiceless,' the masses could only riot in the streets. Indeed, in the early morning of May 1, they marched from the Edsa shrine to the Presidential Palace, destroying millions of pesos of property, resulting in several deaths and scores of wounded until they were dispersed by the police and Palace guards. It is important to note though that while marching to the Palace, the masses chanted slogans. Newspaper reports quote these slogans and in so doing give us a rare chance to actually hear the crowd. On the move, it addresses us with such things as *Nandito na kami, malapit na ang tagumpay*, (We're here, our victory is close at hand!) and *Patalsikin si Gloria! Ibalik si Erap! Nandyan na kami! Maghanda na kayo!* (Get rid of Gloria! Return Erap! We are coming! Get ready!).[33]

Here, the crowd is fueled by the desire to give back to Gloria what they think she's given to them. In exchange for unseating Estrada, they want to unseat her. She took his place, and now they want him to take hers. Through their slogans, the crowd expresses this giving back of a prior taking away. It says: 'We are here, our victory is close at hand;' 'We are coming, you better be ready!' The crowd thereby takes for itself an apocalyptic power. 'We' here has already arrived even as it continues to come. Certain of its arrival, it asks those who hear to be ready. Having arrived, they will settle their debts, collect what is owed to them and thereby put an end to their—the crowd's and the listeners'—waiting. Where the crowd in People Power II clung to the sense of the messianic without a messiah, this other crowd comes as a messianic specter delivered by resentments whose satisfaction could no longer be deferred. It is perhaps for this reason that middle class observers repeatedly referred to it in English as a 'mob,' a 'rabble' made up of 'hordes'. These words imply not only 'savagery' and disordered appearance and speech. As the word 'horde' indicates, the masses were also seen to be irreducibly alien: foreign invaders stealing upon a place they had no place in.[34]

Eschewing a stance of forbearance, this crowd demanded recognition without delay. 'Here we are,' it shouted. 'Be prepared.' For many among the middle class, to hear this crowd was to realise that they were not quite ready to hear them, and that they will always have been unprepared to do so. The masses became suddenly visible in a country where the poor are often seen by the middle class to be unsightly, spoken about and down to because deemed incapable of speaking up for themselves. They are thus acknowledged in order to be dismissed. Marching to the Palace, however, and chanting their slogans, they assumed an apocalyptic agency. They threatened to bring about a day of reckoning that was simultaneously desired and dreaded by those who saw them. In their uncanny visibility, the masses did not so much gain a 'voice' that corresponded to a new social identity. Instead, they communicated an excess of communication that could neither be summed up nor fully accounted for by those who heard them. Unprepared to hear the crowd's demand that they be prepared, the middle class could only regard it as monstrous. Hence the bourgeois calls for the conversion of the masses and their containment by means of 'pity,' 'compassion' and some combination of social programmes and educational reform. But such calls also demanded that those who made up this crowd, one that was now totally other, be put back in their place, removed like garbage from the Edsa shrine and from the periphery of the presidential palace.[35] By the late morning of labor day, the military, spooked earlier by the specter of 'poor people power,' had dispersed the marchers. Their violent outbursts like their abandoned rubber slippers were relegated to the memory of injustices left unanswered, fueling the promise of revenge and feeding the anticipation of more uprisings into the future.

Notes

[My thanks to Pete Lacaba and the contributors to Plaridel, to RayVi Sunico, Tina Cuyugan, Lita Puyat, Karina Bolasco, Jose and David Rafael, Carol Dahl, Chandra Mukerji, Matt Ratto, Paula Chakrabarty, Teresa Caldeira, James Holston, Jean-Paul Dumont, Adi Hastings, and Michael Silverstein for providing me with a variety of sources and insights that proved invaluable for this essay. I am especially grateful to Rosalind Morris and Michael Meeker for offering thoughtful comments on earlier drafts of this essay].

Endnotes

[1] The link between telecommunication technologies and the politics of belief that I pursue here is indebted partly to the work of Jacques Derrida, especially in such writings as '*Faith and Knowledge*: *The Two Sources of "Religion" at the Limits of Reason Alone*,' translated by Sam Weber in Jacques Derrida, *Acts of Religion*, ed. by Gil Anidjar, New York: Routledge, 2002, 42–101; 'Signature Event Context,' in *Margins of Philosophy*, translated by Alan Bass, Chicago: University of Chicago Press, 1982, 307–330; and *The Politics of Friendship*, translated by George Collins, London: Verso, 1977.

[2]See the bundle entitled 'Telefonos, 1885–1891' at the Philippine National Archives, Manila for sketches of a plan to install a telephone system in the city as early as November, 1885. By December 1885, an office of Telephone Communication had been established (*Communicacion Telefonica*) and the first telephone station set up on the same date at Santa Lucia, Manila.

[3] Jose Rizal, 'Por Telefono', Barcelona, 1889. Reprinted in Manila: R. Martinez and Sons, 1959 and in various other anthologies of Rizal's writings. For a more extended discussion of telegraphy and the formation of a wish for a lingua franca among the first generation of nationalists, see Vicente Rafael, 'Translation and Revenge: Castilian and the Origins of Nationalism in the Philippines,' in Doris Sommer, ed., *The Places of History: Regionalism Revisited in Latin America*, Durham: Duke University Press, 199, 214–35.

[4] For an elaboration of other modalities of these telecommunicative fantasies and their role in shaping nationalist consciousness, see Vicente Rafael, *White Love and Other Events in Philippines History*, Durham: Duke University Press, 2000, especially Chapters 4 and 8 on rumour and gossip as populist modes of communication in Philippine history.

[5] For a useful collection of documents and newspaper articles relating to the corruption case against Estrada, see Sheila Coronel, ed., *Investigating Estrada: Millions, Mansions and Mistresses,* Quezon City: Philippine Center for Investigative Journalism, 2000.

[6] The quotations above come respectively from Uli Schmetzer, 'Cellphones Spurred Filipinos,' *Chicago Tribune,* January 24, 2001; Ederic Penaflor Eder, 'Tinig Ng Genertion Txt', *Pinoy Times* February 8, 2001; Malou Mangahas, 'Text Messaging Comes of Age in the Philippines,' *Reuters Technology News*, January 28, 2001.

[7] Wayne Arnold, 'Manila's Talk of the Town is Text Messaging,' *New York Times*, July 5, 2000.

[8] For a succinct historical analysis of the Philippine state, see Benedict Anderson, 'Cacique Democracy in the Philippines, in his book *The Specter of Comparisons*, London: Verso 1998, 192–226. See also John Sidel, *Capital, Coercion, and Crime: Bossism in the Philippines,* Stanford: Stanford University Press, 1999; and Paul D Hutchcroft, *Booty Capitalism: The Politics of Banking in the Philippines*, Ithaca: Cornell University Press, 1998.

[9] The technology for monitoring cellphone use does exist and there is some indication that the Philippine government is beginning to acquire these. It is doubtful, however, that such technology had been available under Estrada. It is also not clear whether the current regime of Gloria Macapagal-Arroyo has begun monitoring or intends to monitor cellphone transmissions.

[10] See Arnold, 'Manila's Talk of the Town is Text Messaging,'; Mangahas, 'Text Messaging Comes of Age in the Philippines,'; Schmetzer, 'Cellphones Spurred Filipinos' Coup.' See also Leah Salterio, 'Text Power in Edsa 2001,' *Philippine Daily Inquirer*, January 22, 2001, (hereafter indicated as PDI); Conrad de Quiros, 'Undiscovered Country,' PDI, February 6, 2001; Michael L. Lim, 'Taming the Cellphone,' PDI, February 6, 2001. There are certain limits to this economic advantage, however. For example, it is expensive to call across networks, so that calling or texting from a Globe phone to a Smart phone is rarely ever done. Indeed, the Department of Transportation and Communication (DOTC) at one point had to intervene in late 1999 to get the two companies to improve inter-connectivity and service as well as lower their costs.

[11] This article was being circulated around the listserves of various NGOs in the Philippines and bore the title 'Pinoy Lifestyle'. I have no knowledge as to the original source of this piece and so it exists in some ways like a forwarded text message. Thanks to Tina Cuyugan (tinacuyugan@mindanao.org <mailto:(tinacuyugan@mindanao.org>) for forwarding this essay to me. All translations are mine unless otherwise indicated.

[12] Wayne Arnold, 'Manila's Talk of the Town is Text Messaging.'

[13] rnrsarreal@aol.com, <mailto:rnrsarreal@aol.com,> In Plaridel, (plaridel_papers@egroups.com), January 25, 2001.

[14] Arnold, 'Manila's Talk of the Town is Text Messaging'; See also Richard Lloyd Parr, untitled article on People Power II and cellphone use in *The Independent*, London, January 23, 2001.

[15] Michael Tan, 'Taming the Cellphone,' Philippine Daily Inquirer, February 6, 2001.

[16] Tan, Ibid.; De Quiros, 'Undiscovered Country,' Philippine Daily Inquirer, February 6, 2001.

[17] Arnold, 'Manila's Talk of the Town is Text Messaging.'

[18] These messages were forwarded by rnrsarreal@aol.com, <mailto:rnrsarreal@aol.com,> to the Plaridel discussion group (plaridel_papers@yahoogroups.com) <mailto:
(plaridel_papers@egroups.com)> January 25, 2001.

[19] Bart Guingona, Plaridel, (plaridel_papers@yahoogroups.com) <mailto:(plaridel_papers@egroups.com)> January 26, 2001. Texting is widely credited with bringing about the rapid convergence of crowds at the EDSA shrine within approximately 75 minutes of the abrupt halt of the Estrada impeachment trial on the evening of January 16. Even prior to Cardinal Sin and former president Cory Aquino's appeal for people to converge at this hollowed site, it has been estimated that over 20,000 people

had already arrived there, perhaps lured by text messages they received. As Danny A Gozo, an employee at Ayala Corporation, points out in his posting on Plaridel, of January 23, 2001 (plaridel_papers@yahoogroups.com) < mailto:(plaridel_papers@egroups.com) > Globe Telecom reported an average of 42 million outgoing messages and around an equal number of incoming ones as well, while Smart Telecom reported over 70 million outgoing and incoming messages texted through their system *per day* during the days of People Power II. He observes enthusiastically that 'the interconnectedness of people, both within the country and outside is a phenomenon unheard of before. It is changing the way that we live!'

[20] Ederic Penaflor Eder, *Pinoy Times,* February 8, 2001. The translation of this text is mine.

[21] I owe this term to James T Siegel, *Fetish Recognition Revolution,* Princeton: Princeton University Press, 1997.

[22] My remarks on Manila's streets were gleaned from the notes and observations I made in the 1990s. On Manila's urban forms, see the excellent essay by Neferti X. Tadiar, 'Manila's New Metropolitan Forms,' in Vicente Rafael, ed., *Discrepant Histories: Translocal Essays on Filipino Cultures,* Philadelphia: Temple University Press, 1995, 285–313. For a lucid portrait of Manila's fantastic street life, see the novel by James Hamilton-Paterson, *The Ghosts of Manila,* New York: Vintage, 1995. Contemporary Philippine cinema which often traverse the divide between rich and poor and acutely explore the spaces of their habitation are excellent primary source materials for the study of Manila's urban forms. For a recent collection of essays on Philippine cinema, see Roland Tolentino, ed., *Geopolitics of the Visible: Essays on Philippine Film Cultures,* Quezon City: Ateneo de Manila University Press, 2000.

[23] I owe this information to David Rafael, former manager of the Glorietta shopping mall in the Ayala Center in Makati.

[24] For a discussion of the historical link between linguistic and social hierarchies, see Vicente Rafael, 'Taglish, or the Phantom Power of the Lingua Franca,' in the book *White Love and Other Events in Filipino History,* Durham: Duke University Press, 2000, 162–189.

[25] Here, I draw from Martin Heidegger, 'The Question Concerning Technology,' in his book *The Question Concerning Technology and other Essays,* translated by William Lovitt, New York: Harper and Row, 1977, 3–35. See also the illuminating commentary by Samuel Weber, 'Upsetting the Setup: Remarks on Heidegger's 'Questing After Technics,' in *Mass Mediauras: Form Technics Media,* Stanford: Stanford University Press, 1996, 55–75. My remarks on the crowd are indebted to Walter Benjamin, *Charles Baudelaire: A Lyric Poet in the Era of High Capitalism*, London: Verso, 1977.

[26] For a discussion of the history of this nationalist fantasy, see the Introduction to Vicente Rafael, *White Love and Other Events in Filipino History,* 1–18. For a comparative approach to the radical potential of nationalist ideas, see Benedict Anderson, *Imagined Communities: Reflections on the Origins and Spread of Nationalism*, London:Verso, rev. ed., 1991.

[27] Flor C., Plaridel listserve (plaridel_papers@yahoogroups.com, <mailto:(plaridel_papers@yahoogroups.com,> January 24, 2001.

[28] 'Flor C' I have subsequently learned, is Flor Caagusan. She was formerly editor of the editorial page of the *Manila Times* and at one point served as the managing editor of *Diliman Review*. I owe this information to the journalist Pete Lacaba. While she would be known to a small group of journalists who are part of the Plaridel discussion group, she would presumably be unknown to the majority of participants in this group. Her anonymity thus remains crucial.

[29] For an elaboration of the notion of *damayan*, see Reynaldo Ileto, *Pasyon and Revolution: Popular Uprisings in the Philippines, 1840–1910*, Quezon City: Ateneo de Manila University Press, 1979. See also the important work of Fenella Cannell on Bikol province, south of Manila, *Power and Intimacy in the Christian Philippines*, Cambridge: Cambridge University Press, 1999.

[30] Jacques Derrida, 'Faith and Knowledge: The Two Sources of 'Religion' at the Limits of Reason Alone,' in *Acts of Religion*, 56–57.

[31] See for example the news reports and opinion columns of the *Philippine Daily Inquirer* from April 26 to May 5, 2001 for coverage of the 'Poor People Power', or as others have referred to it, 'People Power III'. In particular, see the following, 'Estrada Loyalists Overwhelm Cops on Way to Malacanang,' May 2, 2001; Amando Doronilla, 'The State Defends Itself,' and 'Now the Fight Over Semantics,' May 2, 2001 and May 3, 2001 respectively; 'Exchanges on Edsa 3,' May 3, 2001; 'Edsa reclaimed by Edsa II Forces,' May 2, 2001; Blanche Gallardo, 'Tears of Joy for Tears of Sadness,' May 6, 2001. See also Jarius Bondoc, 'Gotcha,' in *Philippine Star*, May 5, 2001; Howie G Severino, 'The Hand that Rocks the Masa,' *Filipinas Magazine*, June 2001, 70–72; Pete Lacaba, 'Edsa Puwersa,' *Pinoy Times*, April 29, 2001.

[32] See for example Conrado de Quiros, 'Lessons,' in *Philippine Daily Inquirer*, May 4, 2001; Walden Bello, 'The May 1st Riot: Birth of Peronism RP Style?' *Philippine Daily Inquirer*, May 7, 2001; La Liga Policy Institute (Quezon City), 'Poor People Power: Preludes and Prospects', as it appears in filipino-studies@yahoogroups.com <mailto:filipino-studies@ yahoogroups.com>, May 6, 2001; Ferdinand Llanes 'Edsa at Mendiola ng Masa,' filipino-studies@yahoogroups.com, <mailto:filipino-studies@ yahoogroups.com,> May 3, 2001.

[33] 'Estrada Loyalists Overwhelm Cops on Way to Malacanang,' *Philippine Daily Inquirer*, May 2, 2001.

[34] 'Horde' comes from the Turkish *ordi, ordu,* camp, and referred to 'troops of Tartar or other nomads dwelling in tents or wagons and moving from place to place for pasturage or for war and plunder,' according to the Oxford English Dictionary.

[35] See 'Edsa Reclaimed by Edsa II Forces,' *Philippine Daily Inquirer*, May 2, 2001 which reports, among other things, how people involved in People Power II 'brought their own towels, sponge and scrubs,' to clean the garbage

that had been left behind by the pro-Estrada crowd, hosing down 'the filth from the ground,' and 'disinfecting,' the Shrine with chlorine. 'They heaped mounds of garbage, sang and danced lustfully over the Edsa shrine marker, rammed a truck into the landscape and directed huge loudspeakers to the shrine door,' fulminated the Shrine rector, Monsignor Soc Villegas.

Vicente Rafael, 'The cellphone and the Crowd,' was first published in *Public Culture*, Volume 15, No. 3, 2003.

PART TWO

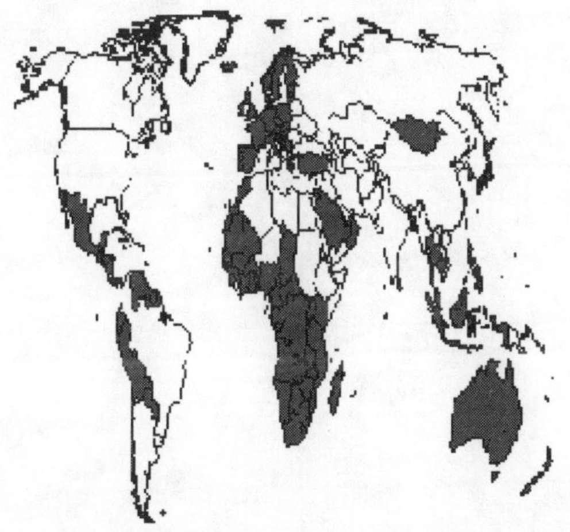

Country Profiles

17. Japan

At the Cutting Edge of the Wireless Information Society

Shinichi Saito

Editors' Note: This fascinating chapter profiles the world's most advanced wireless information society, Japan. Three book reviews in Part III of this book also profile useful learnings from the success of NTT DoCoMo's i-Mode service, in terms of industry strategy, content models and management culture. In March 2003, the number of cellphone subscribers in Japan was 75.7 million and the number of subscribers to cellphone Internet services reached 62.5 million, representing 82.6 per cent of all cellphone users. The cellphone user base exceeded the number of fixed phone subscribers in 2000. The m-commerce industry is expected to cross 3.2 trillion yen by 2006. Over 56 per cent of cellphone Internet users have used m-commerce. The e-Japan Strategy is aimed at making Japan the most advanced IT nation within five years. A growing body of social science research is examining a wide variety of social impact, ranging from gratification effects to digital shoplifting. A key challenge ahead is to reduce the digital divide in areas like access by disabled citizens.

Introduction

In November 2000, the Japanese Diet passed the Basic Law on the Formation of an Advanced Information and Communications Network Society (IT Basic Law) to cope with rapid, worldwide, socio-economic changes. The law came into effect in January 2001. In accordance with the law, the government set up the IT strategic headquarters and established an IT strategy known as the e-Japan

Strategy, aimed at making Japan the most advanced IT nation within five years. To solidify the strategy, IT strategic headquarters released the e-Japan priority policy programme in March 2001. The programme calls for promoting the development of an advanced network infrastructure, including high-speed wireless Internet access systems. In July 2003, e-Japan Strategy II was established, with the e-Japan priority policy programme 2003 announced in August. E-Japan Strategy II calls for creating of a 'ubiquitous network' that enables users to send and receive information anytime and anywhere via an omnipresent computer network. Full development of a wireless access system is considered indispensable in the anticipated ubiquitous network society. What, then, is the current wireless situation in Japan? This essay outlines the unprecedented changes that Japan faces today and the outlook for future wireless environments.

Connectivity

Like many other countries, Japan has experienced a rapid diffusion of Internet use over the past few years. While numerous reports publish estimates of the number of Internet users each year, there is still no consensus. Figures periodically reported by the ministry of public management, home affairs, and posts and telecommunications (MPHPT) and by the Internet Association of Japan (IAJ) are probably cited most often. According to the MPHPT, the number of Internet users exceeded 69.4 million in December 2002, representing 54.5 per cent of the population (MPHPT, 2003). On the other hand, the IAJ estimated 55.3 million Internet users in the same period. Thus, these two most-frequently cited sources have a nearly 14-million person disparity. Differences in survey methods and definitions of Internet users mainly account for the disparity in the two estimates. While the estimated number of Internet users differs depending on the survey organisation, undoubtedly the number has increased dramatically in the past few years. Today, the medium has found its way into many Japanese households.

Broadband

The MPHPT declared 2001 as the 'First Year of Broadband' (broadband usually refers to cable Internet, DSL, FTTH, and wireless LAN). Rapid growth in high-speed Internet access services has characterised the Internet scene in Japan since then. The IAJ estimated that 39.3 per cent of total Internet users used broadband at the end of 2002.

Estimates by the MPHPT were slightly lower than this figure, but reached nearly 30 per cent.

Among broadband Internet access services, wireless LAN service has not yet become widely diffused. However, wireless LAN will become increasingly important, especially in the expected ubiquitous network society. Wireless LAN service in public locations known as 'hotspots'[1] was launched in April 2002. Located at such places as railway stations, hotels, restaurants, and fast food shops, hotspots enable Internet access that is as fast as ADSL service from laptop computers or PDAs. Currently, the main hotspot communications standard is IEEE 802.11b (also known as Wi-Fi), which provides high-speed data communications at speeds up to 11 Mps.

According to the IAJ (2003), about 15 per cent of total Net users access the Internet via laptops, PDAs, or other mobile equipment. Among these users, more than 44 per cent access the Net using laptops connected to cellular telephones (including PHS) as the main access (note that in Japan, people prefer laptops to PDAs)[2] . Only 4.4 per cent use hotspots as a main access. Thus, while hotpots are still at the early stage of diffusion, the number of hotspot access points has rapidly increased from 578 in July 2002 to 1522 in March 2003 (IAJ, 2003)[3] .

Cellphones

The number of subscribers to cellular telephones has also increased dramatically in recent years. According to the MPHPT, the number of cellular telephone subscribers soared from 20.9 million in 1996 to 60.9 million in March 2001 and reached 75.7 million in March 2003. The number of cellphone subscribers exceeded the number of fixed phone subscribers in 2000.

While the number of cellphone subscribers is still expanding, the growth rate has slightly decelerated since 2001. According to a survey conducted by the Nomura Research Institute (NRI) in March 2003, about 79 per cent of individuals aged 15- to 69-years old already have a cellphone. Thus, cellphones are nearing the saturation point. Table 17.1 shows a breakdown of cellphone use by age and gender.

As of August 2003, three carrier groups (the NTT DoCoMo Group, the KDDI Group, and the Vodafone Group [J-PHONE]) provide mobile communications services[4] . The nationwide service area is divided into 10 regions. Each carrier group assigns a regional carrier to serve either in one or multiple regions.

Table 17.1: Percentage of cellular phone users by age and gender

	15–19	20s	30s	40s	50s	60s
Male (March 2003)	80.3	96.9	96.2	92.8	72.6	54.1
Male (March 1997)	19.6	48.9	43.1	40.8	29.1	—
Female (March 2003)	93.9	95.7	87.6	75.3	61.2	40.2
Female (March 1997)	10.7	33.1	18.4	10.2	8.6	—

Source: Nomura Research Institute, Ltd. (2003).

Cellphone Internet

One of the most notable trends in the communications scene in recent years is an explosive growth in Internet access services via cellphones (including PHS)[5] . The first cellphone Internet service (i-Mode service by NTT DoCoMo) started in February 1999. Other carriers began services soon afterwards (ezweb by the KDDI Group and J-Sky by J-PHONE in the same year). The number of cellphone Internet users has rapidly risen since then. According to the MPHPT, the number of subscribers to cellphone Internet services reached 62.5 million in March 2003, representing 82.6 per cent of all cellphone users.

The 2003 NRI survey discussed above showed that more than 90 per cent of people in their teens, 20s, and 30s own Internet-enabled mobile handsets (browser phones), but the corresponding figures for people in their 50s and 60s are much lower (Table 17.2). Furthermore, not all users of Internet-enabled mobile handsets actually access the Internet via the browser phones. As shown in Table 17.2, many younger-generation browser phone users actually use the browser feature. Browser phone owners in their 50s and 60s, however, use the Internet access feature less frequently. Thus, the survey revealed an apparent generation gap in terms of cellphone Internet usage rates. By contrast, there was no clear difference by gender. The World Internet Project (WIP), a survey project conducted in Japan in October 2002, indicated a similar result. Despite such a digital divide by age, cellphone Internet usage, along with the rapid diffusion of broadband access, may represent an early stage of a ubiquitous networked society.

Content and Commerce

According to the MPHPT (2003), cellphone Internet users currently account for 40.2 per cent of the total number of Internet users. In

Table 17.2: Percentage of browser phone owners and subscribers among cellphone users

	15–19	20s	30s	40s	50s	60s
Browser phone owners (Male)	88.5	91.7	89.9	79.1	61.9	52.1
Browser phone users (Male)	80.3	65.4	67.6	50.4	23.0	15.5
Browser phone owners (Female)	96.8	90.2	92.5	76.7	58.0	52.8
Browser phone users (Female)	77.4	69.3	63.4	51.7	25.9	17.0

Source: Nomura Research Institute, Ltd. (2003).

contrast, 82.4 per cent of total Net users use PCs to access the Internet. Some industry analysts, however, predict that the number of cellphone Internet users will exceed that of PC Internet users in the near future.

Three carrier groups (The NTT DoCoMo Group, the KDDI Group, and the Vodafone Group) provide a variety of advanced mobile Internet services, including not only standard e-mail and Web browsing services, but also application downloading services, GPS (global positioning system), picture/video mail services by mobile terminals with built-in digital cameras, and video distribution services. At the end of 2002, more than 3,000 items were available from the official menu for NTT DoCoMo's i-Mode users, and more than 60,000 unofficial Websites could be accessed via i-Mode (Kishida, 2003).

3G (Third-generation)

Although many services are available for cellphone users, only a limited number of users actually use advanced services such as video mail or distribution services, partly because of low data transmission capabilities. Diffusion of third-generation (3G) mobile phones will, however, enable more effective use of these services. As Ishii (2003) noted, the 3G service is 'a high-quality, digital new-generation mobile communications service that enables up to 384 Kbps data communications on a global basis'. NTT DoCoMo launched the world's first 3G mobile communications (W-CDMA service under the brand name FOMA) in October 2001. Two other carrier groups soon followed suit. As of September 2003, KDDI Group's 3G service known as CDMA 2000-1x leads the 3G mobile phone market. While 3G mobile phones are still at the early stage of diffusion, they will soon find their way into mobile phone uses.

M-commerce

Japan is the first and arguably largest mobile e-commerce (hereinafter referred to as m-commerce) market, which was stimulated by the i-Mode service. Although some industry observers initially saw cellphones as unsuitable for commerce, the m-commerce market seems to be growing rapidly. Indeed, the industry is expected to expand to over 3.2 trillion yen by 2006 (*Nihon Keizai Shimbun*, January 1, 2003). With such a rapid diffusion of m-commerce, trade and business publications such as the *Nikkei Sangyo Shimbun*, the *Nikkei Ryutsu Shimbun*, the *Nikkan Kogyo Shimbun*, and the *Nihon Kogyo Shimbun* now frequently provide up-to-date, insightful reports on m-commerce industry trends. In such circumstances, many European software companies have already entered the Japanese market.

A survey by InfoComm Research Inc (2003) indicated that 56.5 per cent of cellphone Internet users have used m-commerce (i.e., purchasing something by mobile terminals). The 2003 NRI survey revealed that about 30 per cent of cellphone users under the age 60 access premium Websites. According to the 2002 WIP Japan survey (2002), however, transaction services (such as mobile banking or stock trading on the Net) are not yet widely used. The survey found user rates of 2.3 per cent for booking tickets, 1.8 per cent for mobile banking, 1.8 per cent for online shopping, and 1.3 per cent for stock trading[6]. It will take some time for these mobile phone transactions to become more common daily activities.

Notwithstanding the currently limited number of users, more banks now offer mobile banking services. As of March 2002, 115 banks offered automatic transfer and balance information services, and six other banks provided balance information services.

Among m-commerce purchases, premium ringtone melodies represented the most-often purchased content (about 82 per cent), followed by premium information (28 per cent), and premium games (22 per cent) (InfoComm Research, 2003). Until recently, cellphone Internet users bought only digital content such as ringtone melodies. Now, however, users have begun to purchase more goods such as CDs, DVDs, and clothes via mobile Internet services.

E-commerce now occupies only a very small portion of commerce as a whole. The potential market for e-commerce is, however, increasing, since the majority of current users are consumers in their 20s and 30s, who are also the heaviest users of cellphones.

Cooperation between Industry and Government

To develop one of the world's most advanced information and communications networks in which wireless access systems play an important role, the e-Japan priority policy programme (including its new 2003 edition) requires private sector initiatives. The Japanese government has indicated its intention to back such industry endeavours.

The e-Japan priority policy programme also stresses the promotion of e-commerce as one of the top-five priority policy areas (m-commerce is considered an important part of the e-commerce market in general). The e-Japan priority policy programme 2003 requires further facilitation of e-commerce. Thus, the government actively embraces and promotes e-commerce and m-commerce.

Research Centres

Recently, many research centres have been established to incubate wireless technologies. For example, in 2003 the Auto-ID Center (http://www.autoidcenter.org/), an international, non-profit organisation with headquarters at the Massachusetts Institute of Technology, opened a new research laboratory at Keio University in Japan. The centre aims to develop a system for automatically identifying anything, anywhere, using radio frequency identification (RFID) technology. RFID, which is expected to be a very important and promising technology over the next decade, enables users to attach wireless ID tags to any item. The tags will allow for automatic access to information, such as the item's location and content, through communications with other network computers.

The government is supporting the development and promotion of RFID. Many companies, along with the government, are trying to apply RFID technology to full commercial uses. For example, as Tagegami (2003) pointed out, 'NTT DoCoMo is among the companies that have begun to study how to integrate the technology into mobile handsets, and innovative development is expected to result in new services that people cannot now imagine.' Tagegami further noted that 'the convergence of RFID technology and mobile-phone services will enable mobile operators to improve their services and will therefore benefit both the operators and their customers'.

The Ubiquitous ID Center (http://www.uidcenter.org/index.html) is another prominent research centre working to establish and spread core technology that can automatically recognise items and eventually realise a ubiquitous computing environment. The Ubiquitous ID Center is working on research that includes the construction of an ID system that can be attached to items (uID: ubiquitous ID), the development of core technology for uID use, and the establishment of core technology for a secure, widespread system of distributing uIDs and the information connected to them.

Culture

University Education

The number of university faculties and departments related to media/communication studies and information science has also rapidly increased in recent years. Many scholars specialising in media/communication studies or information science offer courses that examine wireless media from a variety of perspectives (e.g., wireless LAN technology, economics of wireless media, or social psychology of mobile communications).

Research on Cellphone Use

With a dramatic increase in mobile phone use in the past few years, a wide variety of social impacts (both positive and negative) has emerged. Social scientists have embarked on examining those impacts. For example, researchers have investigated the extent to which cellphone use in general, and cellphone Internet use in particular, has affected users' communication behaviours (e.g., a frequently addressed issue is how cellphone use affects inter-personal relations).

One example of such research is the above-mentioned WIP Japan investigation into the gratification derived from cellphone Websites. According to the study results, more than half of the respondents said they use the system to 'kill time' (69.5 per cent) or to gain information on hobbies and leisure activities (50.3 per cent) and that use of Websites is fun (60.5 per cent). Similarly, a survey conducted by a study group on mobile communications (2002) revealed that as a result of mobile phone usage, most users reported feeling less anxiety, either often or some of the time (84.2 per cent), and that many of their family members feel the same (54.7 per cent). In addition, many reported feeling less irritation (61.4 per cent), because mobile communications allow them to keep in touch whenever they want. Thus, research indicates that cellphones are indispensable for many users, especially younger-generation users.

Negative Impacts

One of the negative impacts of the rapid spread of cellphone use relates to so-called online dating sites. Reports frequently cite the rise in crimes related to such dating sites. Most of the victims are reportedly girls of middle- and high-school age who access dating sites mainly by cellphones. To restrict the use of such dating sites and curb these crimes, the first legislation regulating dating sites took effect in September 2003.

Another negative impact of cellphones is 'digital shoplifting'. Hirai (2003) summarised some positive and negative effects of camera-mounted mobile phones (camera-phones). Along with positive impacts such as stimulating the photo-printing business, negative impacts such as the misuse of camera-phones require special attention. One significant social problem is what is referred to as digital shoplifting. Unfortunately, some people use camera-mounted mobile phones to photograph magazine pages and obtain information without purchasing the publication. As Hirai noted, 'both bookstores and the publishers of magazines are concerned about the lost revenue that results from such use of handset cameras' (2003, p. 2). The new technology will necessitate new etiquette guidelines and new rules.

Recommendations

Overcoming the Digital Divide

Like many other countries, Japan experiences a digital divide. Research has shown that gaps in Internet usage rates by age, gender and education are gradually decreasing. As mentioned earlier, however, wide gaps still exist among different age groups in terms of mobile Internet access. This is just one example of the digital divide that exists in Japan. A digital divide also exists between persons with disabilities and non-disabled persons. Supporting and promoting the use of information and communication technologies among people with disabilities and the elderly is an urgent task. With the e-Japan priority policy programme (including its new 2003 edition), one of the government's main aims is to lessen the digital divide. To reduce the digital divide, we need the united efforts of the government, private enterprises, and universities. Social scientists should continue research on how people use the Internet and wireless media and provide detailed analyses of Internet users. We should also pay more careful attention to issues such as the digital divide and important topics such as strengthening information security.

Resources

List of Useful Websites

http://www.icr.co.jp/index.html (InfoComm Research Inc)
http://www.iajapan.org/index-en.html (Internet Association of Japan)
http://www.kantei.go.jp/foreign/policy/it/index_e.html (IT strategic Headquarters)
http://www.nri.co.jp/english/index.html (Nomura Research Institute)

References

A Study Group on Mobile Communications (2002). *Development of mobile phone use and its impact* (Grants-in-Aid for Scientific Research report).

Hirai, Takao (2003). 'Some positive and negative effects of the use of camera-mounted mobile phones', *nG* (next generation) *Japan*, vol. 2 (no. 8), 1–3.

InfoCom Research Inc. (2003). *A survey on the actual status of broadband and Internet shopping use* (http://www.commerce.or.jp/result/min38/index.html) (in Japanese).

Internet Association of Japan [IAJ] (2003). *Internet White Book 2003* (in Japanese), Impress, Tokyo.

Ishii, Kenichi (2003). 'Diffusion, policy, and use of broadband in Japan', *Trends in Communication*, 11(1), 45–61.

Kishida, Shigeyuki (2003). 'DoCoMo to limit access to mobile dating sites', *nG* (next generation) *Japan*, vol. 2 (no. 4), 1–3.

Ministry of Public Management, Home Affairs, and Posts and Telecommunications (MPHPT) (2003). *Information and Communications in Japan: White Paper 2003* (in Japanese) Gyousei, Tokyo.

Nomura Research Institute (2003). *A survey on information and communications users* (http://www.nri.co.jp/opinion/r_report/clo/index.html) (in Japanese).

The World Internet Project Japan (2002). *Internet usage trends in Japan—Survey Report 2002* (in Japanese), Communications Research Laboratory.

Takegami, Kei (2003). RFID applications increase in Japan, *nG* (next generation) *Japan*, vol. 2 (no. 7), 1–2.

Endnotes

[1] NTT Communications started a wireless broadband Internet access service called HOTSPOT® in May 2002. In this essay, however, I use the general term 'hotspot' to refer to wireless Internet access service in public locations.

[2] Although PDAs are not widely used, newcomers have entered the PDA market one after another in recent years. PDA shipments reached 1.6 million units in 2002 (InfoCom, 2003).

[3] Many communications carriers and IT-related companies have expressed their intention to participate in the hotspot business.

[4] In October 2000, IDO Corporation, DDI Corporation, and KDD Corporation merged into the KDDI Group. The J-PHONE Group is now a member of the Vodafone Group.

[5] Note that cellphone Internet use in this study refers to viewing the Web on a cellphone/PHS display or sending/receiving e-mail via a cellphone/PHS. It does not include Internet access by PCs connected to cellphones.

[6] Percentages represent those who used each type of Website during the month immediately preceding the WIP Japan survey.

18. South Korea

Broadband Pioneer—The Wireless Way of Life

John Lee

Editors' Note: This chapter highlights South Korea's role as a world leader in broadband Internet—wireline and Wi-Fi. The country is also the largest producer of communications equipment in Asia. Broadband usage has 80 per cent household penetration. There will be an estimated 50,000 hotspots nationwide by the end of 2004. Mobile data services for cellphone users are rich in multimedia content. In the summer of 2002 ahead of the 2002 World Cup, the major wireless carriers launched 3G wireless Internet services on the CDMA platform EV-DO, which enabled speeds of up to 2.4 Mbps. M-commerce is alive and well in Korea, with over 20 million online banking accounts and wide usage of wireless as a transaction platform. Korean operators have also launched a platform called 'Wireless Internet Platform for Interoperability' (or WI-PI). Other initiatives have been launched targeting RFID, wireless personal networks (WPAN), ultra wide band (UWB) and intelligent wireless sensor networks. The Korean government has also set up the 'Korea Venture Fund'.

Introduction

The Internet in Korea has rapidly grown into a broadband tidal wave phenomenon, touching all parts of life. This has been driven mainly by the South Korean government's commitment to developing communications technologies and investing in the establishment of an information communications superhighway over the past few years.

As a result, Korea today boasts of one of the world's highest penetration rates in terms of wireless telecommunications, and the world's highest fixed-line broadband household penetration exceeding even larger populations such as the US and Japan. The country is also the largest producer of communications equipment in Asia, driven by the growth of global communications giants such as SK Telecom, Korea Telecom, Samsung Electronics and LG Electronics.

Connectivity

Part of the reason for South Korea's high penetration in fixed line and wireless based broadband has been the government's subsidy for upgrading the nation's fixed line and fibre optics networks, which has resulted in the lowest access costs in the OECD group. Since 1998, the total number of broadband users has been doubling every year and is now estimated at nearly 80 per cent household penetration. This essentially means that the entire nation is 'wired' and 'unwired' on broadband.

Broadband and Wi-Fi

In the years since broadband began to be offered in late 1998, the total number of households with broadband access has grown to over 15 million in 2003 (there are an estimated 18 million households in South Korea, with a total population of 48 million). According to the Korean ministry of information and communications (MOIC), the total number of Internet users topped 32 million people, or 80 per cent of people aged six and over in the country.

As part of this trend, Korea has been the global prototyping ground for different emerging standards of wireless broadband access. Wireless LAN (WLAN) services are well developed in Korea with over 20,000 WLAN hotspots active nationwide in 2003. Major carriers such as Korea Telecom and Hanaro Telecom estimate that this will grow to 50,000 nationwide by the end of 2004. Such WLAN technology has been centred on 802.11x based technologies, but also face competition from other wireless network technologies such as Flarion and Array Comm.

In the case of Flarion, its Flash-OFDM mobile network technology offers similar speeds to a WLAN, but with increased operations, security and bandwidth compared with 802.11x technologies. The Flash-OFDM WLAN technology, otherwise known as 2.3 Ghz 802.20, is being rolled out and tested by Hanaro Telecom, the second-largest fixed line broadband operator in Korea. In the case of Array Comm, a similar 2.3 Ghz spectrum-based technology provider has partnered with Korea Telecom and anticipates Korea will be 'the' testing ground for competing WLAN technologies as the Korean government allocates spectrum at the end of 2003.

Thus far, the uptake of Wi-Fi in Korea has been mixed although boasting the highest penetration of hotspots in the world. A number of startups have rolled out specialised services and subscription programmes in Wi-Fi—particularly focusing around PDA-centric gaming, location-based services and communications, but no compelling commercial model has proven itself other than a number of novel concepts. Despite some early disappointment, many IT companies and the Korean government expect consumer uptake to lag widespread infrastructure availability until 2005, when Wi-Fi accessibility is built into a wider range of devices beyond WLAN card enabled PDAs and PC laptops. As part of this, a number of major consumer electronics companies such as Samsung and the Korean government are actively pushing the development and widespread adoption of WLAN specialised semiconductor and system on chip (SoC) for next generation WLAN/Ethernet and intelligent wireless communications systems.

Handset Technology

The sophistication of handsets in the South Korean wireless market has been astounding. Integration and convergence of multiple consumer technologies such as digital cameras (currently at 1.3 megapixels), MP3 players with stereophonic 64 harmonies, PIM functionality and TV reception services are blurring the lines with other traditionally separate consumer electronics devices. Given the fact that the leaders in Korea are global consumer electronics innovators such as Samsung Electronics and LG Electronics, such convergence of functionality and service is right at the cutting edge. As of July 2003, the newest model by Samsung Electronics is a convergence PDA/phone with 192 megabytes of memory, built-in 1.3 mega-pixel camera and full PDA/PIM functionality based on the Windows operating system, in a smaller footprint than a Palm V PDA. As the shift from 2G to 3G phones accelerates in Korea, so will the handset upgrade cycle and the corresponding sophistication of services tied to the capabilities of the handsets.

Wireless Data Explosion

Wireless Internet data services have emerged as a growth engine for Korea's three major wireless carriers, as demonstrated by their year-on-year revenue and financial performance data. Revenues from SK Telecom, the largest mobile operator in Korea, rose 91 per cent year-on-year pushing wireless data related revenues to $548 million and representing 14 per cent of total company revenues. The trend, however, for voice-related wireless revenues has been on the decline while rates for local and long distance have gone into a free fall.

Content and Commerce

According to the Korean ministry of information and communications, the wireless Internet market in Korea will reach 7 trillion won (approx. $6 billion) by 2007, from its current size in 2003 of 2.6 trillion won.

As in Japan, the three carrier groups (SK Telecom, Korea Telecom and LG Telecom) provide a variety of advanced mobile Internet services, including not only standard e-mail and Web browsing services, but also application downloading services, location-based services (via GPS and base station triangularisation), ringtones, video on demand, picture/video mail services and mobile gaming services. The market is expected to grow even more rapidly in 2004 with the Korean government opening up wireless Internet service network market to all companies in August 2003, opening up competition between Internet portal players such as Yahoo! and Daum versus the i-Mode style mobile directories of the wireless carriers.

3G (Third-generation)

Korea is arguably the most advanced 3G nation in the world, with multiple platforms in prototype or full commercial launch.

In the summer of 2002 ahead of the World Cup football championship, the major wireless carriers launched third generation wireless Internet services on the CDMA platform EV-DO, which enabled speeds of up to 2.4Mbps, enabling large file size transmission such as moving images and video. Since then, the number of EV-DO subscribers reached 1 million in a short eight months and has since been estimated at close to 2 million subscribers which was more than twice the number of 3G subscribers on Japan's DoCoMo IMT-2000 service.

M-commerce

Korea belongs to an elite group of nations (other than Japan) that has been successful in commercial rollouts of 3G-related m-commerce consumer services. Typically, wireless carriers charge either on a per usage basis (in which a single time micro-charge and packet data download provide a double revenue stream) or a small subscription fee added on to a subscriber's monthly bill. Examples of the most popular per usage downloads in order of popularity are mobile game downloads, ringtones, graphics wallpaper (for small colour screens) and media on demand services such as short video and music clips. Subscription-based services that are proving to be very popular are colour ring services (muzak-like answering service), location-based car navigation services (via base station triangularisation) and TV services (e.g., view television channels on your phone while commuting home on a subway).

M-finance

Korea has taken a global lead in the innovation of mobile commerce and payments, with over 20 million online banking accounts and wide usage of wireless as a banking and transaction platform. A number of major banks and telecom carriers in the market have launched multi-channel banking with sub-brands for mobile such as BankOn (LG Telecom-Kookmin Bank), which is the first SIM-based service enabling users to perform full banking services using a mobile phone. This goes beyond simple information checking such as balance checking but, given the confluence of smart card usage in Korea, also allows for full transactions and ATM cash downloads, eliminating the need to visit ATMs. Phones already act as an integrated smart card/phone debit card, which allow users to make payments from their mobile phones directly at points of sale (POS) across the country at coffee shops, restaurants and metro fares.

Capacity, Capital and Cooperation

Cooperation between Industry and Government

To develop a consolidated mobile Internet platform and eliminate overlapping investments, the Korean government and Korea's three mobile service operators (SK Telecom, LG Telecom and Korea Telecom) launched a platform called 'Wireless Internet Platform for

Interoperability' (or WIPI). Previously, all operators made overlapping investments in incompatible network platforms, which made inter-carrier communication problematic. As a result, the hope is that WIPI (currently at its 1.1 version) will not only become a standard in Korea, but a wireless platform standard overseas as Korean wireless carriers export their technology to other countries.

The government has taken a particularly active role in fostering wireless usage in Korea, by offering massive subsidies to wireless carriers to continually upgrade their networks as well as to consumers in the form of handset subsidies to spur faster handset uptake and replacement cycles.

The Korean government has named 10 strategic technologies to develop Korea as a top five global player by 2006, and a top three global player in each by 2010. In addition, the government has launched a number of other initiatives in areas such as radio frequency identification networks (RFID), wireless personal network (WPAN), ultra wide band (UWB), SoC related to next generation WLAN/Ethernet, and intelligent wireless sensor networks. In a number of these areas, Korea already ranks in the top 10 countries in the area and has allocated billions of dollars in subsidies for research and development to spur long-term growth in the industry.

For example, the ministry of information and communications has launched a 'ubitiquous sensor network' project to be launched across the nation by 2008 which will link diversified IT systems and home appliances by wireless technology, and manage information via radio frequency and RFID tags. The government has invested over $100 million and established of an RFID research centre working with global standardisation bodies and technology partners.

Intel recently announced a Korea-based research and development centre concentrating on the development of technologies for digital home and wireless communications. The projects slated to be carried out will include UWB and Wi-Max (Wireless Interoperability for Microwave Access) last mile wireless broadband.

Finally, the government also stimulates the local wireless venture industry indirectly by being the largest limited partner for the Korean venture capital industry. In late 2002, it set up the 'Korea Venture Fund', a $300 million venture capital pool which is distributed annually in a bidding contest, for the purpose of cultivating local startups in the wireless and broadband sectors. Every year, dozens of venture capital firms present proposals for cultivation of the industry, which the MOIC judges based on content and vision. The fund recipients then have to target broadband/wireless early to mid-stage venture capital funds focused exclusively on domestic Korean startups, with fund sizes ranging from $25–40 million using the Korean government as a large anchor investor and soliciting outside strategic investors.

Learnings and Recommendations

Consumer wireless data services have succeeded in countries like Japan and Korea for a number of reasons: offerings of a wide range of content (controlled via a walled garden for content quality but not competing directly), getting the basics right in platform (transport, access, gateway, middleware), and successfully leveraging outside partners (equipment vendors, content providers, strategic partners) and not trying to 'do it all'. But enterprise uptake of these mobile services has yet to take off in Korea; some early success amongst younger demographics have been experienced in mobile games and download of ringtones and images.

Carriers in other parts of the world hoping to emulate Korea's market success can adopt the following key learnings:

- Get the content right: open up the platform to enable a wide range of content, potentially focus on some proprietary content for premium services only.
- Build a broad and open alliance/partner strategy—don't try and 'do it all' across horizontal wireless value chain (i.e., portal, content, transactions).
- Focus on providing the best platform (transport, access and gateway), create high switching barriers for users based on platform use.
- Wait for key network/handset technologies to be in place (i.e., packet v/s circuit switching, handsets enabled for xHTML/Java).
- Be sure to address key consumer market trends to generate network economies. Some content or services Koreans like may not go over well in other markets—for instance, fortune-telling is a major service that culturally may not transfer to the US.

19. Singapore

Powered by InfoComm

Lunita Mendoza

Editors' Note: This chapter clearly profiles the pioneering and facilitative role of the Singapore government in developing the city state as a major infocomm player. As early as the end of 1998, almost half of Singapore's households had mobile phones. Today, there is complete mobile coverage in Singapore. It was one of the first countries in Asia to get an Internet connection. Its wireless operators are active regionally in cellular markets as well as Wi-Fi services. Singapore is also taking advantage of its ethnic and commercial ties with China and India in content and human resources areas. Singapore is consistently rated high in global benchmarking studies of e-readiness and productivity. Its policy machinery is pro-active and in step with trends in media convergence.

Introduction

Despite its small size, the city state of Singapore continues to be a significant test bed and regional hub for major wireless players and media analysts. The number of mobile subscribers grew from 743,000 subscribers in 1997 to 2,859,000 subscribers in 2001 and 3,245,000 in 2002.

Connectivity

SingTel initially had a monopoly on fixed lines and international services until March 31, 2007. However, in May 1996, a duopoly was established. In January 2000, this policy was abolished and the market was fully opened to competition in April. SingTel was partially privatised.

As early as the end of 1998, almost half of Singapore's households had mobile phones. Today, there is complete mobile coverage in Singapore. Coupled with the availability of the pre-paid cards, there are virtually no infrastructure limitations to universal telecommunication access.

Singapore has a well-developed domestic communication network with 27 telephone exchanges. The fixed network has been fully digitalised since 1994. The Singapore One Network for Everyone (ONE) backbone uses ATM technology over the fibre optic cable at speeds of up to 622 Mbps. In 2001, the island had five digital mobile cellular networks (two GSM 900, two GSM 1800 and one CDMA). A significant milestone was crossed in July 2000 when the number of mobile cellular subscribers exceeded fixed telephone lines.

In 2001, Singapore had over 14,000 telephone circuits with international terrestrial links to Malaysia and submarine cable and satellite links to the rest of the world. Investments in regional submarine fibre cable systems include SEA-ME-WE 3 (South East Asia-Middle East-Western Europe 3), China-US and Japan-US.

Singapore was one of the first countries in Asia to get an Internet connection. This was a culmination of many years of connectivity experience with academic networks preceding the commercial Internet. The pioneering network initiatives of NUS were critical, attesting to the importance of the academic sector in developing a sustainable Internet environment.

Among the operators, StarHub participates in three initiatives launched by the Wireless Broadband Alliance to open access and enhance user experience for wireless broadband users in Singapore. It operates wireless hubs at Singapore's Changi Airport and Suntec City convention centres. StarHub wireless broadband users are able to access hotspots from Australia's Telstra, China Netcom and Malaysia's Maxis at no additional charge.

Two additional wireless broadband operators have joined StarHub's local wireless broadband marketing partnership programme. Nautilus Networks and PC Connect will work with wireless players StarHub, 5G and Bluengine to ensure that their hotspots work smoothly with each other.

StarHub and The Coffee Bean & Tea Leaf have announced that all 41 coffee outlets will soon offer wireless broadband services to coffee and tea lovers throughout Singapore. StarHub customers can also engage in instant two-way SMS communication between the Web and their handsets.

SingTel has joined hands with the National Library Board (NLB) to offer Outdoor Wireless Surf in all 23 public libraries in Singapore. SingNet customers can get Wi-Fi access at more than 2,000 locations in 14 countries through its partnership with global roaming carrier, GRIC Communications Inc. SingTel is also planning 3G trials in 2004.

SingTel claims an aggregate mobile subscriber base in the region of 37.4 million (as of June 2003). This represents a 51 per cent year-on-year increase in the total number of mobile subscribers of SingTel, SingTel Optus and SingTel's four regional associates, compared with the previous year's base of 24.8 million customers.

Content

By emerging as the info-communications hub for the Asia-Pacific region, Singapore harnessed high-value inputs such as software design and multimedia content that go into products and services to be delivered around the region.

For example, Singapore is taking advantage of its ethnic and commercial ties to China and India by providing content translated and digitised in multilingual Singapore for delivery over the Internet and IP-based networks in those large potential markets.

Singapore recognises that not all info-communications industries will be present in the city state itself. The manufacture of items like handsets can be done cost-effectively in markets with lower labour and real estate costs but Singapore seeks to be the nerve centre, the centre for management expertise, technological development, and intellectual property.

Among the operators, SingTel offers directory assistance connect and directory assistance SMS services for its mobile customers. SingTel has also launched two new wireless financial services, in partnership with global financial information provider, Standard & Poor's (S&P), and VoIUM GlobalMarkets, a content provisioning partner.

Yahoo! users can exchange SMS messages with mobile subscribers of Globe Telecom in the Philippines and PT Telelomunikasi (Telkomsel) in Indonesia, two of SingTel's regional mobile associates.

In an interesting reflection of our changing infocom environment, operator M1 announced that it would terminate its paging service on December 31, 2003, more than six years after its launch. In markets like Singapore, pagers are slowly being phased out as cellphones become more ubiquitous and data-enabled. Cellphones also offer more interesting content in areas like entertainment: for instance, M1, in collaboration with MediaCorp, offers customers mobile content by popular TV personalities such as Gurmit Singh and Zoe Tay.

Capacity

Singapore's workforce reflects a strong focus on the infocomm sector including wireless media. It has consistently been rated as a leader in productivity and e-readiness. Singapore's infocomm sector directly employs more than 50,000 people, with the sector being one of the most successful generators of jobs over the last decade.

Singapore: Leading in the Information Society

2002 World Competitiveness Index: Ranked 5th in the world—1st in APAC—by IMD
2001/2 Global Competitive Index: Ranked 4th in the world—2nd in APAC—by World Economic Forum
2002/3 Global Information Technology Report: Ranked 3rd in the world—1st in APAC—by World Economic Forum
2002 e-readiness Ranking: 11th in the world—2nd in APAC—by Economist Intelligence Unit

Another 50,000 infocomm professionals are employed by the user industries (IDA, Infocomm Manpower Survey, December 2002). All these industries are now influenced by the global trends of digitalisation and convergence, resulting in new patterns of competition and co-operation between firms throughout the sector.

Cooperation

Singapore has a comprehensive set of government-to-government agreements in place that facilitates market access and growth of trade and commerce. The country has a recognised legal and business environment, which means that it is a preferred location for business. The addition attraction offered by the Singapore government that will come to fruition over time highlights the fact that it will take more than a sterilised order to attract business. This comes in the wake of surrounding countries also coming up to par with great infrastructure and as good government.

New trade agreements will continue to underpin a connected island, attempting to provide the country with the broadest band connectivity possible to other major economic centres of the world.

The Singapore government is involved in ICT in three ways. It is an avid user in its own right. It is also a big provider of electronic governance services to the public. Finally, it is a big promoter of ICT.

As a User

The Singapore government was an early adopter. In 1981, it introduced a Civil Service computerisation programme to train all public employees in the use of Information Technology. In 1998, the Ministry of Defence launched the world's first Internet-based government procurement system. Today, the Singaporean government is recognised as one of the most computerised in the world. Back in 2000, the government was using a system that integrated financial and procurement systems of government ministries and agencies, allowing users to browse supplier catalogues and purchase products.

As a Provider

It has been a while now since the government started promoting electronic services delivery with the objective of creating an e-based society. All ministries have a Web presence, integrated through the government's central portal. Very soon, all government over-the-counter procedures should be online. One example of a successful application is the online submission of income tax forms.

As a Promoter

In the 1980s, the government turned its attention to IT, beginning with 1981 Civil Service Computerisation Programme. In 1986, a committee on national computerisation was formed to create a national IT plan. In 1992, the IT2000 plan was published, calling for the construction of a broadband infrastructure, common network services like directories and billing, experiments with applications, forging international strategic alliances with industry leaders in Japan, Europe and the US; and establishing a policy and legal framework on issues like data protection, privacy, copyright and intellectual property rights and the admissibility of computer-imaged documents in court.

Policymaking Culture

Singapore's governing political culture emphasises a consensus-building approach that involves both public sector leadership and private sector input in marshalling national resources and achieving national goals.

One of the most striking developments has been the rise in activities of the ICT sector-specific regulatory agency over the last decade. By the end of 2000, there were 102 such agencies around the world operating separately from telecommunications service providers and 52 of them operating separately from communications-related government ministries.

The International Telecommunication Union's (ITU) Sector Reform Unit (SRU) selected Singapore as one of five SRU regulatory case studies conducted in 2001. This was due to the fact that in 1992, Singapore became one of the first countries in the world to create a sector-specific telecommunications regulatory body, the Telecommunication Authority of Singapore (TAS).

Singapore was also one of the first countries to create a regulatory body specifically to respond to the convergence of telecommunications and information technologies (computing). It did so in late 1999, through the merger of TAS with the country's former National Computer Board (NCB), creating the Info-Communications Development Authority of Singapore (IDA). In the short time since it was established, IDA has been lauded for developing a fair and impartial framework for competition, the code of practice for competition in the provision of telecommunications services (the Code), based on the principles of technological neutrality and asymmetric regulation between dominant and non-dominant licensees.

In addition, in a process balancing the need for public consultation with quick market entry, IDA approved a reference interconnection offer (RIO) that allows any licensee to establish fair terms and conditions for interconnection with the incumbent fixed-line operator. Together, the Code and RIO serve as valuable models for countries seeking to implement full competition.

In addition to Singapore's role as a regulatory pioneer, the nature of IDA's mandate is a rare combination of regulatory duties and roles in promoting and developing the country's communications capabilities and industries. IDA represents Singapore's comprehensive effort not just to regulate ICT industries, but also maintain it as an info-communications hub. In addition, the government holds indirect equity stakes in major ICT market players and provides grants and subsidies to companies engaging in telecommunication innovation or pioneering new technologies or services.

Vision of Info-Communications

The government of Singapore adopted the term *info-communications* to convey the concept of computers, content, and transmission as a converging whole. Info-communications is the banner under which Singapore is building the country's digital future. The Singaporean government was not content merely to foster the development of the info-communications market. Rather, it wanted Singapore to be a regional hub for a variety of inter-related and complementary info-communications industries, believing that if it took a leadership role in this sector it would serve as a multiplier for the economy as a whole.

In an effort to attract foreign investment, Singapore recognised that multinational corporations would gravitate to locations where they could benefit from sophisticated communications technologies and a trained labour force to support them. It believed that local businesses would also benefit from an integrated package that enabled them to find, generate, process, and send information in an efficient and seamless way.

Singapore has also implemented its InfoComm 21 Masterplan, a government policy initiative that is designed to transform the country into a thriving and prosperous Internet economy by 2010. The initiative, much of which is co-ordinated and implemented by the IDA, is something of a conceptual umbrella for Singapore's planned approach to industrial development in this sector.

The major pillars of the plan include implementing and fostering competition in Singapore's telecommunications services market, developing information technology training and educational competencies, and making the Internet an integral part of the nation's industry and society.

The structural components (mostly working units within IDA) that have been put in place to pursue the government's goals illustrate several hallmarks of the government's approach to ICT issues, as noted by the ITU:

- The government can and should act as a catalyst of market change and growth;
- Private sector and public sector forces can work in tandem to achieve goals;

- Singapore must attempt to be competitive in most, if not all, ICT industries;
- Industry regulation is just one tool, among several, that the government can use to establish market conditions conducive to growth and competitiveness.

Capital: Singapore Government the Biggest Investor

The Singapore Infocomm industry revenue grew by 5.1 per cent from $30.65 billion in 2001 to $32.17 billion in 2002, according to the annual survey on infocomm industry for 2002. The companies surveyed also forecast continued growth for the next two years—4.5 per cent for 2003 and 7.5 per cent for 2004.

Since the founding of the country in the early 1960s, the government has continued to provide the catalyst of initial capital investment to spark and nourish critical industries. This policy, which has earned the government the moniker of Singapore Inc., is based on the concern that the relatively small size of Singapore's domestic market might not provide sufficient market-based incentives for the growth of healthy industries.

The government provides direct financial support, predominantly through two methods: the holding of indirect equity stakes often amounting to controlling interests in domestic operators; and the provision of grants and subsidies to companies engaging in telecom innovation or pioneering new technologies or services. Since 1974, the government has utilised a holding company to channel its capital investments into the Singapore economy.

Temasek Holdings Ltd., which is wholly owned by the government, has substantial holdings in most economic sectors, including transportation, energy, banking, shipping, diversified energy, real estate, and communications. Altogether, the government, through Temasek, holds some S$ 70 billion in assets, according to ITU sources, through equity stakes and control of most of Singapore's major communications-related conglomerates Media Corp., Singapore Technologies, SembCorp, and Keppel Group.

Temasek Holdings retains significant, often controlling, equity stakes in each of Singapore's major domestic telecommunications operators. It owns more than 75 per cent of incumbent operator Singapore Telecommunications (SingTel) directly. Through Media Corp. and Singapore Technologies, it controlled more than 70 per cent of exclusive cable television operator Singapore Cable Vision (SCV)—bought over by StarHub—and, through Singapore

Technologies, it also controls StarHub Communications, the country's second fixed-line and third mobile operator.

Through SembCorp and Media Corp., it owns more than 50 per cent of Pacific Internet, a dial-up ISP, and through Keppel, it owns 35 per cent of Mobile One (M1), Singapore's first competitive mobile network operator.

The government's stated aim is to invest strategically, but to not hold equity beyond the point when the enterprise becomes self-sustaining. The goal of strategic disengagement has not always been easy to achieve. The government's continuing investment in SingTel, for example, has raised questions in foreign markets concerning investments by the incumbent operator.

In addition to these government holdings, IDA, has a budget of roughly S$ 600 million from Parliament to fund various grant and incentive programmes to the private sector. Perhaps the best example of how Singapore's government has supported info-communications is found in the creation of Singapore ONE (One Network for Everyone).

Singapore ONE was born during the late 1990s as part of the National Computer Board (NCB) IT2000 master plan to enhance Singapore's information technology and broadband infrastructure. It consists of a broadband, ATM-based core network (known as 1-Net) that provides multimedia content and applications to homes and businesses at transmission speeds of up to 622 Mbps, and a comprehensive development programme to promote connection and use of the network, including electronic commerce applications.

TAS (now IDA) was one of the initial members of the consortium, 1-Net Singapore Pte Ltd, which owned the network. IDA held its equity interest through a wholly-owned subsidiary, Singapore Communications Investments Pte Ltd., which initially owned 40 per cent of 1-Net. SingTel and SCV split the remaining 60 per cent. Subsequently, the IDA subsidiary divested 30 per cent of 1-Net's equity to Pacific Internet and CyberWay (later purchased by StarHub). On September 27, 2000, IDA announced the sale of 1-Net to Media Corp., the nation's largest broadcaster (and itself a government-linked company). IDA is now out of the broadband infrastructure market.

Resources

1. Statistics Singapore
2. Infocomm Development Authority (Singapore)
3. Wireless World magazine
4. International Telecommunications Union

20. Australia

Early Adopter of Wireless

Michael Sainsbury

Editors' Note: Culturally, Australians are early adopters of new media technologies. There are approximately 15 million cellphone users in Australia (more than 70 per cent penetration), far surpassing the number of landlines (10 million). There are six mobile phone networks, among whom Hutchison has launched '3', the country's first 3G mobile network. Australia's vast size also has political ramifications in terms of requirements for wireless services in remote rural areas. Content and marketing services companies as well as trade associations like the Mobile Industry Advertising Association are addressing mobile marketing strategies. An innovative urban hotzone, CitiLan, has been rolled out in Adelaide. M-commerce is still extremely nascent in Australia. The government has made more than $4 billion selling mobile spectrum since 1997. The mobile telecommunications industry now contributes more than A$5 billion par year to the Australian economy.

Connectivity

There are six mobile phone networks in Australia: nationwide players like Telstra (GSM and CDMA), Optus (GSM) and Vodafone (GSM). In some metropolitan areas, there are additional players: Hutchison with W-CDMA (service '3' in Sydney, Melbourne, Brisbane, Perth and Adelaide only) and CDMA ('Orange' in Sydney and Melbourne only). Outside these areas, '3' roams onto Vodafone and Orange onto Telstra.

There are approximately 15 million cellphone users in Australia, far surpassing the number of landlines (10 million). Mobiles phones range from A$99 to A$1,000 (for smart phones). Industry revenues are expected to have reached over A$9 billion for 2003.

Tariffs are perceived to be somewhat expensive—pre-paid starts at A$1 per minute although high-end plans of A$50 per month (or more) provide much cheaper calls. In comparison, Australia has fixed line local calls at a cost of 20 cents anywhere in the country.

SMS messaging is extremely popular, growing at 50 per cent per year. In October 2003, more than 400 million SMS messages were sent across Australia's mobile networks at an average price of A$0.20 each. Several hundred Wi-Fi hotspots also exist.

Industry Dynamics

In 2003, the Australian mobile telephony and data market had one of its biggest growth years even though penetration had crossed the 70 per cent mark. The intensively competitive environment had led to a lull in 2002 as network service providers paused to re-think their strategies before leaping into the unknown world of next generation networks.

Then, in 2003, two major events happened which jolted the market back into life. The biggest mobile network, Telstra, recanted on a year-long strategy of removing the handset subsidies which are very popular in Australia. The move had allowed its competitors to steal market share but also had the effect of dampening the market. By the end of 2002, it had become clear that handset subsidies—which often allowed users brand new, state of the art mobile phones at minimal cost—were still the most effective way to sign up customers for contracts of up to two years.

In Australia, contracted customers spend about three times as much as those using a pre-paid service, and are much less likely to change service providers. So by the busy Christmas period in 2002, Telstra had begun to use subsidies again to prop up its fading market share. By the New Year, its programme was in full swing, during 2003, it regained its ascendancy in adding the most new customers—and most importantly the most new contracted customers.

The other major event was the launch, of '3', the country's first 3G mobile network, built as W-CDMA at a cost of A$3 billion by Hutchison Telecommunications Australia. It was the third of group's multi-country launches, coming only two months after the initial launches in the United Kingdom and Italy.

The combination of aggressive behaviour from Telstra and Optus over the past few years and poor management at Vodafone, which created the cost base of a much larger player, pushed the number three player into deep financial trouble. When its new chief Grahame Maher took the reins he was charged with slicing the company's costs in half. He proceeded to do this by cutting back staff numbers, selling off the companies wholly-owned retail chain and outsourcing a number of key areas. Vodafone's biggest problem is that is losing valuable contract customers to its better-funded rivals. It has about 19 per cent market share, but as little as 11 per cent of the Australia industry's revenues.

Vodafone has also to deal with the Optus joint venture with Richard Branson's Virgin group, Virgin Mobile, which has gained some traction at the bottom end of the market. The strategy allows Optus to focus on higher-end customers yet still get Virgin traffic on its network.

Australia's vast size also has political ramifications. The coalition arrangement between the conservative Liberal Party and the rural-based National Party gives the people who live in rural and regional Australia—known locally as 'the bush' —a strong voice at government level. One of the key issues in the past three election campaigns has been the level of infrastructure and services for the bush. Of particular concern were the provision and level of telecommunications services to the 14 per cent of Australians who live outside the dozen or so major cities.

Content, Community and Commerce

Telstra and Optus launched WAP services in late 1999—but like elsewhere (as in Europe) it was something of a flop. The introduction of GPRS two years later did not do much to improve this.

In 2003, a number of more sophisticated offerings were launched. Vodafone brought its GPRS-based Vodafone Live service, complete with branded camera phone handsets by Sharp, to the market. Youth market offerings concentrated on MMS and games. The service was heading towards 100,000 users by the end of 2003.

Telstra rolled out a similar service, called Loop, on its CDMA 1XRTT network but the lack of camera-enabled handsets saw slow take-up—It then explored the possibility of bringing its features onto its GPRS network.

In April, Australia was the third country in the world to see the launch of Hutchison's '3' network, although it was plagued by early version handsets and the lack of a range of handsets (only NEC and Motorola supplied units in 2003). Hutchison's business plan calls for A$3 billion to be spent by 2005 on the network and it has hit the market using cheap voice as a lure for customers. It is the only network capable of offering video phone calls but all other operators believe they can run all other current applications on their 2.5G networks.

Premium SMS services also become available in 2003. They are emerging as a popular tool for advertisers, particularly fast moving consumer goods and reality television programmes.

Legion Interactive is an interactive communications specialist with a strong focus on media and consumer promotions. It provides interactive channels that allow corporates to effectively communicate with their customers, and direct marketing via interactive devices. Legion Interactive facilitates interactive communications through Interactive Voice Response (IVR), Short Messages Services (SMS), Multi-media Message Services (MMS), Interactive Voice Messaging (IVM), e-mail, the Internet, Wireless Application Protocol (WAP) and interactive television.

BlueSkyFrog, launched in 1999, was the first company in Australia to offer consumers cross-network SMS messaging as well as mobile icons, ringtones, picture messages and outbound voice messages. Members of BlueSkyFrog can download content via the Web and Interactive Voice Response (IVR). At the end of 2003, BlueSkyFrog attracted over 3.6 million page impressions per month, and had 811,795 customers to receive e-mail and mobile marketing messages—making it one of the most desirable channels for companies adding e-mail and mobile messages to their marketing mix.

Television network Channel Seven pioneered an SMS loyalty points system to great effect during the 2003 Rugby World Cup, which was held in Australia.

Intelematics Australia is the leading provider of mobile information, safety and security services for Australian motorists providing in-vehicle assistance and in-vehicle communication services to organisations looking to offer added value to their motor vehicle customers. It is a joint venture company of NRMA Motoring & Services and RACV, two of Australia's foremost motoring club organisations.

M-commerce is still extremely nascent in Australia despite a number of attempts to kick-start it. Several large Australian banks have experimented with mobile banking but have since discontinued the service.

Capacity, Capital and Cooperation

Most development in the Australian mobile industry is being driven by the networks and their vendor partners. Optus has a close alliance with Nokia which provided some of its GSM network and has signed a deal to provide Optus with a W-CMDA network which saw its first trials in 2003. The two companies have a project called FutureLab.

Ericsson, which supplies networks to Telstra (GSM), Vodafone and Hutchison (W-CDMA), is also a long time driver for mobile content and applications. Each year it holds a competition called Frontier for mobile application developers across Australia.

Telstra has an experimental 3G and Wi-Fi network in Adelaide, Australia's fifth biggest city with a population of over 1 million people, called mNet—in conjunction with Alcatel, Motorola, Cisco Systems, software developers Chimo and Medical Communications Associates.

In September 2003, mNet launched the CitiLan Wi-Fi network, which provides wireless Internet access throughout the Adelaide CBD and North Adelaide, creating an easy-access, low-cost 'urban hotzone' covering entire city streets.

The network is the result of an innovative collaboration between Adelaide City Council, service providers Internode/Agile and AirNet, network hardware leader Cisco, the South Australian government and mobile broadband consortium m.Net Corporation. ISPs AirNet and Internode will provide services to casual users and existing customers. Visitors will be able to pick up an instruction brochure from tourist information centres and businesses and register at the CitiLan Website.

The Australian Mobile Telecommunications Association—which consists of service providers, Australian subsidiaries of global mobile and wireless equipment suppliers, and software houses—hosts an annual mid-year conference which is a highlight of the local mobile calendar.

Mobile advertising is being driven through the Mobile Industry Advertising Association, a mix of network carriers, advertising agencies, content providers and media companies. It held its second annual awards last July.

Much of the hardware market for mobile phones in Australia is dominated by European and Asian players. Only one major local Australian company, Voxson, makes mobile phones, software and parts for GSM and GRPS handsets but it had an 'extremely disappointing' financial year from June 2002–May 2003 (the Australian financial year), according to chairman Ray Weekes. It posted a loss of almost A$10 million on sales of only A$25 million and restructured its operations in June 2003.

Culture

Outside Scandinavia, Australia was one of the first countries in the world to see a fast and continuous take-up of cell phones. Culturally, Australians are early adopters of new media technologies. Mobile phone subscriber numbers passed those of fixed-line phones by 2001 and penetration is expected to have crossed 70 per cent at the end of 2003.

The set price of local calls in Australia—fixed by the government every few years—has seen less substitution of fixed services by mobiles than in some other territories. Almost every home in Australia has access to some sort of fixed telephone service, mainly copper—although 25 per cent, or 2 million, Australian homes have access to fibre-optic cable.

The advent of cross carrier text messaging in 2000 saw SMS services begin to explode and, in the middle of 2003, cross-carrier MMS was available between all networks. The high number of operators for the Australian market means that the market is highly competitive. Almost every day most newspapers, particularly in city areas, feature stories of the ups and downs and experiments and successes of the mobile lifestyle.

Capital

With the mobiles sector having become very competitive in Australia, the Australian government has made more than A$4 billion selling mobile spectrum since 1997. A report released in 2003 by the Australian Mobile Telecommunications Association (AMTA) showed the mobile telecommunications industry now contributes more than A$5 billion per year to the Australian economy.

'This is more than either the automotive vehicle component manufacturing sector or the accounting services sector,' says AMTA CEO Graham Chalker. AMTA engaged the Allen Consulting Group to undertake a research project, which found that the industry makes a substantial and growing contribution to the wider telecommunications industry and the Australian economy. In 2002–03, Industry Gross Product was A$5.16 billion.

In 2001–02 more than 32,000 people worked in the industry, which accounts for more than a third of total telecommunications employment and more than A$1 billion in wages. This was almost a 50 per cent increase on the number of people who worked in the industry in 1996–97.

'The mobiles industry has become a key part of the Australian economy and a significant employer,' says Chalker.

In 2003, the industry is expected to have invested A$1.4 billion in capital to support new networks and expansions and upgrades to existing networks. This is in addition to regular operating expenditures. Since 1997, capital expenditure on new networks has exceeded A$8 billion, the report found.

'These investments underpin expansions to mobile network coverage that make mobile services available to more people in more places as well as supporting the development of new services which Australians are embracing,' says Chalker.

Recommendations

The biggest question facing Australia's mobile and wireless industry is 3G. With Hutchison's launch in mid-2003, the exiting operators are holding their breath. All three say they are ready and willing but simple economics dictate that Australia cannot bear 4G networks. Already one GSM network—the ill-fated One. Tel—collapsed in 2001, making others wary of repeating the mistake of over building.

We will surely see an eventual rationalisation of the industry. Telstra and Optus have moved well ahead of the other operators and many believe that Vodafone must do a deal with Hutchison if it is to survive in Australia.

In late 2003, two other fixed wireless services—Unwired and Personal Broadband—were preparing to launch in Sydney and then other major metropolitan areas. The Australian experience shows that they will need to partner with at least one of Telstra or Optus to leverage their large exiting customer bases. Wi-Fi is still only a very niche and unprofitable business in Australia. The spread of the country and its metropolitan cities means Wi-Fi may not be as suited to Australia.

The proliferation of mobile and wireless technologies along with a wealthy and mobile-savvy population means Australia is a market ripe for the right applications. Mobile network providers have become more open in the past few years to partnering with third parties as they are keen to improve average customer revenues by driving users to new mobile data applications.

Other technologies are also emerging, and the high bandwidth fixed wireless technology 802.16 is seen by many in Australia to be an ideal answer for those areas which cannot get fixed line broadband via ADSL or cable.

Resources

Australian Mobile Telecommunications Association: www. amta.org.au
Telstra www.telstra.com
Optus www.optus.com.au
Vodafone www.vodafone.com.au
Hutchison www.hutchison.com.au

21. India

The 'Mobile Party' Begins, but Wi-Fi Languishes

Madanmohan Rao

Editors' Note: India is now the world's fastest growing mobile market; CDMA and GSM subscriptions are finally taking off, and the traditional regulatory mess regarding licensing seems to have been finally sorted out. With a rich content base, a wide range of premium SMS services have been launched. Gartner telecom analysts expect that by the end of 2004, India will have more mobile subscribers than fixed lines. Once the market stabilises, there will be only four major cellular operators left in the field. Interesting initiatives have been launched to bridge the digital divide via wireless access, but much more scalability is called for. The market for Wi-Fi is also only opening up slowly. Research funding for wireless technologies is pouring into Indian R&D labs, numerous wireless startups have emerged, and outsourcing of enterprise wireless application development is playing to India's strengths as a software powerhouse. (Coverage of India's WLL market and regulatory environment are also covered in greater detail in previous chapters.)

Introduction

India's telecom policy of 1999, which allowed cellular operators to migrate to a revenue-based licence fee regime, led to the first drastic lowering of GSM tariffs. The launch of CDMA-based services in 2003 truly sparked off a boom and heated competition among the GSM operators, now making India the fastest growing cellphone market in the world.

'The mobile party has just begun,' says Dilip Mehta, managing director of RPG Cellular Services. With a subscriber base which crossed 30 million in February 2004, India is adding over 1.5 million subscribers a month. The year 2003 witnessed the start of the mobile explosion: In January 2003, India had just 10 million mobile subscribers. Gartner telecom analysts expect that by the end of 2004, India will have more mobile phones than fixed lines. Once the market stabilises, there will be only four major cellular operators left in the field. Despite the rapid growth in urban mobile users and SMS content services, rural India has not yet been brought into the fold, and the market for Wi-Fi is only slowly opening up.

Mobile Users

In its early years in the mid-1990s, the cellphone was largely for the privileged sector of India, due to high device costs and tariff rates. All this changed in the last four years with new players, tough competition and high-profile advertising campaigns. The cellphone has become not just a white collar tool but a blue collar business accessory as well used by employees ranging from real estate agents and chauffeurs to plumbers and even some domestic maids. A key driver of the mobile market is the youth segment for whom the cellphone has become, variously, a status symbol, security blanket, and fashion statement. Accordingly, the cellular operators have launched different kinds of branding campaigns targeted at these various groups.

For instance, Spice Telecom, Karnataka's oldest wireless service provider, has tied up with the state's department of posts to market its 'Uth' pre-paid card and recharge coupons for the youth segment. A total of 77 post offices have been identified to participate in this cross-promotion in Bangalore and Mysore. Idea Cellular has tied up with Discovery Channel for a 13-week series on ideas, and a six-episode series CNBC called 'The Power Of An Idea.'

Airtel has tied up with MTV to launch the Airtel-MTV mobile SIM card. Escotel Mobile's partner is Yahoo! India for the EscotelYouth mobile card, with tariffs and services specially designed for the youth segment.

T.N. Prabhu, managing director of the 8888 service for SMS content and downloads at the Indiatimes portal, classifies users of the service into four personality types: community activist, info-seeker, fun monger, and buyer.

For many users, the cellphone is actually a substitute for (rather than a complement to) a landline—but to be fair, the landline sector

Table 21.1: Cellular coverage areas and operators in India

Circle	Covered areas	GSM Operators	CDMA Operators
Metros	Cities: Delhi, Mumbai, Chennai, Kolkata	Bharti Cellular, Hutchison Essar, MTNL, Idea, BPL Mobile, RPG Cellular, BSNL	Reliance Infocomm, Tata Teleservices
A Circle	States: Maharashtra, Gujarat, Andhra Pradesh, Karnataka, Tamil Nadu	BPL Cellular, Idea, Bharti Cellular, Bharti Mobile, BSNL, Fascel, SpiceComm, Aircel	Reliance Infocomm, Reliance Telecom, Tata Teleservices
B Circle	States: Kerala, Punjab, Haryana, Uttar Pradesh, Rajasthan, Madhya Pradesh, Bengal	Escotel, BPL Cellular, Bharti Cellular, Bharti Mobile, BSNL, SpiceComm, Aircel, Digilink, Hexacom, Reliance Telecom	HFCL Infotel, Shyam Telelink, Reliance Infocomm
C Circle	States: Himachal Pradesh, Bihar, Orissa, Assam, Jammu & Kashmir, North Eastern states	Bharti Telenet, Reliance Telecom BSNL, Hexacom	Reliance Infocomm

Source: Adapted from *Voice and Data* magazine.

is also likely to experience high growth. 'A family will continue to have a landline, but it will also have several mobile phones,' says MTNL Mumbai chief R.L. Dubey. The landline operators are also busy rolling out broadband Internet access in many Indian cities.

Market Forecasts

Given India's relatively late mobile boom as compared with the rest of Asia, there is still plenty of growth potential in basic voice and data services. The Cellular Operators Association of India (COAI) expects the region to challenge China as the world's largest wireless market in the next few years. 'India is the fastest-growing mobile market in the world today,' says Jorma Ollila, Nokia's chairman and chief executive.

The Yankee Group's Global Subscriber Forecast Report projects that the Indian market will have 3.5 per cent cellular penetration rate by end-2005, up from 2.7 per cent in 2003. India and China will be the main drivers in the Asia-Pacific region, with total mobile telecommunication service revenues rising to $245 billion in 2006 from $107 billion in 2001, according to the Yankee Group.

'We expect the Indian mobile market to cross 50 million subscribers by December 2004 (from April 2005 earlier) and 100 million subscribers by August 2006 (from January 2007 earlier). We expect the Indian mobile market penetration to hit 20 per cent by April 2010,' say Lehman Brothers analysts.

'The big hard-working Indian elephant has finally hit the dance floor and is going to make it rumble and shake. The Chinese dragon will have to watch its tail,' says Vijay Chopra, director telecoms and media at Dutch bank ABN Amro in Singapore.

India is playing 'catch-up' with China, according to Suresh Goyal, head of investments of the Indian arm of global banking giant HSBC Holdings. Investments to the tune of $9 billion could be expected as the cellphone user base doubles from an estimated 50 million in 2004 to 100 million in 2005, according to Vikram Mehmi, head of mobile service operator IDEA Cellular. Several Indian telcos have already had successful initial public offerings.

Industry Dynamics

The Indian telecom industry witnessed unprecedented growth in the year 2003, according to the Telecom Regulatory Authority of India (TRAI). During the year, around 17.5 million mobile subscribers were added (more than triple the additions of 2002) bringing the total in December 2003 to 28.20 million against 42.0 million fixed lines subscribers. The combined subscribers base of basic and mobile subscriber stood at around 70.5 million at the end of 2003, for a population of over a billion. India formally achieved a teledensity (number of phone connections per 100 people) of 7 at the end of December 2003, 15 months ahead of the target set in the National Telecom Policy 1999.

As compared with markets like China with two just two mobile operators, India has a more fragmented market with many more players. India had 14 private operators when it opened up its mobile telephony sector in the mid-1990s. Since then, four have become dominant players—the Bharti group (partnered with Singapore Telecom), Essar (partnered with Hong Kong-based Hutchison), the Sterling group, and the Tatas.

The number of operators per circle is six in India, as opposed to two in China, three in Thailand, five in Malaysia and three in Indonesia. Geographically, the Circle A states—Maharashtra, Gujarat,

Andhra Pradesh, Karnataka and Tamil Nadu—have the highest concentration of mobile users with over 8 million subscribers. Metro circles have 7.5 million mobile users. Delhi, with around 3 million mobile users, is the largest circle followed by Mumbai, with 2.7 million users. The number of wireless subscribers in Delhi is more than the number of fixed-line users.

The Bangalore-based BPL group, which has interests in consumer goods, also provides mobile services in Mumbai through a joint venture with France Telecom. Sterling group's Aircel (which itself is a one-circle player) has bought out RPG's Chennai circle, and Escotel has been acquired by Idea Cellular.

Digging beneath the overall subscriber numbers for the nation, interesting patterns have emerged. For instance, the cities of Chennai and Kolkata have not been as robust as expected in terms of GSM penetration and uptake, whereas the states of Kerala and Punjab have exceeded expectations.

Research firm Gartner cautions Indian operators to focus not just on expanding their subscriber base, but on profits as well. Gartner analysts also urge more pro-active roles by Indian regulators in guarding against predatory pricing in the market, keeping tariffs at sustainable levels, ensuring interconnection agreements and unbundling of services.

At the end of 2003, Reliance was the largest mobile player with a subscriber base of 6.24 million (5.57 million CDMA and 0.67 million GSM), followed by Bharti at 5.5 million (GSM), BSNL (4.8 million) and Hutch (4.1 million). The Reliance group is famous for its claim that mobile telephony would work in India only if phone calls became cheaper than postcards.

The lead players will continue to jockey for dominance in the months and years to come. In addition to Orange, Bharti, BPL, Reliance and Tata, the two state-run fixed-line providers BSNL and MTNL in Delhi and Mumbai have also become quick movers in the market.

The Reliance group shook up the Indian wireless market in 2003 with its high-profile advertising campaigns and cut-throat prices. The Reliance CDMA service uses CDMA2000 1x technology with data rates of up to 144 Kbps.

Reliance Infocomm has launched pre-paid services on Samsung and Motorola CDMA handsets in a bid to woo pre-paid subscribers (75 per cent of existing mobile users) of GSM services. Its low cost STD and ISD voice services have been a huge hit in India. RIM handsets are available through 15,000 retail outlets across the country and recharge vouchers are stocked in over 70,000 retail outlets. Some rough patches have been encountered, however, in bill generation and dispatch.

Reliance Infocomm, as part of its pre-paid services, has tied up with the US-based Comverse Technologies for the pre-paid card system. The Comverse pre-paid system supports more than 30 different languages and currencies, and can be configured to run in all major wireless network environments including Reliance's IS-826 IN standard.

The Bharti groups aims at achieving 100 million subscribers by 2006. With Ericsson as a vendor, Bharti Tele-Ventures has launched EDGE services. Its Airtel service has introduced international GPRS roaming with SingTel in Singapore and Globe in Philippines. To increase its reach and profile, Hutchison Essar (known in Mumbai as Orange) has launched a service called mobile Hutch Shops, consisting of 30 vans equipped with sales and billing units.

All the mobile operators are way below the TRAI benchmarks on various aspects of user satisfaction, according to the 3rd Annual Mobile Users' Satisfaction Survey conducted by *Voice & Data* magazine and research firm IDC India in October 2003. The survey covered aspects like network quality, signal strength, echoes, cross-connection, SMS delivery delays, customer care, and billing disputes. BPL led with an overall satisfaction score of 79.8 per cent, followed by Idea Cellular and Hutch. Nokia leads in GSM handset share, followed way behind by Samsung, Panasonic, Motorola, Sony-Ericsson and Siemens. Korea's LG leads in CDMA handset share, followed closely by Samsung and Nokia. About 10 per cent of users send more than 10 SMS messages per day, and 16 per cent of users receive more than 10 SMS messages per day.

Vendors with telecom gear for wireless operators include Nokia, Lucent Technologies Inc, Alcatel, Nortel and Motorola. Chinese mobile phone manufacturer Bird—with an estimated 15 per cent market share in China—has announced its entry in the Indian market with four GSM handset models priced between Rs 5,000 and Rs 12,000.

Finland's Nokia, the world's largest maker of mobile phones, has announced plans to increase its market share in India and boost software offshoring from the country. In the handset market, Nokia accounts for an estimated third of market share. Nokia has rolled out a basic series called the 1100 for fast-growing markets such as China, India and Russia. The phone will work on GSM 900/1800 networks. Nokia is struggling to keep its global market share at around 40 per cent.

According to Nokia CEO Jorma Ollila, 80 per cent of the increase in global mobile subscribers in the next few years will come from new growth markets. Nokia predicts the global base of mobile subscribers would rise to at least 1.6 billion by 2005 and to 2 billion by 2008, up from the current 1.2 billion. Mobile services now reach less than 20 per cent of the world's population and some 4 billion people are without telephone service of any kind. 'We expect that for an increasing number of these people, a mobile will be their first and only phone, and within a few years there will be a large number of mobile users who have never seen a fixed telephone,' Ollila says.

Tariffs and Revenue Growth

Competition has lowered tariffs in the country to some of the lowest in the world, with a 300-minute basket costing the Indian customer $16 as against $21 in China or $40 in Malaysia, according to a report from analyst firm Crisil. Indian subscribers thus currently enjoy some of the cheapest rates in the world.

According to TRAI figures, nearly 70 per cent of the domestic mobile subscriber base was pre-paid in 2002–03. The minimum effective local call charges in cellular services have declined from Rs 14.51 per minute (for 400 minutes of usage per month) in March 1995 to Re 0.77 per minute in September 2003 (or just 5 per cent of 1995 levels, and less than half of 2003 levels).

Revenue from value-added services such as roaming and SMS has doubled from just 20 per cent of the total revenues of cellular operators in March 2002 to nearly 40 per cent in June 2003, according to a TRAI study in late 2003. Revenue from voice traffic declined to 60 per cent at the end of June 2003, down from 80 per cent in March 2002. The decrease is attributed to a 69 per cent drop in call charges. Prepaid card users were more frequent users of SMS and other data services while the post-paid subscribers used roaming more. In addition, international SMS charges more than doubled from Rs 2 in 2002 to Rs 5 in 2003.

Mobile Data Traffic: SMS, Games and Other Downloads

Given India's relatively late mobile boom as compared to the rest of Asia, there is still plenty of growth potential in basic voice and data services. Outbound SMS traffic grew by 200.8 per cent year-on-year to reach 7.39 billion messages in 2003. This is equivalent to 35.8 outbound SMS messages per cellular connection per month. This level still lags those of markets like China, which recorded about 55.9 outbound SMS messages per cellular connection per month, and the Philippines which recorded a staggering 189.4 outbound SMS messages per cellular connection per month in 2003.

A number of creative SMS services are proliferating in India, including a matrimonials service called 'Fifty-Fifty' for matchmaking prospective brides and grooms. Yahoo! India has tied up with a number of operators for messenger and e-mail services delivered via SMS. Rediff.com has launched Rediff Mobile Search, a service that will enable mobile subscribers to search for ringtones based on movie names.

The Indian Railways offers ticket confirmation via SMS for users who sign up for the notification service at the Website. The Delhi police have introduced an SMS-based enquiry service for aspiring purchasers of used cars, in the form of a helpline to check on the legal status of the vehicle. The Kerala High Court sends SMS alerts to lawyers about upcoming cases, thus saving them the inconvenience of having to hang around the courts endlessly.

For the highly-competitive examinations ranging from the SSC board in Karnataka to the CBSE Board across the country, anxious students and parents—tens of thousands of them —are receiving exam scores via SMS.

In February 2002, IndiaTimes—the portal of the *Times of India* publishing group—launched an infotainment service with the shortcode 8888. It is available to GSM users as well as CDMA networks like Tata Indicom, and receives 1.8 million requests every day for SMS horoscopes, dating, cricket, jokes and other branded content.

Cricket contributes more than 60 per cent of the SMS traffic in India, outside of the person-to-person messaging category, according to Chennai-based Badri Seshadri, CTO of Web and wireless cricket information company CricInfo (www.cricinfo.com). Cellphone services currently offered include live text content for SMS, live text content for IVR operation through text-to-speech conversion, audio content, SMS alerts and WAP content (Rao, 2003). CricInfo provides live international scores to numerous operators in India and has relationships with mobile operators and intermediaries in the UK, Australia, Pakistan and Sri Lanka.

Tapping into India's cricket craze, operators like Hutch offer Java-based mobile games including an exclusive Rahul Dravid game. 'An average of 120,000 mobile games were downloaded every month by Hutch users in early 2004. This is a 50 times increase from an average of 2,200 mobile game downloads in February 2003,' says Hutch corporate marketing VP, Harit Nagpal. Games can be downloaded for a price of Rs 50 to Rs 99.

As of early 2004, Reliance IndiaMobile subscribers sent an estimated 150 million messages every month; the SMS service has now been extended to users roaming in dozens of countries abroad.

Mobilefirst, an alliance of leading cellular operators BPL Mobile, Escotel, and Spice Telecom, along with Idea Cellular and TV channel Aaj Tak, launched a mobile version of the tambola game. SMS-based lottery services also have a lot of potential: the offline lottery business in India is estimated to be worth over Rs 50,000 crore a year ($10 billion) annually, and cellphones can tap into segments unreached by the PC-based Internet or V-SAT terminals.

Enterprise software company, Mensamind—which launched a blogging site called mensablogs.com—has tied up with portal IndiaInfo.com to offer mobile multimedia blogging services to IndiaInfo users in India and abroad.

Idea Cellular has tied up with IMI Mobile to source multimedia content. IMI Mobile owns IPR for movies and music videos. Nokia has also tied up with Indian Performing Rights Society (IPRS) to create ringtones which cost consumers Rs 10 per download. It also has tieups for downloadable screensavers with Bollywood production houses like Harry Baweja, Rajshri Pictures Pvt. Ltd., and R.S. Entertainment.

Indian festivals like Diwali, Durga Puja, Id and Navratri are also convenient occasions to host SMS gaming, contests, greetings and downloads. Reliance IndiaMobile reported 10 million Navratri-specific downloads in 2003, including garba event listings and recipes. And over 50,000 users responded to an instant poll on: 'Up to what time should playing loud music be allowed during Navratri?'

On the regulatory front, TRAI has permitted SMS on fixed and WLL telephones. The WLL-based handsets already have the SMS facility; most fixed-line instruments would need to be replaced with SMS-enabled handsets. The Bharti group has recently launched an SMS-enabled fixed-line telephone set.

'The Internet has changed the way we live our lives and the way we do things,' says Reliance Infocomm's president (applications and solutions group) Mahesh Prasad. Its R-World Java-based service allows Internet access and MMS, which can be launched on the handsets via the pre-configured R-World button. 'Quite obviously, only mobile voice will not be enough,' says Prasad.

Services offered range from news and video streaming to festival updates and examination results. The Reliance Application Platform

(RAP) is used for services like mobile banking for account holders in banks like HDFC Bank.

While SMS has been an excellent medium for messaging and instant forwarding of news, challenges have also arisen with the equal potential for spreading rumours. During the bomb blasts in Mumbai in August 2003, news—as well as rumours—travelled quickly through the city's networks. SMS congestion was also reported in many parts of the network as traffic tripled for some operators.

On a regional basis, MTV Networks Asia has announced the launch of a brand new business division, MTV Mobile, targeted at youth users in nine countries across the region—China, Hong Kong, India, Indonesia, Malaysia, the Philippines, Singapore, Taiwan and Thailand. The services like wallpapers, picture messaging, games, VJ ringtones, celebrity content and even 'VJ Flirt' will be available in English, Chinese (simplified and traditional), Bahasa Indonesia and Thai.

Mobile Marketing

Convenience, speed and low cost have made SMS a growing marketing medium in India, thanks to a range of SMS marketing firms like Mobile2Win. Ad agencies like O&M Interactive are looking at using the mobile space to offer value-added services to clients, using mobiles as customer-response devices.

Kellogg's has launched an SMS-based health tips programme on nutrition to increase awareness about the ingredients of its products like Kellogg's Cornflakes. Kinetic Engineering Limited tied up with Yahoo to generate awareness for its Nova moped via an SMS promotion campaign for subscribers of Airtel, Escotel, RPG and Oasis. A database of over 50,000 active users was generated in the process, says Sulajja Firodia Motwani, managing director of Kinetic.

Mobile2Win: Promoting SMS Marketing in India and China

Mobile2Win (www.Mobile2Win.com, or 'M2W') is an innovative wireless marketing solutions company in Asia. A joint venture between Siemens, Contests2win and Softbank, it owns and operates the 8558 shortcode, and has a combined potential reach of over 300 million mobile consumers in India and China. 'We work closely with brands and advertisers to conduct SMS marketing campaigns for them—be it through the pull mechanism of piggybacking on their media or by sending out SMS marketing messages to opt-in users,' says Rajiv Hiranandani, vice president at Mobile2Win and former head of sales at Yahoo India.

(*Box Continued*)

(*Box Continued*)

On the gaming front, Mobile2Win rolls out more than 20 games per month, deployed across all the major operators in India and China as well as across more than 40 operators worldwide. Among Indian operators, the Java games experience more than 150,000 downloads per month on Hutch and around 80,000 per month on Airtel.

Mobile2Win is the wireless partner for Sony Entertainment TV in India. This includes running promotions and show updates on a separate shortcode acquired for them, 2525. M2W also handles the backend for Club Nokia in China and the complete frontend and backend for Siemens mobile world site in India.

As of early 2004, Mobile2Win conducted over 48 brand campaigns on SMS in India and more than 50 in China. The company also helps marketers building 'wireless clubs' via SMS—similar to e-mail databases. 'Our 8558 tie up with China Mobile and China Unicom can cover 100 per cent of the Chinese mobile market,' says Hiranandani.

Its China successes have included a 'Guess tomorrow's temperature' SMS campaign advertised during TV commercials of Coke in Shanghai. It drew over 4.5 million SMS messages in one week in the summer of 2003 and also signed up 700,000 users for an opt-in promotional campaign of Coke.

'We conducted similar SMS campaigns advertised via 20 million traymats at McDonald's in China and gave away free fries and Coke for some of the users who signed up,' says Hiranandani. In Mumbai, Mobile2Win provides SMS-driven billboard advertisement services, where potential advertisers can SMS special codes printed on billboards at various city locations to publicity firms to receive information on exact location, pricing and availability.

He predicts that India could have over 70 million cellular subscribers by mid-2005. 'Around two to three million of these will own Java-enabled phones. There is a huge potential in the Java Game download space with the operators. Also, more and more TV channels are using SMS and premium content to increase interactivity and stickiness for their shows,' Hiranandani observes.

It will even be possible to pay parking meter charges via SMS, once parking meters become widespread in China and India. He recommends that operators (especially in India) start reverse billing so that application developer revenue share and the overall SMS market can grow.

'Do not view mobile technology and marketing as a temporary fad—it is here to stay. The growth in numbers in India and China have proved that! Therefore using technology, value-added services and solutions that will aid your product to reach out to the consumers must be looked at,' Hiranadani advises marketers and advertisers.

Mobile2Win has tied up with SMART and ZED Sonera of the Philippines to distribute its Java gaming content. Operator SMART has a 60 per cent market share in the Philippines, and ZED Sonera is a leading game distributor with clients like Yahoo in the UK and Germany. Mobile2Win's SMS campaigns have been deployed by Coke, Sprite and the Matrix movie in China.

Virgin Atlantic, in keeping with its image of 'excitement', has tied up with Mobile2Win to promote its Delhi-London flight route via SMS-based promotions and free-ticket offers. The launch of Wrangler's new store in Mumbai was accompanied by an SMS competition in conjunction with Mobile2Win; winners received m-coupons which could be redeemed at the store.

(*Box Continued*)

(*Box Continued*)

CNN and Mobile2Win launched a promotion for the news programme 'Your World Today' via an SMS contest; one winner was picked for one new question each week. Other clients signed up for SMS-based marketing solutions in India include Standard Chartered Bank, Bacardi, Domino's and Fosters. Lufthansa also conducted an SMS-based auction with Mobile2Win. Sony TV's promotion for the movie *Lagaan* drew 55,000 SMS responses in eight hours.

Mobile2Win held live and interactive wireless promotions in China's cinema halls for the movie *Matrix Revolutions*: the first person to respond via SMS when the word 'Revolutions' fully appeared on the screen would receive a prize and be notified in the theatre itself. A related 30-day campaign let mobile users in China type in 'MR' and SMS it to the shortcode 8558 for a chance to win cinema tickets and T-shirts. Similar campaigns were conducted for the Mandarin movie *Turn Left Turn Right.*

The SMS services brand from the Indiatimes portal— 8888—has launched promotions for clients like Levi's, Temptations, Kwality Walls, Close up Lemon, Blaupunkt, Marico, McDowell and Mountain Dew. ABN AMRO and Aviva Life Insurance promote their financial services via SMS thanks to a tieup with ActiveMedia Technology.

Jet Airway's 'Jet Escape' vacations are promoted via SMS as well. Zee TV has used SMS to promote its new programme, 'Jeena Isi Ka Naam Hai'. Pepsi let subscribers send questions via SMS to cricket players.

'It is not just well-off consumers who are making the most of cellphones. We are witnessing remarkable pickup by autorickshaw drivers, sandwich makers, betel-leaf vendors and even fishermen,' says Sanjoy Mohanty from wireless operator Escotel. The business community and upper middle-class consumer segment uses post-paid services, and the lower middle-class uses the pre-paid facility. A sandwich maker near a Mumbai college reportedly experienced a dramatic rise in business after being able to receive SMS messages from students placing orders just before the lunch-time rush hour.

There are numerous ways in which companies and carriers are innovating, says Mohanty: yellow pages services, beauty tips, cookery information and text-to-speech services.

Permission marketing campaigns are proliferating. Coke has launched a promotional campaign called SMS2Win, where users can send the serial number printed on the bottle sticker via e-mail or SMS. A reply of win or loss is sent back to the user by SMS as well. Within the organisations, salesmen from companies like Coke are now using cellphones instead of order salesbooks to check out outstanding payments of vendors.

Successful mobile marketing must follow the '6 Is'—immediacy, imagination, interactivity, informativity, integration, and individuality, recommends S.P.S. Shukla of Reliance India Mobile Marketing (RIMM), part of the Reliance group.

Neeraj Roy, managing director of virtual marketing company Hungama, observes that the new trends in mobile marketing are 'Mobile Adver-Games' and position-based interactive communication. Hungama has recently designed mobile marketing campaigns for companies like Electrolux, HDFC Bank and Cartoon Network.

These are early years for mobile marketing in India and adequate attention must be paid to ensure accurate measurement of user data as well as standards of privacy for individual users, cautions Andre Nair of WPP Media.

SMS has also helped television programmes become more interactive. A number of popular TV programmes are weaving in SMS for polling, games and quizzes. A news programme, which asked users to vote on whether a particular minister should be allowed to stay in office, drew 130,000 SMS votes. 'IVR (interactive voice recognition) services also have a lot of potential in India,' says Dilip Venkataraman, head of brand management at Zee News.

Escotel and New Delhi Television (NDTV) offer a news service by SMS in English and voice in English and Hindi called 'NDTV Wireless' for users who dial 6388. The TV quiz show 'Kaun Bange Crorepati?' ('Who wants to be a millionaire?') launched a play-along service via SMS (by dialing 7827), and experienced an SMS centre (SMSC) overload at launch with a deluge of over 500,0000 messages in just seconds.

TV channel Discovery India is weaving an SMS response feature into programmes like women's show 'Berman & Berman' to gain an insight into opinions of viewers who SMS their opinions to the number 3636.

One obstacle which may come in the way of future SMS growth is the fact that cellphones in India have only English keypads. Vernacular SMS services in Indian languages like Hindi, Tamil and Malayalam are available only through coded picture messaging technology (CPMT) and pre-configured messages.

'The SMS growth rates will taper off once most English speakers become cellphone users. The challenge for us will be to creatively devise keypads and interfaces which can accommodate inputs in Indian languages,' advises Ajit Balakrishnan, CEO of Indian portal Rediff.com.

India's cellular revenue is slated to touch Rs 366 billion by year-end 2007, up from Rs 69 billion in year 2002, according to Gartner. Data, which currently contributes Rs 3.6 billion towards India's cellular operators' revenue, is slated to expand to Rs 66 billion, or 17 per cent of total revenues, by 2007.

The Regulatory Tangle: Finally Getting it Right?

India's telecom policy environment has been quite turbulent, but the recent single-licence policy for mobile and fixed-line phone services is expected to trigger off changes in the investment and M&A front. Limited mobility subscribers have been re-classified as fully mobile players under the unified access service licence.

Some fixed-line operators were earlier offering limited mobility services using the CDMA system, but are now ready with fully mobile services. Cellular firms have complained that the terms of the unified licences are stacked against them as they allow fixed-line operators to enter the booming wireless market at a very low cost.

Disputes have arisen as the Telecom Regulatory Authority of India (TRAI) has on occasion accused cellular operators of using the revised interconnect usage charges (IUC) as an excuse to raise tariffs.

In January 2004, Arun Shourie, then Indian communications minister, announced that all existing multiple access radio relay systems would be replaced with wireless in local loop technology by the next financial year. He abolished custom duty of up to 15 per cent on infrastructure equipment in an effort to promote the rollout of both GSM and CDMA networks.

Per-minute charges for mobile calls have dropped from Rs 16 in 1996 to Rs 2 in 2004. 'These are fantastic gains for Indian customers within so short a period as six years,' says telecom veteran T.H. Chowdary, IT advisor to the Government of Andhra Pradesh (Keniston and Kumar, 2004).

'What is more significant is that while the UK and EU took 14 years from 1984 to 1998 for across-the-board competition in every sector of telecommunications network and services, India is experiencing it within half the time', observes Chowdary. But the continuous legal battles between regulators and operators are getting tiresome for industry players, investors and consumers. Chowdary calls for better education of lawmakers and more self-governance and lobbying among consumer associations in the mobile sector.

TRAI has finally got many things right with its unified licensing policy, says Ashok Jhunjhunwala, professor at IIT-Madras and a telecom pioneer (Keniston and Kumar, 2004). But the regular should have better managed the controversy over WLL (IS-95 CDMA) or limited mobility services—and ought to do more to promote rural telecom access and ensure that the solutions that are provided can support bit rates of at least 28.8 Kbps for Internet access.

'Wireless technology is a key component for providing connectivity to rural areas. Frequency charges should be made zero for a rural service provider (RSP) for the first 10 years in order to rapidly enhance telecom and Internet services in rural areas,' he advises.

Wi-Fi: A Laggard Market in India

Low awareness of Wi-Fi benefits and high cost of installation are main barriers to adoption, says a report prepared by the Indian Market Research Bureau (IMRB). Companies seeking to offer Wi-Fi need to start the journey from the top of the segment—IT, telecom and banking and financial services segment—while looking at Wi-Fi as a networking solution for both wide local area network (WLAN) and telecommunications. Low laptop penetration is possibly the cause of low level of Wi-Fi usage among other segments, says the report.

On the Wi-Fi front, key Indian ISPs such as those owned by the Tatas, the Bharti group and Satyam Infoway are commencing operations at Wi-Fi hotspots in India. Satyam Infoway (Sify) has launched a Wi-Fi service in the airports of New Delhi and Chennai for Rs 60 per hour; Wi-Fi cards can also be rented.

The Bharti group, which already has broadband DSL offerings, will augment this service with Wi-Fi boxes for Rs 3,000. The package is targeted at residential users; other targets include conference and trade show venues like Pragati Maidan in New Delhi. India's version of Starbucks—the Barista café chain—has 15 Wi-Fi-enabled locations in Mumbai and Delhi, thanks to a tieup with Tata Teleservices, which has a stake in the coffee chain. The Tatas also plan to target distributor networks and warehouses.

WLAN diffusion seems to be mainly in educational institutes, hospitality companies and large corporates, claims *Voice&Data* magazine. Campuses like those of the International School of Business in Hyderabad and the Indian Institute of Information Technology in Bangalore already offer Wi-Fi access to students. Intel has begun marketing PCs with its wireless Internet chip.

Data Access is working on seamless migration from Wi-Fi to GSM-based Internet coverage. Systems integrators active in the business include Tulip IT Services and Convergent Data. Some hotels in India like Le Meridien and the Taj hotels already have Wi-Fi installations. Offices of companies like Microsoft and Punjab National Bank are Wi-Fi enabled. Cisco leads in market share of WLAN products, followed by D-Link and more distantly by Ericsson and Proxim.

In early 2004, the number of hotspots in India was at best a few hundred. Government rules also do not encourage wide-range Wi-Fi deployment. Business models for revenue share are still being worked out. Yet, growth is expected to mushroom next year. Cisco expects that the fastest growing markets in the Asia-Pacific region for Wi-Fi will be China, India and the Philippines.

Wireless in the Last Mile: Overcoming the Digital Divide?

It is a story that is regularly making its rounds in mobile marketing conferences in India and has even been picked up by *The Economist* magazine: the creative use of cellphones by fishermen in the south Indian state of Kerala.

Fisherman in Kerala are using mobile phones to dynamically reconfigure distribution chains while they are at sea, depending on which seafood market offers the best prices for the catch. This is a classic example of m-services at 'the moment of value', according to Peter Keen and Ron Mackintosh, authors of 'The Freedom Economy.'

Some fishermen report that their profits in the peak fishing season have doubled since they can contact multiple ports and seafood markets and identify the best sale prices for the catch of the day. Routes, fishing locations and even the composition of the catch itself can be planned and coordinated while the fishing boats are still at sea.

A wide section of Kerala's coastal strip, stretching for about 600 kilometre, is covered by two operators: BPL Mobile Cellular and Escotel Mobile Communications. Wireless services are available 25 kilometre out to sea, and in some cases twice as far. Twenty per cent of BPL Mobile subscribers and 8 per cent of Escotel subscribers in Kerala are reportedly from the fishing community, together accounting for close to 100,000 fishermen as of late 2003. Both operators have even launched special camps to train fishermen on how to use SMS and are devising financing schemes especially for fisherfolk.

Those who have been using cellphones for creative distribution chain planning over the past two years have already been able to improve their living standards. Earnings for some have doubled from Rs 2,500 a week to Rs 5,000 a week (46 Indian rupees = 1 US dollar).

Add to this the safety factor—through weather warnings and emergency calls—and the fishermen of the Kerala State Fishing Boat Operators' Association are even arranging for loans and financing for acquisition of new cellphones!

Other forms of wireless connectivity based on WLL and proprietary radio standards have been used in rural initiatives like the Gyandoot project in Madhya Pradesh (Rajora, 2002) and the Village Knowledge Centres (VKCs) of the M.S. Swaminathan Research Foundation. The Gyandoot model involved WLL-based connectivity for kiosks operated by local entrepreneurs, providing fee-based information services like land records and crop prices (see the chapter on WLL earlier in this book for more such case studies). The VKCs involved a cluster of village information centres connected via a hybrid wired and wireless network with VHF duplex radio devices and dial-up telephone lines; self-help groups provided online access to healthcare, educational and vocational content (Senthilkumar and Arunachalam, 2002).

The Indian IT Powerhouse: Software and Services for Wireless Technology Companies

From *Wired* magazine and the *Wall Street Journal* to *Business Week* and the *Financial Times*, the growth—and perceived threat to Western jobs—of Indian outsourcing has been receiving high profile coverage in the global media.

The R&D outsourcing market for IT in India is projected to grow from $1.3 billion in 2003 to $9.1 billion in 2010 at a compounded annual growth rate (CAGR) of 32.05 per cent, according to research firm Frost & Sullivan, which undertook the study for the department of IT. The R&D outsourcing market for telecom in India—including wireless—is slated to grow from $700 million in 2003 to $4.1 billion in 2010 at a CAGR of 28.73 per cent.

A Merrill Lynch report on core technology research and development services in India identifies Indian companies which are focusing not just on IT budgets of their clients but the R&D budgets in areas like ASIC design (targeted by Wipro and HCL Technologies), VoIP (Hughes Software), and protocol stacks (Sasken). Domestically, however, Indian IT companies spend far less a percentage of sales revenues on R&D (0.6 per cent on average) as compared with US firms (14.8 per cent on average) (Rao, 2002).

Research funding for wireless technologies is already pouring in to the Indian Institutes of Technology and central research labs. A number of startups along the entire wireless value chain are sprouting in high-tech hubs like Bangalore.

For instance, Bangalore-headquartered Jataayu Software offers MMS, SMS and WAP platforms for Tata Teleservices' CDMA network, called Tata Indicom. An estimated 80 per cent of the WAP, MMS and SMS data in India is delivered through its gateways, and the company recently exhibited at the ITU Cairo 2004 conference to strike alliances with African wireless operators. Venture capital firms like JumpStartUp and Acer Ventures have already funded a range of wireless startups in India, or US-based startups with significant design and development in India.

JumpStartup has invested in three companies in the wireless/mobile space: July Systems (building the next generation service creation, delivery and management platform for wireless data carriers), Meru Networks (building an enterprise class, standards compliant voice/data WLAN infrastructure) and Hellosoft (building unified, DSP-less, communication stacks for VoIP/WLAN/GSM-GPRS solutions).

The three companies are based in Silicon Valley with very significant India based operations. The companies have raised between $10 million and $30 million. 'Trend lines indicate that Wi-Fi is going to be the dominant mode of communication. However, there are issues relating to security, standards, business models and costs. Wi-Max could become interesting over a 5 to 10 year period in some countries in Asia, especially with Intel's push into low-cost silicon for such solutions,' predicts Sanjay Anandram, Bangalore-based founding principal of JumpStartUp.

The carriers are the ones who have to figure out the appropriate business models, e.g., an integrated Wi-Fi-cellular solution. They will have to figure out how to deal with issues like roaming, billing, customised service offerings, security, customer care, content management and digital rights management. 'There are opportunities for companies to create solutions in this space. There's a huge opportunity for a low-cost low bit-rate unified VoWLAN phone with GSM-GPRS capabilities for the SOHO/SME segment,' says Anandram.

Acer Technology Ventures (ATV), a venture capital business unit of Taiwan's Acer Group, has a wide range of investments in wireless space: AirMagnet provides wireless network management solutions; Apherma provides audio solutions for cellular phones; Atrua offers solutions for wireless security; Centrality Communications and IRF Semiconductor develop IC components for handheld devices. ATV has offices in Santa Clara (US), Taipei, Shanghai, Beijing and

Bangalore. Based in Bangalore, Samir Kumar, VP of Acer Technology Ventures, also identifies China Linktone and Spreadtrum as notable Acer investments in China, and HelloSoft and July Systems in India.

'The balance of technology is shifting from the West towards the East. If India and China use their market potential skilfully, their companies could become primary players in the telecom space,' says Ashok Jhunjhunwala, professor at IIT-Madras and a telecom pioneer (Keniston and Kumar, 2004). Companies based in the US lead in heavy-traffic optical networking, routers, switches and last-mile access equipment. Chinese and Indian companies can become leading players in software and systems integration with promising niches in device and software product as well.

References

Keniston, Kenneth & Deepak Kumar, (2004). *IT Experience in India: Bridging the Digital Divide*, Sage Publications, New Delhi.

Rajora, Rajesh (2002). *Bridging the Digital Divide: Gyandoot, the Model for Community Networks*, Tata McGraw-Hill, New Delhi.

Rao, Madanmohan (2003). *The Asia-Pacific Internet Handbook, Episode V: News Media and New Media*, Tata McGraw-Hill, New Delhi.

Rao, Madanmohan (2002). *The Asia-Pacific Internet Handbook, Episode IV: Emerging Powerhouses*, Tata McGraw-Hill, New Delhi.

Senthilkumaran, S. & Subbiah Arunachalam (2002). 'Expanding the Village Knowledge Centres in Pondicherry', Regional Development Dialogue, United Nations Centre for Regional Development (UNCRD), Autumn issue.

22. China
Mobile Superpower

Lunita Mendoza

Editors' Note: This chapter outlines the mesmerising potential of China's mobile market—while also highlighting the IPR and political challenges that are opening up. China has the world's largest mobile subscriber base and second largest Internet population. Wireless subscribers outnumber wireline users. The markets are growing for products and services ranging from used cellphones to SMS gaming. A total of 220 billion SMS messages were sent in 2003. Chinese government officials are concerned about the use of SMS to spread rumours and political messages. The WLAN market may cross $900 million by 2007. A number of home-grown wireless players are also venturing into foreign markets. Controversies are emerging as China flexes its muscles in terms of its own standards and requirements for wireless security, digital content and 3G protocols. Still, China is viewed as a powerhouse that will benefit Asia as a whole and will have repercussions for wireless players around the world.

Introduction

With China surpassing the rest of the world to become the nation with the largest population of mobile subscribers, Chinese mobile operators are moving aggressively to tap into this community with next generation networks and the wireless data services that they support to drive future growth.

From major conference rooms in Asia to coffeshop tables, China is certainly one of the hottest topics of discussion. China's economy grew at an official 9.1 per cent in 2003—its best performance in seven years. At the same time, a political buzz surrounds developments regarding its frosty relationship with Taiwan, and the opening of the six-nation Central Asian security group—led by China and Russia—to fight Islamic extremism. China-watchers continue to churn out books about the future of the Middle Kingdom: a book called '*Jesus Comes to China*' even claims that Chinese Christians are highly pro-American, and so a Christianised China will be more likely to see eye-to-eye with the US on many international issues!

The Mobile Market: Dazzling Growth

By the end of October 2003, the number of telephone subscribers in China reached 512 million. Mobile phone subscribers exceeded fixed-line subscribers for the first time as well, with the rapid rise of the domestic communication manufacturing industry. Communication product sales increased by 33 per cent in 2003 compared to with 2002, with a 111 per cent rise in profits.

Export by Shanghai Bell Alcatel in the first 10 months of 2003 amounted to RMB 1.12 billion, an increase of 54 per cent from 2002. Huawei's profit totalled RMB 2.4 billion, while ZTE's sales income has enjoyed a consistent 50 per cent rise. The growth of Panda, Amoi, and Qiaoxing has also been considerable, with sales income growing by over 80 per cent each.

Cellphone sales for 2004 are expected to reach 70–90 million. China's cellphone penetration rate is about 20 per cent, with more than two-thirds of them subscribing to China Mobile. But ARPU has been falling with the increase in competition.

Product differentiation is also transforming the Chinese handset market. Gem-encrusted handsets are reportedly in demand in China, especially handsets like TCL Mobile's diamond-studded mobile phones. TCL has also sold millions of phones with imitation jewellery.

Used cellphones, on the other hand, are favoured in markets like rural areas. Gartner estimates that 40 per cent of the 62 million new mobile accounts in China in 2003 came from people using second-hand cellphones. In terms of overall market share, 10 to 15 per cent of all cellphones sold in China may be second-hand (eventually reaching 30 per cent, according to BNP Paribas Peregrine), mirroring trends in other developing countries like India and Indonesia and some parts of eastern Europe.

On the 3G side, Qualcomm China says that it expects to see more than one 3G standard rolled-out. 'We will continue to provide our CDMA expertise and industry-leading technology solutions to support China's 3G rollout,' says Jing Wang, chairman of Qualcomm China and Qualcomm's Asia Pacific Strategic Committee. For Jing, China is the world's largest wireless market and also home to some of the most exciting companies in the 3G wireless industry globally.

SMS Traffic and Services: Business Potential but Political Challenges

In China's 'thumb economy' SMS traffic yielded a revenue of $333 million to operators and content players in 2003 and is expected to increase to $533 million in 2004.

Shanghai Telecom Corp, the city's biggest fixed-line phone operator, has a content subsidiary called Shanghai Online Information Network Co Ltd which launched an SMS service called 'Know Everything about China'. Subscribers can receive 3–5 SMS messages every day about Chinese vocabulary, history and customs. The Chinese newspaper *Shanghai Morning Post* uses SMS for raffle services.

The Star Group's Star China TV entertainment company uses SMS in its Mandarin language channels for programmes like reality shows and dating. Channel [V] Chinese Music Awards has received more than 2 million SMS messages while voting for best songs and singers.

With 282 million mobile subscribers as of February 2004, concern is also growing over SMS spam and the ministry of information industry (MII) is working on a regulation to better supervise and standardise the SMS market. Regulation of SMS is now necessary, says Chen Jinqiao, director of the Chinese Academy of Telecommunications Research under the MII.

China's MII is trying to facilitate SMS interconnection between the fixed and mobile networks. MII statistics reveal that in the first 10 months of 2003, the volume of SMS in China reached 108.7 billion messages, 14.4 billion more than that in all of 2002. It is estimated that the market scale of mobile phone SMS services in China's mainland will be worth RMB 5.4 billion in 2003, and hopefully surpass RMB 10 billion in 2004.

Leading Chinese portals Sina, Sohu, and NetEase are now riding a new investment wave thanks in part to profits from SMS services. At the time of writing, the market value for Sina and NetEase is well over $1.5 billion; for Sohu it is more than $1 billion. Even though operator China Mobile takes about an 80 per cent cut on SMS-delivered news and information, all the portals have earned profits. Sohu has tried to push multimedia messaging; its CEO Charles

Zhang climbed part of Mount Everest to promote the subscriber service, beaming regular updates of his trek to customers! 'We wanted to find a way to get money, and then SMS exploded,' remarks Zhang.

Chinese portals NetEase, Sina and Sohu are now reporting increasing profits thanks to billable SMS services for news alerts, information, cartoons and games. At 1.5 yuan (US 20 cents) per message, the huge volume of cellphone users and Internet traffic is enough to generate decent profits. In 2002, China Mobile, the country's biggest wireless operator, introduced a micro-payment system which lets portals share in revenues for wireless content services. China Mobile users sent 80 billion SMS messages in 2002, up from 15.9 billion in 2001.

Traffic began to increase dramatically when China made its debut appearance in the soccer World Cup last year. Future SMS revenues could arise from online dating services and stock trading. With 59 million Internet users at the end of 2002, China is the world's second-largest online population, ahead of Japan and behind the US. Another 27 million Chinese are expected to be online by 2004.

SMS services are already driving M&A activity in China. Sina.com has announced intentions to buy Crillion Corp, a provider of services for mobile phone users, for up to $125 million in cash and stock. Shenzhen-based Crillion provides job referrals via SMS.

At the same time, political observers have been noting how China is actively embracing wireless and Internet markets, but is also grappling with the political challenges of online access and mass mobilisation that the Web and wireless allow. Chinese government officials had to track and crack down on rumours about SARS being spread via SMS. Tracking features have also been reportedly used to patrol 'objectionable' SMS traffic (e.g., about the banned movement Falun Gong) before the 16th Party Congress. Such tracking services are available from a number of vendors, including Huawei, ZTE and Nokia.

PDAs and Other Handheld PC Devices

Results from IDC's Asia-Pacific Handheld Device Tracker reveal that the Asia-Pacific (excluding Japan) handheld market decreased 29 per cent year-on-year to 1.59 million units in 2003. Although many countries in the region grew over the course of the year, China's market fell as buyers shifted their attention from low-end handhelds to other products such as digital cameras and mobile phones.

'The sheer size of China's market dragged down the entire region as locally-produced low-end handhelds fell out of favour there,' says Manny Lopez, Personal Systems senior research analyst, IDC Asia-Pacific. 'But the silver lining is that more sophisticated products are continuing to sell across the region. Devices that go beyond just basic organiser features, such as multimedia and Wi-Fi, are helping to stir interest in the market.'

But HP has managed to buck the trend with its strong product line-up. HP held the top position in all four quarters of 2003, overtaking local PRC vendor Hi-Tech Wealth as the number one vendor in Asia-Pacific (excluding Japan).

PalmOne was able to move up one notch to second place by maintaining its performance in the region while the Chinese vendors dropped in rankings. Hi-Tech Wealth, GSL, and Minren recorded declines in shipments in 2003 as Chinese buyers continued shifting away from the low-end handheld devices that had initially made them successful in 2002.

WLANs

With 70–80 million Internet users expected by the end of 2003, the total value of the Chinese WLAN market at the end of 2002 was estimated to be around $14 million, a growth of over 180 per cent over 2001. It is estimated that by the end of 2003, investments will have doubled to around $30 million. Analysts expect growth will slow with the gradual maturity of the WLAN equipment market and with 3G diverting some of the available investment resources. WLAN growth is forecast to flatten to about 50 per cent within the next two years and reach a total value of over $900 million by 2007 out of a total telecom investment projected at over $2 billion.

With over 90 hotspots in Beijing alone, WLAN nomadic connectivity is becoming common as the number of hotspots increase in campuses, airports, shopping centres, hotels, medical facilities, factories and other public areas like Starbucks and Häägen Daz. Reality continues to bite seeing how McDonald's restaurants recently terminated their free Internet trial in China over issues involving bureaucratic red tape. Services have yet to be resumed.

In December 2003, Shanghai Telecom began implementing outdoor coverage of wireless broadband in Lujiazui, Nanjing Road, Huaihai Road and Xintiandi. Wi-Fi's huge success in South Korea has attracted the attention of players like China Telecom. Shanghai

Telecom's Wi-Fi challenge to the GPRS Internet access of Shanghai Mobile should be a competition to watch.

In November 2003, Calypso Wireless Communication Corp. and China Telecom reached an agreement worth $500 million over three years for Wi-Fi services and VoIP mobile phones.

At the same time, however, US businesses have expressed alarm about newly-issued regulations that appear to require equipment makers to deal with their Chinese competitors to obtain access to the encryption standards required for wireless networks. All manufacturers and importers of wireless network equipment used in China, were given a six-month grace period to comply with the regulations even though products contracted, imported or produced prior to the effective date of the ban. Critics of the new policy say the move appears to be a blatant attempt to boost the domestic industry. Chinese manufacturers lag behind in wireless technology, and the government could also be concerned about having to follow, or even pay patent royalties on, standards set overseas. China recently released its own standard for DVD technology, EVD, as part of the drive to wean itself away from reliance on foreign standards.

Table 22.1: China Telecom market for fixed line, mobile, broadband and fixed access investment from 2001 to 2003

Year	Fixed lines (million)	Mobile (million)	Broadband (million)	% Growth (fixed)	% Growth (mobile)	% Growth (broad-band)	Fixed Access Investment (RMB billion)	% Growth (fixed access)
2001	179.00	144.81	1.64	23%	70%	–	264.18	14%
2002	214.00	207.00	4.43	20%	43%	170%	203.46	–23%
2003	250.00	242.00	7.90	17%	17%	78%	207.00	2%

Source: Krone Asia.

The China Market: Global Players

Despite rampant concerns about piracy of intellectual property, China is still one of the most popular destinations for business expansion in sectors like mobile and Internet.

In November 2003, Siemens announced that its sales in China reached RMB 30.1 billion that financial year, increasing by 8 per cent compared with the financial year of 2002. Ernst H. Behrens, president and CEO of Siemens (China), maintains that China is an attractive market and will play an increasingly important role in the global service of Siemens in the future.

Jan Malm, president of Ericsson (China), says his company has 26 representative offices, 10 joint ventures and four fully-invested companies in China. Ericsson has a nearly 35 per cent market share in China's GSM market and 40 per cent of its GPRS market. It is one of the two MMS solution providers for China Mobile. The company also has an approximate 15 per cent market share of China's CDMA business.

Fujitsu has also established Fujitsu (China) Information System Co., Ltd based in Beijing, for a range of IT services.

Next Generation Technology Markets

An In-Stat report predicts that China's mobile market will grow from 206.75 million subscribers in 2002 to 402.28 million subscribers by 2007, a CAGR of 10.5 per cent for the forecast period. Over the next five years, China's wireless services market will see dramatic development and transformation with the move from today's voice-centric, price competitive model to one characterised by a more complex mixture of voice, wireless data and multimedia services. At the low end of the market, improved network efficiency and capacity will drive operators to continue to target and grab the all-important first-time subscribers. The result will be massive growth in the number of mobile subscribers.

Table 22.2: Chinese carriers and vendors involved in 3G trials

Carrier	Vendors
China Telecom	Samsung, Lucent, Huawei, Siemens, Alcatel, Guangzhou P&T
China Netcom	ZTE, Nokia, UTStarcom, Lucent
China Mobile	Huawei, Ericsson, ZTE, Motorola
China Unicom	Huawei, Ericsson, Motorola, Nortel, Dailan Huanyu
China Satcom	ZTE, Capitel
China Railcom	ZTE, Eastcom

Source: Lehman Brothers.

During the forecast period, 2G subscribers will still represent the majority of the market, but 2.5G and 3G subscribers, mostly driven by increasing corporate wireless implementations and improved device functionality, will grow as well. The 2.5/3G market will be dominated by high-end subscribers, as an increasing number of customers connect to corporate networks and usage levels continue to approach levels seen in the mobile office paradigm.

Dealflow: Mobile Business Booms in China

A quick scan of the trade press in recent months shows a hot and heavy flow of deals between Chinese operators, local partners and foreign vendors. At the same, Chinese players flush with success in the local market are foraying abroad as well. For instance, Zi Corporation has signed a licensing agreement with UTStarcom China for text input technology for both the Chinese and English languages. Beijing-based Oztime, Zi Corporation's Chinese e-Learning subsidiary, has signed a major contract with China Telecom Corporation Limited to license its e-Learning products and services. Netgear has entered into an agreement with the digital home business unit of Legend Group Limited, the largest IT enterprise in China. Infineon Technologies has established its latest IC Design Center and new subsidiary—Infineon Technologies Xi'an Co. Ltd—in Xi'an. China Mobile recently awarded Siemens a contract for Anhui, Shanghai and Zhejiang provinces worth around 30 million euros.

Inventel supplies telephone manufacturers in Southern China with the electronic design and software for DECT standard cordless telephones. In 2003, 6 million telephones were sold integrating Inventel technology, accounting for 15 per cent of the world market. The market for DECT telephones should grow at a rate of 15 per cent per annum till 2008.

Table 22.3: China's electronic information product exports, January to October 2003

Product	Export (10 thousand units)	Export volume (USD 10 thousand)	Growth rate of export volume
Telephone	15280.7	131877.7	11.8%
Program-controlled switchboard	12.8	32950.9	72.6%
Handset (vehicle) wireless phone	7329.2	543554.6	42%
Mobile communication BS	1.57	39450.2	12.8%
PC	297.7	179878.5	68.3%
Printer	3074.9	334669.8	59.1%
Monitor	4996.6	732851.5	62.2%
Colour TV set	1741.6	197154.9	18.9%
Laser disk video player	9082.1	398475.4	45.7%
Electronic component		963144.4	26%
IC	889426.7	475117.9	156.5%

Source: Hong Kong Trade Development Council.

At the same time, many Chinese enterprises have expanded outside the mainland; there are 22 Chinese companies listed in Singapore. The wholly-owned subsidiary of Accord Customer Care Solutions Limited (ACCS), Distribution Management Solutions Pte. Ltd. (DMSPL), has clinched six new contracts with several mobile phone distributors and AMS providers in Asia-Pacific. ACCS provides an integrated range of services in the management of after-sales customer care to 24 brands, namely, Alcatel, Audiovox, Bird, CECT, Daxian, Haier, Hyundai, Ikomo, Kyocera, LG, Motorola, NEC, Nokia, O_2, Panasonic, Philips, Psion, Sagem, Samsung, Sendo, Sharp, Siemens, Sony Ericsson, and TCL.

Cellstar was named as exclusive distributor of Bird Phones. Bird is currently the leading phone manufacturer in China. Formerly a maker of pagers, the company claims to have sold 7 million units in China, unseating Motorola. The Chinese company reported $1 billion in revenue in 2002, and is expecting to double it in 2003. While China remains its major market, the company is aggressively rolling out its products in Southeast Asia where the demand for low-cost handsets is high. The company was founded in 1992, and was awarded a manufacturing license by the Chinese government in 1999. The company is currently publicly listed on the Shanghai Stock Exchange.

McGordons is the distributor of Panda mobile phones outside China. Panda customises both the phone it self, consumer boxes, display, special customer software and selection of colours. Other telecom companies from China with an overseas presence include Amoi Electronics Company from Shanghai. Chinese mobile phone manufacturer Bird—with an estimated 15 per cent market share in China—has announced its entry into the Indian market with four GSM handset models.

Challenges: Intellectual Property Rights

From movies to mobiles, safeguarding IPR issues in China has been an uphill struggle. Blockbuster, the world's largest video rental chain, is shutting shop in Hong Kong and scrapping plans to expand into mainland China due to piracy concerns. Though China officially has been vocal in condemning copyright violations, scores of shops in leading cities such as Beijing sell pirated video discs—often for less than $1—without fear of being closed down. To be fair, a significant number of Westerners are also seen gleefully purchasing pirated VCDs in China.

'China: A Powerhouse that is Going to Benefit Asia as a Whole'

An interview with Khoong Hock Yun, Assistant Chief Executive of IDA, Singapore

Several concerns arose during Singapore's mission to China in 2003 that encompassed issues like IPR infringements and the hesitation to trust in a country that was notorious for 'hit and run' debacles. Till today, Singaporeans and other foreigners alike continue to tell tales of how they got burnt on the Silk Road to riches. Nevertheless, in his speech at one of China's biggest wireless exhibitions in Beijing, assistant chief executive of the IDA Khoong Hock Yun stressed the issue of trust in relation to business dealings.

Singapore would more likely be a bridge between the western and eastern world, being an important value-added player. Some observers wonder when China will come to a stage where concerns on basic business sanctity will no longer be a worry. 'It took Singapore 20—30 years to get where we are today. It may take China one generation to also become different,' observes Khoong.

'You really need to be in the market, you can't keep travelling there and going back—to sense the market, the opportunities, the feedback, the benefits, and to develop a platform for other companies to come together.

'On the telecoms side, obviously it will take some time for the Chinese market to deregulate or to liberalise. The Chinese are looking to various countries to learn more about what are the pitfalls and benefits of liberalisation and how to manage the transition process. I think in Singapore, even though we liberalised in 2000, we are still in the transition process; the market doesn't become perfect overnight.

'The growth of China will have a positive impact for Singapore, just like the growth of the US economy has always been of benefit to us. But having an Asian powerhouse is going to benefit Asia as a whole. There will be some products that will make more sense to the Asian market and having a large powerhouse like China helps give that balance. There are good things that come from the West; we would like to see good things as well that come from the Asian market.'

Source: Lunita Mendoza

In the realm of mobile content protection, China's first mobile phone ring infringement lawsuit saw the Music Copyright Society of China (MCSC) suing TCL Mobile Communication Co. Ltd., and Beijing Digital Electronic Communication Technology Co. Ltd. in December 2003. TCL Mobile paid RMB 2.5 million to MCSC to compensate for its economic losses.

C.K. Leong, regional managing director of Krone Asia, takes the zen approach by driving home the importance of quality to operators in China via educational sessions. 'For the copycats there are absolutely no guarantees on their manufacturing processes. In some instances, these are done in dodgy factories or with poor quality materials. The end result is very often an inferior product that will fail over time—or when pushed to the limit,' says to Leong. 'The copyrights are getting more and more emphasised by the China government especially with the entry into the WTO. Consequently, it would be safe to say that there are efforts by their government to do something—compared to some years ago.' Obviously, the efforts have yet to yield real results. Leong says Krone is seeking legal action for the infringements produced by the local manufactures.

In a report by the *Beijing Review*, Christian Murck, chairman of the American Chamber of Commerce in China, observes unexplained delays in granting foreign firms licences made available to domestic competitors and a pattern of delay and restricted access is emerging.

In a report on global barriers to trade, the US Trade Representative's Office, weighed in on the dispute in early 2004 where China granted free access to the new wireless encryption standard—called WAPI—to only 11 Chinese telecom-equipment makers. Many foreign vendors were nervous of being shut out of the Chinese market for the fast-growing wireless data network technology. In April 2004, however, the Chinese government decided to delay adoption of the new wireless Internet security standards, but analysts feel that such issues and tactics will arise again further down the road.

The Chinese decision to back down temporarily came with a quid pro quo: the US has agreed to reconsider the ban on exports to China of high-tech products that could have military use. Such give-and-take is expected to become a standard routine as China continues to flex its muscles that have developed became of its market size and technological prowess in the mobile and wireless Internet arenas.

23. Thailand

GSM Competition and Penetration

Srikanya Mongkonsiri

Editors' Note: This chapter provides an overview of Thailand's mobile media market. There are four wireless operators in Thailand: AIS, DTAC, Orange and Hutch, accounting for approximately 21 million cellular users. SMS messaging and downloads are popular among young generations, especially teenage girls; MMS and GPRS services are picking up as well. But operators have suffered declining ARPU and there is a perception that because the current prime minister is the owner of AIS, fair treatment to every operator is impossible. Wi-Fi access is provided by companies like Internet Thailand Public Company Limited (INET). The hotspot front is yet to pick up momentum, other than in coffeeshop chains like Starbucks.

Overview

There are four wireless operators in Thailand: AIS, DTAC, Orange and Hutch. The Thai market is dominated by AIS (Advanced Info Service Plc.), founded by Prime Minister Thaksin Shinawatra and has a 60 per cent market share. There are approximately 21 million cellular users in Thailand. The wireless market can be described as hyper-competitive in terms of media spending and consumer promotion campaigns, to gain new subscribers and to prevent existing subscribers from moving to other competitors.

Typical cost of cellphones are around 6,000 baht with PDAs at around 10,000 baht. SMS messages are popular among young generations, especially teenage girls. Most Thai teens love to send SMS,

MMS or melodies to one another, with downloads priced at 15 baht for a one-time download. SMS is being seen as a little outdated when compared to MMS, pictures or animated pictures. Phones with cameras are seen as trendy.

The Thai wireless market today is very much like Europe in 1998, where operators suffered declining growth rates and average revenues per user even as penetration increased. Unfortunately, there is a perception that the telecom operators will not get fair treatment as the current prime minister in the owner of AIS.

Fixed Line Services

Thailand's telecommunications services have improved substantially over the past few years, especially in Bangkok, which now has an abundance of fixed lines for offices and residences. The Bangkok metro area is served by two fixed-line operators: the Telephone Organisation of Thailand (TOT), a state enterprise, and Telecom Asia, a private company. The latter can usually install a phone within a few days, the former somewhat longer. Local calls are not timed and carry a fixed tariff of 3 baht per call. Areas outside Bangkok are served by the TOT and the private company TT&T Public Co., Ltd. The fixed line network has a total capacity of 7.75 million lines, of which about 6.02 million are in use. In addition, 218,968 public telephones are available nationwide.

Table 23.1: Fixed line density

Conventional Services			
Year	Nation	Density Lines/100	BKK/P
1992	3,164,125	5.5	4.85:1
1993	3,769,842	6.4	2.27:1
1994	4,473,248	7.5	2.11:1
1995	5,242,091	8.6	1.94:1
1996	6,119,906	9.9	1.74:1
1997	7,087,026	11.4	1.53:1
1998	8,143,103	12.9	1.33:1
1999	9,318,344	14.6	1.16:1
2000	10,605,920	16.4	0.98:1
2001	12,022,571	18.4	0.85:1

Note: BKK/P is the ratio between greater Bangkok and provinces.
Source: TOT.

International calls are handled by the Communications Authority of Thailand (CAT), which provides direct access to just about any country. International rates have been falling, but remain higher than neighbouring countries. International phone cards are widely available in tourist areas.

It is estimated that about 10 per cent of the population (62 millions) have access to telephone lines. By 2005, the fixed line proportion is expected to be 6.6 person per 1 line. Internet usage in Thailand has increased from about 3 million people in 2002 to 6 million in 2003.

Wireless Services

Mobile phones are easily available at competitive prices from service providers using GSM digital networks. A wide range of service packages are available with many prices and options from which to choose. The number of mobile subscribers has soared from a less than 2 million in 1997 to 12.6 million in June 2002.

AIS entered the age of non-voice application services in 2000, with the launch of mobileLIFE service with SIM-based transactions. GPRS services were launched in 2001, MMS in 2002, and WAP portal—wap.mobilelife.co.th—in 2003. AIS claims to have 12–13 million subscribers, with 10.3 million pre-paid users and around 2.18 million post-paid users. Each month AIS reportedly acquires approximately 120–180,000 new subscribers. EDGE services (Enhanced Data Rates for GSM Evolution) are expected in the near future, in cooperation with four major network suppliers including Ericsson, Huawei, Nokia and Siemens.

DTAC is a joint venture between Total Access Communications (TAC) and Telenor Mobile, the Norwegian telco with regional operations in Malaysia, Singapore and Bangladesh. Telenor holds a 40 per cent stake, both directly and indirectly, in TAC, following a $750 million investment made in 2000. Telenor is ranked the 12th largest mobile operator in the world, with 27 million subscribers globally. It is the market leader in Norway, Bangladesh, Ukraine, Montenegro and Moscow. DTAC claims to have 6.2 million subscribers, and an acquisition rate of approximately 110–130,000 new pre-paid phone users per month.

CP Orange, previously called Wireless Communication Service (WCA), is the joint venture between Orange SA, Telecom Asia and CP Group. TA Orange claims to have nearly 2 million subscribers. CP is the founder of TelecomAsia Corp (TA). Together they are the major shareholders in TA Orange, whose marketing slogan 'The

Future is Bright, The Future is Orange' can be seen in billboards across much of Asia.

Newcomer Hutch—Hong Kong conglomerate Hutchison Whampoa Ltd.—launched its mobile phone service in Thailand on February 27, 2003. The company, 65 per cent owned by Hutchison Wireless Multimedia Holding and 35 per cent by CAT, uses a CDMA (code-division multiple access) network covering Bangkok and 25 of Thailand's 76 provinces. CAT operates the network while Hutchison is responsible for marketing the service.

Hutch uses CDMA2000 1X high-speed multimedia services which include talk, games and karaoke. One of its most successful promotional campaigns is 'Buy One, Get One Free,' expected to draw over 200,000 new users. Currently Hutch claims to have 500,000 subscribers. Perhaps the biggest drawbacks till date are that there are only two CDMA handsets available here, few roaming destinations overseas, and is priced higher than its rivals.

WLANs and Hotpots

Internet Thailand Public Company Limited (INET), the leading Internet and e-commerce infrastructure provider in Thailand, offers Wireless Zone, a Wi-Fi access service throughout all leading office buildings that are connected with INET's MetroLAN. INET has teamed up with IBM, Intel and Astra for this service in leading areas including Sukhumvit, Ploenchit, Silom and Sathorn.

The hotspot front is yet to pick up momentum. The Starbucks coffee chain has introduced the Wi-Fi service in some of its Bangkok branches (e.g., Thonglor, Emporium). CS Loxinfo has announced that it would introduce Wi-Fi in major IT stores and shopping malls, while some of the bigger hotels in Bangkok have also started offering such services. DTAC is also looking at offering some of its business customers 802.11b Wi-Fi services.

Content Services

Both Thai and English news service alerts are available for service subscribers who have to pay a monthly service to the operator. The service cost around 300 baht per month for each feed (Thai news or English news). Stock price of the day alerts are also being promoted, as are discount campaigns (e.g., for beauty salons). SMS is used as a tool for TV game shows and TV drama to draw viewers to interact with games and storyplots.

Notable content providers in Thailand are BECi, Samart Info Media (e.g., traffic reports) and Shin Corps (e.g., photo services like Photo Mail). In June 2003, Yahoo!2SMS introduced Yahoo Instant Messenger in Thailand. The service allows PC users to send SMS messages to mobile phone users from both Thailand and Singapore and be able to receive replies instantly. The service is provided through collaboration between Yahoo! Southeast Asia and mobile providers SingTel, AIS and DTAC. With Yahoo!2SMS, users of Yahoo Instant Messenger worldwide can expand their buddy lists with phone numbers of their contacts. Sending a message through Yahoo! Instant Messenger is free for both sender and recipient while SingTel and AIS subscribers have to pay an SMS charge for a return SMS-to-IM message. In Thailand, various instant messenger-to-SMS services are limited to one-way traffic, supporting communication from PCs only. Messages also lack tags to indicate the name of the sender. With Yahoo!'s Y2SMS, mobile subscribers are presented with the sender's Instant Messenger username and can respond via SMS.

DTAC recently lowered the price of its GPRS service with a 999-baht package that has no restrictions on the time spent online or the data downloaded. DTAC has also introduced a 'GPRS hotline' number to make it easier for new users to set up their equipment. DTAC reports 20 per cent revenue growth in non-voice services in the first five months of launch and increases in both GPRS (from 7.6 GB to 18.5 GB from January to May 2002) and multimedia messaging service (MMS) traffic (about 4,000 per day in 2003).

Even the government is getting in on the SMS wave. In October 2003, the government sent mobile phone users text messages on the progress of APEC submit held in Bangkok.

24. Malaysia
Emerging Regional Test-hub for Advanced Wireless Services

Lunita Mendoza

Editors' Note: Though Malaysia is behind the high-adoption curves of Asian markets like South Korea or Singapore, it is way ahead in mobile penetration and infrastructure as compared with many other developing nations of the world. Rising affluence, lower handset and service charges, large youth segment, market deregulation and low import tariffs are among the key drivers for growth in this market. Challenges remain in maintaining quality of service, and in managing the rocky transition to 3G. M-commerce in Malaysia is still very much in its infancy. Hotspots have emerged in some urban pockets but have yet to see significant usage. The key drivers for private WLAN uptake in Malaysia will be its high population density, high mobility and presence of intelligent office buildings. Innovative measures like the Most Creative Local Mobility Application awards help spur software development on the wireless front. Global players like Alcatel and Ericsson have invested millions to develop Malaysia as a hub for 3G business in southeast Asia. This chapter also provides informative insights and perspectives into Malaysia's wireless industry.

Overview

Malaysia's mobile market has seen remarkable growth in the aftermath of the Asian financial crisis in 1997. There were 7.7 million mobile phone subscribers at the end of 2001 (according to the Strategic Intelligence Research group). In 2004, 50 per cent of the 25 million people in Malaysia now have a mobile telephone service. This means Malaysia has the second highest mobile penetration in South East Asia. Rising affluence, lower handset and service charges and—as with many countries gripped by mobile mania—its large youth segment are the key drivers for growth in this market.

In October 2003, the Malaysian enterprise telephony market grew 8.8 per cent from 2002 levels, registering $43.3 million in revenue. Mobile carriers and phone vendors know that there is still significant room for growth in the Malaysian market, given that mobile adoption falls behind other Asia-Pacific markets like Singapore and Hong Kong.

SMS (short message service) has reached a critical mass in Malaysia, and coupled with high Internet and PC penetration rates, sets the stage for rapid adoption of higher-order wireless data services. GPRS (general packet radio services) will be the next wave. The three major carriers already offer GPRS services nationwide, with one carrier even offering GPRS to the large segment (85 per cent) of pre-paid users.

While Malaysia is ahead in terms of infrastructure among developing countries, the industry has been beset by service performance difficulties such as poor reception. This has led to the establishment of the Malaysian Communications and Multimedia Commission (MCMC) in 1998 and government involvement in the industry to ensure that carriers are able to maintain quality of service.

This, coupled with the massive infrastructure investment that carriers would need to make for rollout of 3G services, led Strategic Intelligence Research to project that the number of 3G players will decline from five to three over the next 12 months, despite optimistic projections about the growth in mobile services revenues.

The 3G players have actually declined from five to two. Out of the five companies that bidded for the three 3G spectra, Telekom Malaysia and Maxis Communications emerged winners. Celcom Malaysia, Time dotCom subsidiary TimeSat and E-Touch failed in their tenders. As the use of data services grows, these players are expected to increasingly leverage on their fixed-line infrastructure to provide converged network services to their end-users.

Adoption of wireless LAN (WLAN) technology is becoming evident, with sales of WLAN equipment sales reaching RM 40 million in 2002. WLAN adoption, however, is taking place mainly as an extension of, or as an alternative to, wired LANs in private corporate and residential sites. Public access WLAN or hotspots have emerged sporadically across Kuala Lumpur and Cyberjaya but has yet to see significant usage.

Table 24.1: Malaysia snapshot

	Malaysia
Population	25 million (approx.) as of September 2003
Land Area (sq. km)	329,000
PRIVATE WLAN	
Financial	Significant barriers to WLAN adoption
Government	Limited potential for WLAN usage
Healthcare	Limited potential for WLAN usage
Manufacturing	High adoption and/or good potential for WLAN usage
Retail	High adoption and/or good potential for WLAN usage
Services (include conferences)	Limited potential for WLAN usage
PUBLIC ACCESS HOTSPOTS	
Launched in	2001
Number of public hotspots	> 128 (as of October 2003)

Source: Author

Given its broad geographical spread and topographical diversity, universal access to mobile communication services in Malaysia can indeed be a challenge. The Malaysian government has taken the approach that deregulation and foreign participation will help develop the telecommunications industry. With deregulation, the industry has changed from a monopolistic regime centred on Telekom Malaysia to the current more transparent and liberalised structure promoting increased competition. Malaysian carriers are also leveraging partnerships with global leaders like British Telecom and Korea Telecom in preparation for the rollout of next generation services and applications.

Both the Malaysian government and carriers have taken a cautious approach to the introduction of 3G, but it is estimated that RM 5 to 7 billion will be spent over two years to build the 3 network. There is also a prevailing view that existing technology (in particular, GPRS) is adequate for Malaysia's current needs.

The MCMC assigned two of the three available 3G spectrum blocks in July 2002. Local carriers have indicated that initial launch dates would be pushed back to 2004 at the earliest, cutting expensive 3G licences and the extensive infrastructure upgrades required for a 3G rollout, coupled with the current global economic slowdown.

What is clear, however, is that Malaysian carriers will keep in check the costs of providing 3G services, a lesson learnt from the debt-heavy carriers in Europe and South Korea. The MCMC made much of the need to avoid duplication of infrastructure by network sharing and made it a requirement for facility operators to accommodate virtual network operators, which would compete with the physical network operators based on services and applications.

Connectivity

Malaysia has more mobile phones than fixed lines (as of 2000) with much more growth in sight. This trend has translated into healthy revenue growth for mobile carriers as a whole. Demand for mobile services has in fact far exceeded that for fixed lines. The introduction of the 'calling party pays' billing and settlement standard, and the removal of import duty on handsets have also helped.

Table 24.2: Mobile subscribers in Malaysia (1998–2002)

Year	Subscribers	Penetration rate
1998	2,200,000	10.1%
1999	3,107,000	14.2%
2000	5,709,000	24.5%
2001	7,685,000	34.6%
2002	8,500,000	55%

Source: Various

GPRS is already available nationwide after being initially offered in the Klang Valley, Penang and Prai. DiGi launched its GPRS in July 2002 and is now available nationwide to all its customers with no monthly fees. The charges for this service are based on the 'pay as you use' pricing model. With the availability of GPRS on its network, DiGi is also the first carrier in Asia to offer GPRS as a pre-paid service.

GPRS services are expected to significantly enrich mobile user experience by enabling the provision of MMS and WAP applications as well as faster wireless Internet access (from 56 to 114 Kbps).

This will provide mobile users the ability to access content and services such as KLSE stock information, business news and lifestyle information. The availability of GPRS will also see the revival of high-speed data Web-based services that are particularly suited to WAP. With its always-on feature and transmission speeds of up to 115 Kbps, GPRS will effectively address the complaints of slow transmission of WAP services over the circuit-switched GSM networks.

Telekom Malaysia, DiGi and Maxis are looking to work with content and application developers to jointly stimulate and create applications. Interest has been shown in the development of mobile/wireless application solutions, multimedia conferencing systems and mobile phone management systems.

On the WLAN front, the top three WLAN adopters in Asia by take-up rates as well as potential have been Japan, Korea and Hong Kong. Malaysia follows this lead group. Global network equipment and wireless vendors have targeted the Malaysian market since early 2000. Malaysia only saw the introduction of WLAN in 2001, but already Malaysian telcos and ISPs regard WLAN as one of their most important businesses for the next five years for verticals like retail and manufacturing. Banks have been reluctant to embrace WLANs because of security concerns, but non-banking subjects such as insurance providers are starting to show interest.

Table 24.3: Mobile services revenues (1998–2000)

Year	Revenue (US$ million)
1998	1,173
1999	1,353
2000	1,633

Source: Global mobile

Prominent network equipment and wireless vendors have targeted Malaysia as an early adopter of WLANs since early 2000. Equipment makers like 3Com see Malaysia as an ideal market given its dominant SME base and are promoting WLANs as a cheaper and simpler means of achieving network connectivity for cost-conscious SMEs.

Malaysia's public WLAN access market is in the nascent stages of development. Hotspot penetration is growing, but remains low. More than 80 per cent of these hotspots located within the Kuala Lumpur area due to MCMC initiatives concentrated on Kuala Lumpur and Cyberjaya. The majority of WLAN service revenues will come from business users in airports, business hotels and exhibition centres.

Currently, the key offering by WLAN players in Malaysia has been wireless access, rather than mobility. Therefore, ownership of the network or hotspot will be a critical success factor for revenues. Consequently, hotspot owners (such as hotels and conference centres), fixed operators and IT specialists will have an advantage because they have access to the customer and the technological capability to provision networks. The key success factor will be the ability to reach and deploy a WLAN network at relevant hotspots for corporate users, e.g., corporate intranet access, usage by hotels.

Wireless operators can also prosper from WLAN hotspot services as they can produce the strongest synergy. DiGi is working with Green Packet (Malaysia) to develop and deploy roaming solutions for wireless and broadband services in Malaysia. It is expected the other operators will leverage their brand and billing relations and follow suit in the near future. There is also the potential to provide seamless movement from the cellular wide area network (WAN) to the WLAN environment. The entities that develop the most compelling service propositions will gain a larger share of the pie.

The Putra World Trade Centre (PWTC) remains the venue of choice for many trade events—over a thousand a year—and has transformed part of its facility into a hotspot based on Airzed's WLAN technology. TM Net has launched its hotel broadband solution at Pan Pacific Hotel Kuala Lumpur, with other hotels set to follow. Maxis and three other Wireless Broadband Alliance's founding members—Australia's Telstra, China Netcom and Singapore's StarHub—have kicked off a pilot trial for wireless broadband roaming across their respective countries.

The key drivers for private WLAN uptake in Malaysia are its high population density, high mobility, and presence of intelligent office buildings. The growing popularity of mobile devices (laptops and PDAs) are other drivers. The SOHO/home WLAN market has seen tremendous growth, growing to over 40 per cent of the total WLAN market in the first quarter of 2002.

On the 3G front, the Malaysian Communications and Multimedia Commission (MCMC) announced in May 2002 that it would award three 3G licences, but subsequently decided in July that only two companies had met the necessary criteria: Telekom Malaysia and Maxis subsidiary UMTS (Malaysia).

State-controlled Telekom Malaysia is the country's leading fixed-line operator, while Maxis Communications is the number one mobile provider. It is estimated that each company will spend between RM 2.5 and 3.5 billion over the next two years in building their 3G networks.

That said, the government has been very cautious in its approach to 3G. There is a view that existing technology (in particular, GPRS) is adequate for Malaysia's current needs and there is no need to rush the introduction of 3G, which would provide higher bandwidths that enable faster Internet access as well as the ability to download music and video to mobile phones.

The MCMC has also made much of the need to avoid duplication of infrastructure and has made it a requirement for the 3G network operators to facilitate virtual network operators. (This policy resulted in DiGi withdrawing its bid for a 3G licence, citing strategic plans to operate instead as a virtual network operator.)

Content

Today's 2G services like voice and SMS are mainly used to facilitate peer-to-peer communications. For 3G, the value proposition lies more in the applications, services and content that can be accessed, be it in the B2B or B2C context. Hence, partnerships between carriers, application and content developers and device vendors (especially since they help determine the user interface) are a critical factor for a successful 3G rollout, since these enable the provision of such offerings. This may spell certain success for virtual network operators to end up as big winners in Malaysia's 3G market.

For consumers, the experience of SMS highlights the demand for creating and sending large amounts of user-generated content. Consumer-oriented devices should look attractive and offer users superior-messaging capabilities based on colour screens and always-on connectivity. With the added bandwidth and processing capability of GPRS devices, content and services should also be developed for this audience. Carriers are likely to enter into partnerships with game publishers and developers, and content owners and media companies to address their needs.

Already, Time dotCom has signed a memorandum of understanding with Japan's NTT Data Corporation to collaborate on 3G content and applications and also entered into other deals with British Telecommunications subsidiary BTexact Technologies as well as US-based Mobileum. Telekom Malaysia aligned itself with South Korea's incumbent Korea Telecom.

M-commerce: Where are We Today?

'M-commerce is a competitive necessity, but not necessarily a commerce strategy'
—A conversation with Stephanie Wong of IDC

With a population of over 23 million and 7.3 million mobile subscribers (Strategic Intelligence Research cites 7.7 million subscribers) as of end-2001, there is definitely a market opportunity for m-commerce in Malaysia.

(*Box Continued*)

(*Box continued*)

M-commerce refers to all purchases of products and services that occur across a mobile data platform, such as GPRS, SMS, HSCSD, 3G, or just the basic 9.6Kbps mobile data platform. Voice-recognition services that interact with the user and the Internet, either on the handset or network, may be used to complete the purchase and are included in this definition as long as the wireless network is used. The difference between e-commerce and m-commerce lies not simply in the fixed and wireless aspects, but also in the potential to capture opportunities in a different perspective. On top of the e-transaction, the m-commerce platform could also be a medium for micro-payments for public transportation, tolls, parking fees, and even for snacks from a vending machine. This could be related to the mobile wallet concept.

Most of the attention on m-commerce focuses on the consumer market at present. The most common m-commerce activities nowadays tend to be more information-based rather than transaction-based. Examples include access to stock quotes, online banking information, flight schedules, weather forecasts, travel directions, and dining guides.

M-commerce in Malaysia is still very much in its infancy. Currently, the only available 'real' m-commerce services in addition to content services are the purchase of Coke from selected vending machines with that functionality. This service is provided via a partnership between F&N Coca-Cola in Malaysia and Celcom. Named the 'ring-a-Coke' initiative, this was the pioneer m-commerce project launched in the country. The user is only billed for the purchase of the drink but not the phone charges (a toll-free number is used). The mobile operator derives revenue from this transaction by getting 10 per cent of the transaction value for each successful transaction made.

M-commerce for physical goods faces a similar problem as e-commerce, in logistics and risk of fraud. Malaysian users tend to be more conservative in their purchasing habits, need to 'touch and feel' a product, and have less trust in the capabilities of m-commerce.

IDC's view on m-commerce is such that it is a competitive necessity, but not necessarily a commerce strategy. With that view in mind, these are IDC's recommendations for implementing mobile wireless applications:

- **Identify your customer's need, and target your application** For example, a wireless travel application that notifies a customer of a cancelled flight via SMS, and then offering alternative flights on which the person can re-book on the spot, is a true benefit that can only be achieved through a mobile wireless application.
- **Do not expect your mobile device to be a replacement for a PC** PDAs and mobile phones with Internet browsers (WAP/GPRS/3G) are not replacements for PCs. The applications are different. A mobile device is not a wireless Web. Reading news is a PC application; reading headlines is a mobile device application.
- **Do not confuse your mobile wireless strategy with your e-commerce strategy** Providing wireless access to information for your customers will give your customers another option for self-service to information, which may initiate a transaction. However, most of the resulting transactions may not be made from the mobile device. Consequently, the return on investment in wireless applications should be measured in your company's overall revenues. If you are considering developing a wireless application, do it because it is part of your company's wireless strategy, but do not tie it directly to your e-commerce.

Policy and Regulation

With deregulation over the years, the industry has changed from a monopolistic regime centred on Telekom Malaysia to the current more transparent and liberalised structure promoting increased competition. Within this context, the government is actively pushing telecom providers to undertake measures that will enhance their competitiveness. In this regard, improved quality and diversity of services at reasonable rates are key priorities.

Before 1946, the post and telegraph department was responsible for providing all telecommunications services throughout Malaysia. In 1946, the telecommunications department was established in Peninsular Malaysia. This department was merged with the telecommunications department in Sabah and Sarawak in 1968 to form Jabatan Telekomunikasi Malaysia (JTM) (which translates to department of telecommunications, Malaysia).

In 1987, the operation of Malaysia's telecommunication services was transferred from JTM to Syarikat Telekom Malaysia (STM) which became a public-listed company in 1990. It assumed its present name, Telekom Malaysia, the following year. However, to ensure Telekom Malaysia's compliance with national policy, the government retained a majority shareholding in the company.

Thus, it was Telekom Malaysia's predecessor that introduced mobile services to the country in 1985, followed by Celcom in 1989. Since then, the government has granted a number of licences to private sector operators in an effort to develop the country's telecommunications industry and infrastructure.

Malaysian Communications and Multimedia Commission (MCMC)

In 1998, the Communications and Multimedia Bill was passed, replacing the Telecommunications Act, 1950 and Broadcasting Act, 1988. Under the new legislation, telecommunication licences would no longer be technology-specific, but service-specific to reflect the convergence of technology.

As required by the legislation, the Malaysian Communications and Multimedia Commission (MCMC) was set up in November 1998 as the regulatory body charged with overseeing the converging industries of broadcasting, information technology and telecommunications. The MCMC is also responsible for developing national telecommunications policy and for the licensing of telecommunications operators.

MCMC requires that all mobile phone operators submit quarterly reports in a bid to benchmark and raise the standard of cellular services. Significantly, the government has said that if it were found that a mobile operator's network is not able to properly handle its existing subscribers, the company would not be allowed to accept any new subscribers.

In a 2003 report published by Asia-Pacific Research Group, Digi Telecom ranks first overall in Malaysia's customer satisfaction index, excelling in the categories of customer service and data quality. Customers of Malaysia's four leading cellular companies rated providers based on 12 metrics, including billing, access to customer service and call quality.

Capacity

A number of measures are being taken to increase Malaysia's human resource capacity in the wireless infrastructure and services world. For instance, Maxis Multimedia Sdn Bhd (Maxis Multimedia) launched its Maxis Developer Programme (MDP), which aims to develop local mobile application and content developers by assisting them to design, develop, test, and market their mobile applications and services. Through MDP, Maxis Multimedia will nurture, foster and promote innovative mobile application developers; assist participants to access financing and start-up capital; provide training support; and provide support for developers to access markets and commercially deploy their services.

Maxis also sponsors awards along with the Malaysian National Computer Confederation and the ministry of energy, communications and multimedia for categories like Excellent ICT Teacher award, Excellent ICT Student award, Local Community ICT Champion and the school Website competition. Recently, the most creative local mobility application awards went to AMICS application and mobileBlog.

Mobile services provider MacroKiosk has developed a platform that enables its clients to offer m-commerce services, like mobile payment and mobile banking, through 364 mobile operators in 115 countries. This network connects to 95 SMS centres through 30 nodes owned and operated by MacroKiosk around the world. The company is 70 per cent owned by KLSE-listed Gold IS Berhad. MacroKiosk has also entered into a strategic agreement with P2 Digital Sdn Bhd, a local software developer, to power Mail2Phone, an SMS service that enables users to access their POP e-mail account through the mobile phone.

Alcatel announced that it will invest more than RM 200 million over the next five years to develop Malaysia as the hub of its 3G mobile business in Southeast Asia. The new investment would be primarily to provide support and expertise in the 3G/mobile business area, and the development of new mobile applications, particularly for 3G communications, with local partners in Malaysia. The creation of Malaysia as the mobile business hub will see Alcatel's mobile experts to be based in Malaysia to increase by more than 50 per cent, by the end of 2004.

Ericsson invested RM 80 million to establish a regional hub in the MSC which houses a flagship showcase of 'live' networks showing the evolution of mobile telecommunications standards from GSM to 3rd Generation (3G) technologies. Ericsson@MSC is the headquarters to some 500 employees of Ericsson Malaysia, out of which more than half support the company's operations across Asia-Pacific.

Cooperation: Partnerships Seen as Key to 3G Implementation

More than network infrastructure, a successful 3G rollout depends on a concerted effort between carriers, equipment vendors, system integrators and content and application developers. In this regard, both Maxis and Telekom Malaysia have indicated that they are looking to establish partnerships for the delivery of 3G-based solutions.

Partnering with local content and applications providers, Alcatel has set up a 3G Reality Centre in Kuala Lumpur to provide Malaysia's mobile industry with a window to the latest services that are being developed. This network will be a key asset for mobile operators as they look to develop a complete service offering and profitable business model for next-generation mobile services. The centre's strategic partners include Fujitsu, which is Alcatel's long-term 3G-infrastructure partner. It is also looking at partnering with other companies that can add value towards the success of 3G in Malaysia and elsewhere in Southeast Asia.

Capital

Recognising that foreign participation is necessary to inject the capital investment, technological expertise, and experience required to develop the market, the government has moved to encourage foreign investment in the industry. This extended to having stakes in domestic carriers. Foreign investors can buy up to 61 per cent of domestic carriers, subject to certain conditions including lowering their stakes to 49 per cent within five years.

According to IDC Malaysia's Top 10 predictions for the local IT market in 2003, the wireless sector is set to drive the telecommunications market with wireless services revenue enjoying an 18 per cent growth over 2002. Data services are expected to gain ground as more mobile data services are rolled out. Broadband specific content and applications will catalyse uptake of the service, but promises little revenue for providers.

At this juncture, the telecommunications market is sufficient to support all three players. The key battleground will be the high-end market, which offers the best yield.

Maxis has the highest average revenue per subscriber (ARPU) of $29, and the lowest churn rate (around 2 per cent); it has been one of the fastest growing companies in the country with EBITDA (earnings before interest, taxes, depreciation and amortisation) growth in excess of 35 per cent per annum since the Asian crisis.

The takeover of Time (believed to have been triggered by the Telekom-TRI merger) will provide Maxis with extra bandwidth for its mobile operations (it has the lowest domestic allocation of spectrum: 2×10 MHz), a dual band network through the amalgamation of its GSM 900 network with Time's GSM 1800, as well as an additional 600,000 customers.

In the longer term, the Telekom-TRI union may yield greater economies of scale and pack more financial and infrastructure punch. Prior to acquiring TRI, Telekom with its mobile services on three different platforms had lost momentum in the market despite its incumbent carrier status. Its GSM division has only a 16 per cent market share, as opposed to Maxis with 29 per cent.

However, the combination of Telekom and TRI will see a unified subscriber base of over 3 million customers and a leap in market share to approximately 40 per cent, relegating Maxis to the number two slot in the local GSM market. Already, there is speculation that cash-rich Telekom, which has a near monopoly of the fixed line business, may make a bid for Time's prized 5,200 km (3,231 miles) fibre-optic trunk and its fixed-line business.

Pre-paid customers account for nearly 85 per cent of all mobile subscribers in Malaysia. Despite the dominance of pre-paid customers in Malaysia, DiGi is the only operator to have made GPRS services available to this important group.

With pre-paid users set to remain a substantial proportion of mobile customers over the next five years, Telekom Malaysia and

Maxis should also focus on increasing the ARPU from their pre-paid customer base by extending and enhancing the range of pre-paid services they offer.

The main problem is that the delivery of packet-based data and content services to pre-paid users presents technical challenges for mobile operators via upgrades and customer care.

Broadband penetration nationwide was very low at only 0.44 per cent, compared with 48 per cent in neighbouring Singapore and 60 per cent in South Korea. (Source: Malaysia's Energy, Communications and Multimedia Minister Lim Keng Yaik)

According to a Strategic Intelligence survey of Malaysian enterprises, always-on e-mail is the strongest requirement for business users. In terms of vertical business segments, the manufacturing sector displayed the highest adoption rate for GPRS at 35 per cent.

Given sufficient efforts at consumer education, carriers can expect a ready market for 3G services. There are already nearly 400,000 subscribers in Malaysia who pay RM 40–75 a month for mobile data applications, making Malaysia one of the most attractive markets for high-bandwidth services in Southeast Asia.

Handset vendors will be among the main beneficiaries as users switch to 3G. Malaysian manufacturers have already demonstrated the ability to produce a steady stream of new models that offer advanced features like colour displays and small size at low costs. The presence of a large home market to serve as a launch pad and demand driver for new models will help to further establish Malaysia's telecommunications equipment manufacturing industry.

Conclusion

To sum up, the mobile market in Malaysia continues to grow increasingly competitive. The mobile landscape is also seen to be changing as the market matures. Consolidation among mobile operators is unavoidable as voice packages becomes cheaper and mobile data services are introduced to compensate this loss of revenue.

25. Nepal

Wireless Destination

Gaurab Raj Upadhaya

Editor's Note: This chapter makes the clear-cut case for the crucial role that wireless communications and mobile devices can play in a mountainous country like Nepal. Wireless technologies form the backbone of all major telecommunications services in the country. The private sector is becoming increasingly involved in small parts of the telecommunications environment, but the monopolistic environment continues to be a constraint. From WLL and Web-to-SMS gateways to Wi-Fi and VSATs, the success of telecommunications in Nepal hinges heavily on wireless.

Introduction

Nepal is nestled in the vast and towering Himalayan mountain range. People come here to get away from the urban rush from around the world: Sydney to Seattle, New Delhi to New York, Cape Town to Cape Verde. But, once in Nepal, people do realise that telecommunications services in parts of urban Nepal are comparable with the rest of the world. A pioneering initiative even brought e-mail access to the Everest Base Camp! With its enormous geographic and topographical diversity, Nepal is a country where wireless communications is more useful and vital than in other parts of the world. Thus far, wireless technologies form the backbone of all major telecommunications services in the country.

Telecom Background

In spite of the poor economy, the available telecom infrastructure in parts of urban Nepal is comparabale with systems in many developed countries. Nepal entered the era of telecommunications over four decades ago. In 1955, with the establishment of a 300-line manual crossbar (CB) exchange, the general public was provided with access to telephones. The Department of Telecommunication in 1959 and Nepal Telecommunication Corporation (NTC) in 1975 were established. NTC is the only provider of telecom facilities (basic as well as international links) and provides about 310,000 subscriber lines including mobile through 130 exchanges established throughout the country. The installed capacity of all these exchanges is 295,827 lines and there are 277,684 in the waiting. Telephone facilities are now available in all districts of Nepal and 33 per cent of the VDCs are connected. The national penetration per 100 populations is 1.14 per cent.

Unfortunately, Nepal was not able to escape the plague of the monopoly telecom operator. NTC had been the sole operator of all telecommunications services upto 1994. It was only in 1994, when private entrepreneurs took initiatives to e-mail and Web access services in Nepal, that the private sector became actively involved in the telecommunications sector.

National Wireless Backbone

The entire telecommunications backbone operated by NTC in Nepal is based on microwave links between repeater stations and exchanges. Except for the Kathmandu Valley and a few other telecom exchanges in the terai (flat) regions of the country, all telephone exchanges are connected via microwave links.

International connectivity to the rest of the world is also via Intelsat—a satellite-based system. There is a direct microwave link to Bangladesh, and fibre optics connection to India.

GSM Cellular Mobile

GSM cellular mobile phone is available in selected cities. There are two variations of the service. The regular 'NTC Mobile' and the prepaid 'Namaste' service. The number of regular subscribers is around 25,000.

The pre-paid service was introduced in mid-August, 2003 and, in the two months since, about 25,000 people signed up for the service. This has also meant that the quality of service for the entire mobile network has gone down, especially in some pockets in Kathmandu valley where having a GSM phone means nothing.

While NTC introduced the long demanded pre-paid service, they also have started to charge for SMS—which was hitherto free. The charge is Rs 1 per message ($1= Rs 73), which is still affordable.

ITNTI, a service company based on Kathmandu recently launched www.nepalonline.com, a paid SMS gateway. But this has been done without any special arrangement with the mobile company, leading to higher operational cost hampering the company's ability to scale up rapidly.

Bitarak.com, another company in Kathmandu, also started an SMS poll service for their magazine. The company used the mobile channel to send advertisements. However, with the introduction of a charge for SMS, the poll service is not as popoular.

NTC has been experimenting with different tariffs. The trend has been to gradually decrease charges. However, NTC still charges for receiving calls during office hours. The general expectation is that the prices will come down further, but the lack of any competition means that NTC is not in a big hurry.

Wireless Internet

There are currently 22 licensed ISPs in Nepal. Of these, roughly 16 are in operation, 12 in the Kathmandu Valley. The larger ISPs also have operations outside Kathmandu Valley.

NTC also operates its own ISP service and private ISPs have a difficult time getting leased lines and other access. This has led to the ISPs building their own network based on wireless technologies. All larger ISPs use wireless technology to provide corporate access to the Internet. ISPs use equipment in the 2.4 GHz frequency range and variations of 802.11b technology to provide these services. But with the spectrum getting crowded in Kathmandu, some ISPs have started using 5.8 GHz equipment. Some ISPs are also using proprietary vendor frequencies equipment.

In late 2002, a new entrant into the ISP field—WebSurfer Nepal—launched itself as an wireless ISP. It used 2.4 GHz equipment to create many micro-cells inside Kathmandu Valley to provide wireless broadband connectivity to clients. This has proven to be more feasible, and other ISPs are also thinking of building their own wireless networks.

Wireless as an edge connectivity has not seen as much use. In January 2003, during the annual meeting of SANOG (South Asian Network Operators Group), Wi-Fi access was provided for the first time at the conference centre. Some hotels are exploring the use of Wi-Fi, but nothing has materialised as yet. With very few people actually carrying laptops in the first place, it will take a long time before Wi-Fi becomes a common phenomenon in Kathmandu—though some cyerber cafés in Kathmandu's tourist mecca Thamel are interested in trying it out.

Outside Kathmandu Valley, V-SATs are used by ISPs to connect to their POPs in other cities. There, too, wireless technologies are used to provide corporate access to clients demanding 24×7 access.

Wireless Local Loop

United Telecom, a joint venture of Indian operators and Nepali businesses, was granted the licence to provide Wireless Local Loop using CDMA technology in Nepal. The company finally started operation in August 2003. This is currently limited to Kathmandu Valley, but even here, it is expected to provide relief to thousands of people who have been on NTC's waiting list for telephones. Currently, the phones do not support data rates above 9.6 Kbps, but the company plans to upgrade this to allow users to connect to the Internet through the WLL phones.

The phone usage rates are similar to NTC's tariff, but the WLL phone will prove to be popular with thousands of families who live as tenants in Kathmandu Valley. However, the users are not allowed to roam between different CDMA cells.

Other Wireless Initiatives

V-SATs were liberalised in 1999 and this enabled private parties to provide data services. Other data services have also been opened up to the private sector, such as radio paging, trunk mobile, and WLL. Radio paging proved to be a popular service when it was introduced about five years ago. However, with the introduction of the pre-paid mobile service, paging companies have started to lose business and we will definitely see consolidation in the paging market.

Policy

The telecommunications acts and relevant regulations for Nepal are accessible at www.nta.gov.np. The frequency management cell at the ministry of information and communications is responsible for allocating and regulating frequencies. But the ill-equipped cell is unable to regulate effectively. The regulation on use of the ISM band is not clear, but ISPs have been using it for many years.

A new telecom policy is expected in 2004 which will address the issues related to the use of these bands and also introduce the technology-independent licensing regime, thereby making wireless Internet more widespread.

26. Bangladesh

Mobile Telephony = Social Good + Good Business

Sayeedur Rahman

Editor's Note: The story of Grameen Telephone in Bangladesh is an inspiring showcase of the fruits of mobile telephony services powered by local citizens and funded by micro-finance programmes. The CityCell service provider is also a mobile pioneer in south Asia. Grameen's travelling 'phone ladies' provide villagers with a vital shared link to services such as calls to hospitals and to relatives both in the country and abroad, in a nations with the lowest number of phones in South Asia. This chapter concludes that there is also a lot of potential for the next wave of communication services based on Wi-Fi, if regulatory obstacles can be overcome—a point also echoed in the last book review in Part III.

Telecommunications has not been on the Bangladesh government's priority list for some time. Available resources are directed at other more basic infrastructure requirements. However, there are signs that the members of the country's planning commission are increasingly motivated to support telecommunications as they view this as a necessity rather than a luxury.

Growth in Bangladesh's mobile telephone sector, which had a humble beginning in the early 1990s, has really picked up pace in the past few years, aided by higher subscriber volumes, lower tariffs and falling handset prices. Bangladesh had more than 1.5 million subscribers by mid-2003. This represented an annual growth of 100 per cent over the previous year. Analysts predict that the market will continue to grow strongly, moving at a compounded annual rate of more than 75 per cent, Bangladesh has a market now for 13 million mobile phones. There is tremendous potential for wireless Internet in Bangladesh as the telephony infrastructure is relatively poor.

Pacific Bangladesh Telecom Limited (CityCell) has consistently set the benchmark of the mobile telecom industry in Bangladesh: it was the first operator to launch mobile phones in South Asia, the first-ever private wireless network in Bangladesh, and the first to adopt CDMA technology in the sub-region (Table 26.1). Since the inception of mobile phones in Bangladesh in the early 1990s when it used to cost $1,500 to get a mobile connection, the rate has gone down to $100, while the lowest per minute tariff is about 0.01 cents.

In November 1996, licences for cellular mobile telephone were issued to Grameen Phone, Telecom Malaysia International Bhd Ltd. and Sheba Telecom Ltd. All these are joint venture companies between Bangladeshi enterprises and foreign partners. Bangladesh Telegraph and Telephone Board (BTTB) a government monopoly, operates 800,000 fixed line phones. In the next five years, demand for fixed lines is expected to touch 3.5 million and experts BTTB will be hard pressed to deliver this service.

Table 26.1: Cellphone networks in Bangladesh

Network	System	Subscriber	Frequency
Pacific Bangladesh Telecom -CityCell	CDMA	215,000	800
Grameen Phone	GSM	1,000,000	900
Sheba Telecom	GSM	65,000	900
Aktel	GSM	220,000	900

Source: Authors, with inputs from news median.

Recently, a lot of attention has been focused on the use of wireless networks, and in particular Wi-Fi, as a relatively low-cost way of getting high-speed network access to rural areas in a less-developed country like Bangladesh. Wi-Fi is not the only wireless networking technology, of course. Packet radio, microwave links and 3G phone networks also have a lot of potential in this context. But a growing number of companies and market analysts have started touting Wi-Fi as the Next Big Thing, the focus for a second-generation Internet-style boom.

Learn Foundation, a non-profit charity set up in 1997, has worked to bring computers and the Internet to isolated rural areas of Bangladesh, using wireless technology. The Learn Foundation has already built radio towers in seven villages in the region and the

Mobile Telephony in Bangladesh: Good Business

In the early 1990s, when Iqbal Quadir, founder of Grameen Phone, was looking for investors to back his idea for a mobile phone network in Bangladesh, he said he was turned down by an executive at a cellphone company in New York who told him, 'We're not the Red Cross'.

By the end of 2001, Quadir showed how Third World ventures can be profitable—while also providing a useful service—when Grameen Phone Ltd. made $27 million in pre-tax profit. It returned that profit after just five years—far sooner than many First World startups!

target is to establish a broadband network in a 2,500 square kilometre (965 square mile) area.

Grameen Phone offers data services through SMS and Wireless Application Protocol (WAP). Web-to-cellphone gateways are offered by Grameen Phone and CityCell. Two leading news dailies provide SMS news alerts.

In a novel experiment, Grameen 'phone ladies' provide villagers with a vital link to services such as calls to hospitals and to relatives both in the country and abroad, in a nation with the lowest number of phones in South Asia. Villagers flock to the village phone ladies to use a shared mobile phone, paying for calls by the minute. Grameen phone ladies expect to earn more than $1,000 a year, a far cry from the annual average income of $380. The Grameen scheme has been hailed as a successful example of introducing technology and communication services to the poor. The mobile has literally changed many village phone ladies' lives. At present, 32,000 village phones are in 52 districts; 50,000 Bangladeshi women are making a living as Grameen phone ladies.

But there are unfortunately very few local companies, startups and venture capital companies targeting hardware/software development for wireless devices and networks in Bangladesh. One of the notable ones is PraDeshta, a private company that specialises in systems integration and network services. Recently, a few universities have taken the initiative to offer courses in wireless media.

As Bangladesh's wireline telephony infrastructure is relatively poor, it is the mobile phone that is creating a huge economic and socio-cultural impact especially among poor people in the villages. Rural areas are benefiting greatly from the mobile phone in Bangladesh.

Resources

http://news.bbc.co.uk/2/hi/technology/2309595.stm
http://news.bbc.co.uk/2/hi/technology/2290540.stm
http://news.bbc.co.uk/2/hi/technology/2254231.stm
http://www.cellular-news.com/coverage/bangladesh.shtml
http://www.shebatel.com
http://www.aktel.com
http://www.grameenphone.com
http://www.citycell.com

PART THREE

Book Reviews

All Book Reviews are by Madanmohan Rao

27. i-Mode Strategy

Takeshi Natsuno, *i-Mode Strategy*. John Wiley (translated from the Japanese language edition published by Nikkei BP), 2003, 160 pp.

The success story of Japan's wireless Internet service, i-Mode, is one of the most unusual business and media developments in the world—and its creators reveal a fine understanding of Internet dynamics and complex systems theory, so vital in harnessing and creating new media phenomena in the networked global economy of the 21st century.

Takeshi Natsuno is managing director of NTT DoCoMo's i-Mode, the world's largest wireless Internet service. After a Wharton MBA, he joined an Internet startup called Hypernet. Natsuno has been featured in numerous articles in Japan and abroad, including a *Business Week* cover story.

While other books such as Mari Matsunaga's *The i-Mode Affair* describe the journey of the company's success, Natsuno describes the *seamanship* that made it possible. The six chapters cover basic business concepts, practice, alliance strategy, effects, success and future plans. The material includes six pages of colour plates and a useful chronology of i-Mode.

'Information technology business grow far more than expected—or they do not grow at all. Never do they grow little by little along predictable lines,' begins Natsuno.

As of August 2002, i-Mode had 34 million subscribers, more than a quarter of Japan's entire population and 80 per cent of DoCoMo users. More than half of these users downloaded Java applications, constituting the largest Java community in the world. Users included not just youth but niche communities (e.g., surfers) and older age groups (e.g., stock traders) as well. Churn rates among users also reduced due to data services. Lucrative alliances and markets were created for content and service players, as well as replacement phones and accessories.

And those who used data services also made about 20 per cent more mobile phone calls than those who did not. This is due in part to features which allow users to directly call numbers displayed on their mobile phones. 'This is not possible using the Internet from a PC. It takes a mobile phone—a device that combines Internet access with voice communication functions,' explains Natsuno.

In its early days, i-Mode was 'born with a sense of crisis'—wireless operator NTT DoCoMo wanted to tackle the flattening growth rate of voice-based wireless communication services by launching a new layer of mobile data services. Instead of just getting new subscribers, the game was to also get new usage and frequent re-use out of existing subscribers (the 'second S-curve').

Natsuno draws heavily on the successes and failures of the various business models of the wired Internet. 'We applied an Internet way of thinking, not a telecom way of thinking, to everything from selection of technologies to proposing a business plan and acting on it,' says Natsuno. Successful learnings from the Internet world include interoperability among multiple device platforms (e.g., Java), seamless connections, participation of user communities, networks of alliances, standardised formats for content (e.g., HTML, MIDI, GIF), and strong focus on usability and utility (e.g., AOL).

Natsuno also identifies some of the challenges of the wired Internet model of business: difficulty in attracting enough paying subscribers, maintaining multiple databases of users and behaviour, hosting multiple application servers, creating convenient and affordable payment gateways, imprecise targeting of advertising messages, and lack of sustainable business models for many small content and service players.

Another significant source of foundational direction for Natsuno came from the inter-disciplinary field of complex systems theory (applicable to weather patterns, organisational behaviour and national evolution), which yielded concepts like interacting components, positive feedback, increasing returns, aggregate behaviours, self-organising activity, emergence, and overall cascading directionality. This leads to phenomena like virtuous cycles and diminishing returns, which explains why different information technologies may succeed or fail, and in unpredictable ways.

The PC-based Internet in Japan was in a 'negative feedback loop' in its early years, with neither adequate content nor users; the mobile phone would have to kickstart a positive feedback loop, according to Natsuno. He signed on 67 content and service pioneers in a number of categories: banking, airline information, hotel reservations, news, weather, recipes, book sales, games and directories. These were grouped into information, e-commerce, databases and entertainment.

Choosing Internet standards like HTML and GIF helped content providers easily re-purpose existing content from the Web, onto i-Mode's transport protocol (TL) and mobile packet communications protocol (PDC-P, based on TDMA).

The choice of partners was driven with a view for emergence and self-organisation. Examples of self-organisation in the i-Mode world include the provision of mobile banking services to consumers, the choice of a leading bank (Sumitomo) as an authoritative trend-setting exemplar, and viral word-of-mouth endorsements from users. The evolution of cartoon characters from Bandai and the creation of the Digital Street search engine (paralleling the wired Internet) are good examples of emergence from individual components.

A balance also had to be struck between too much control of content (e.g., direct ownership) and un-mediated access to content (leading to inefficient services): hence the creation of official and unofficial partner sites. Successful content play also requires a mix of local and universal content. Pokemon and Disney are good examples of 'borderless content,' according to Natsuno.

Content providers were motivated to provide fresh, in-depth, beneficial and continually improving offerings. i-Mode provided integrated billing for official partners, in exchange for 9 per cent of revenues as fees. Nationwide and regional meetings of content providers and editors helped improve the efficiency and quality of content proposals.

i-Mode also stayed on course with changes in user preferences. For instance, users first preferred lighter phones, but then moved to heavier ones which offered better content visibility and ease of use.

Alliance strategies included partners willing to be innovative and share risk, like Sun (for personalisation and other services via Java), Sony (for technology development, e.g., Playstations with i-Mode features), Dentsu (for mobile advertising via the joint venture D2), AOL (for e-mail gateways) and Lawson (for offline services like printing and purchase pickups via the e-Convenience joint venture).

Value for providers of free sites derived from the ability to keep in touch with existing customers (e.g., airlines) and branding opportunities (e.g., recipe sites of Ajinomoto). Popular corporate usage include e-mail, courier applications, inventory monitoring and update, and map displays for service location.

Further on down the road, Natsuno feels that the convergence path of mobile phones adding on computing features is more natural and likely to succeed than PCs which strip away some features to become wireless PDAs.

In the radar sights of i-Mode are 3G services, e-money and an international strategy. The success of 3G will depend on a delicate balance between the quality of services offered and the cost to the subscriber. E-money can target micro-transactions and other consumer purchases; this will depend on upgradations to POS registers via infrared or Bluetooth standards. Internationalisation efforts are focused on Hong Kong, Netherlands, Belgium and Germany.

'I want to develop the complex system that has formed in Japan into a global system. Global emergence, self-organisation and positive feedback: those are my goals,' Natsuno concludes.

Book review by Madanmohan Rao

28. The Mobile Internet

Jeffrey Lee Funk, *The Mobile Internet: How Japan dialed up and the West disconnected; What Japan's experience tells us about the mobile Internet*. Bermuda: ISI Publications, 2001, 200 pp.

How does the mobile Internet differ from the fixed-line Internet? How did the mobile Internet become so phenomenally successful in Japan? What new business models, content strategies, and alliances does the mobile Internet unleash? And what can other countries do to get around their initial blunders on the mobile Internet front, for current and upcoming generations of wireless technologies?

These are perhaps the most intriguing and pressing questions facing wireless Internet markets around the world, and this book by Jeffrey Lee Funk provides a comprehensive treatment of the business dynamics underlying the success of NTT DoCoMo's i-Mode service, along with comparisons and advice for Western markets.

The book provides insights drawn from 150 interviews with 60 market innovators, and combines a historical treatment of the mobile Internet explosion in Japan along with a strategic template of what other markets can learn from Japan.

Jeff Funk is an associate professor of business at Kobe University's Research Institute for Economics and Business Administration; he first came to Japan in 1985. His previous books include *Competition Between and Within Standards: The Case of Mobile Phones*.

The material is divided into 10 chapters, and is copiously illustrated with figures and charts and packed with references and market research statistics.

In a nutshell, the wireless Internet exploded in Japan due to a superb positive feedback loop between the initially chosen content, mobile device capabilities, phone prices, packet networks, business models, and user targeting. At first the users were young and the services provided were simple; then the user base expanded, content became more complex, and the devices much richer—which is not how the US and Europe are approaching this market. In fact, WAP seems to have generally failed in the West, but is working well with Japanese carriers.

The successful players in Japan's mobile Internet value chain include carriers (NTT DoCoMo, KDDI, J-Phone); content and service providers in areas like finance (Daiwa, DLJDirect), entertainment (Bandai, Dwango), horoscopes (Index), ringtones (Giga), concert tickets (Lawson, Pia), navigation (Toshiba, Matsushita), music information and sales (Tsutaya), employment (Recruit), restaurant information (Guru Navi), travel (Open Door), virtual shopping (Rakuten, NetPrice), books (Kinokuniya), portals (Digital Street, Yahoo Japan), m-payment (Bit Cash, NTT, Japan Net Bank); advertisers (ValueClick, D2C); and dozens of phone manufacturers.

The key lessons, according to Funk, are to view the mobile Internet and fixed-line PC-based Internet as complements to one another and not substitutes; to begin with simple content and applications (unlike complex ones, which is the mistake made in the US and Europe); to grow the mainstream market with young users first and then business users; to first increase reach of the medium and then richness of the medium; to convert subscribers into heavy users and then paying users; and to innovate in the areas of new content partnerships and m-payment services.

'The reason why the mobile service providers must create such a comprehensive business model for the mobile Internet is that the service provider plays a much more important role in the mobile than in the fixed-line Internet in creating the necessary positive feedback,' argues Funk.

This must happen quickly since the mobile service providers make the most money (in Japan almost seven times that of all content providers put together). In addition to airtime and packet charges, revenues are derived from content subscriptions, ads, ticketing, entertainment, messaging, product/service sales, coupons, and transaction commissions. People are willing to pay for content if it saves time or kills time.

If the device and content pricing is right, young people quickly gravitate to the medium since they are more mobile, place less emphasis on richness of content, and are interested in the expressive power of customised ringtones, screensavers and animations—which may not appeal as much to older audiences.

Japanese phones can be priced as low as $50, and have full colour, Java capabilities, and polyphonic sound.

'The US success in the fixed-line Internet has blinded many Americans, and to a lesser extent Europeans, to the possibilities inherent in the mobile Internet,' says Funk; it is a mistake to look at the mobile Internet through the filters of fixed-line Internet users.

One chapter documents the explosion of Japan's wireless Internet market. In May 2001 alone, service providers made $700 million from monthly and packet/air time charges on the mobile Internet, while content providers made almost $100 million.

NTT DoCoMo, the first to start the positive feedback loop (before KDDI's EZ Web and J-Phone's J-Sky), launched the i-Mode service in February 1999. It had a simple menu, and charged 9 per cent handling fees to its handpicked 'official' partners for content accessed from this menu. A large variety of cheaper handsets with colour screens and the launch of simple entertainment content marked the second phase of i-Mode's growth, with 350,000 subscribers signing on each week in late 2000. The third phase in explosive growth was marked by the increase in unofficial sites, advertising on these sites, newly launched trade magazines, favourable press, and clicks-and-bricks hybrid services for content creation (e.g., PhotoNet's 10,000 locations for uploading user content) and merchandise delivery (e.g., picking up tickets at Lawson's convenience stores).

Entertainment and e-mail were the two killer apps; followed by news and weather. Much content is delivered as e-mail, with multimedia plug-ins. Interesting experiments on the ad front include quiz competitions on newly launched products, to test ad reach and recall.

Japanese service providers subsidise handsets, and pay much higher activation commissions to cellphone retailers than in the West; the West also prices mobile Internet phones in its markets as high-end phones.

Japanese players also exploit cross-media synergies between wireless and PC Internet (e.g., financial planning services), TV (ringtones based on theme songs), print (catalogues with abbreviated codes for placing orders on cellphones), and convenience stores (shopping, demand aggregation via viral marketing).

Upcoming trends to watch include phones with still/video cameras, higher bandwidth, car navigation systems, Intranet usage (e.g., Tokyo Gas employees accessing maps), and location-based services.

Challenges in Japan, though, arise in avoiding spam, reducing the lagtime for registration of official sites, opening up i-Mode's payment gateway to all providers, and ensuring more linkages between official and unofficial sites.

Western players must avoid these mistakes and create a new wheel of positive feedback, Funk urges. 'Firms often focus on complex technological solutions when they could obtain better results with a simpler approach. They often focus too much on existing users, even when the technology is more appropriate for new and different users. Firms also modularly modify their systems when an entirely new system, including new business models, is needed. Finally, many mobile service providers are trying to create closed systems when today's economy requires openness,' he concludes.

Book review by Madanmohan Rao

29. DoCoMo: Japan's Wireless Tsunami

John Beck and Mitchell Wade, *DoCoMo: Japan's Wireless Tsunami: How One Mobile Telecom Created a New Market and Became a Global Force*. New York: AMACOM, 2003, 240 pp.

DoCoMo, the 'first dramatic success of the wireless age,' is also a fascinating story of the creation of world-beating innovation inside a famously conservative parent, Nippon Telephone and Telegraph (NTT).

'Largely unknown outside Japan, it has created more wealth in recent years than almost any other company in the world,' the authors begin. DoCoMo is ahead of the West in technology, financials and market strength. But a challenging question is: Will it dominate the new economy as Sony and Toyota dominated the old one? While that question may not be easily answered, there are still clear lessons from its success for business strategists and new media analysts around the world.

Mitchell Wade is a strategist at Rand for the last 10 years, COO of Your North Star, and consultant to Accenture. John Beck is director of research at Accenture's Institute for Strategic Change and co-author of 'The Attention Economy.'

The real story behind DoCoMo's success is that it is all about feelings: passion, love, acknowledging and harnessing disparity, impatience, luck, fun and strength, according to Beck and Wade. It went from zero to over $30 billion in revenues, without acquisitions, in just a few years—in a floundering economy and in a product category that few users understood.

DoCoMo succeeded because the right people in the company had the right mix of powerful, human emotions—and succeeded in skilfully managing them. 'A company that understands the power of human passions, and manages those passions in its customers, employees and leaders, will create value faster than its competitors,' the authors advise business readers.

The authors also paint informative profiles of the management of DoCoMo, drawing on an environment of post-War reconstruction, a burning desire to learn and catch-up with the West, passion for quality, and the hope of actually being ahead in some key domains. The authors also dispel some common myths and misconceptions about Japanese work culture.

i-Mode sparked a 'love affair' among its users, by tapping into deeply felt needs, desires and wishes. Given smaller house sizes in Japan and less opportunities for private voice conversations, i-Mode's data communication services helped provide the anonymity and comfort for particularly sensitive exchanges among users. In the workplace, it enabled e-mail communication for those who did not have adequate access to PCs or workstations. It has enabled more dynamic and flexible scheduling of meetings. News reading habits are also changing.

In innovation diffusion theory, there are five kinds of users: innovators, early adopters, early majority, late majority and laggards. Challenges can arise in moving from early adopters to the early majority, who need not want to learn or hear from the early adopters. And sometimes the early adopters are not the key kinds of consumers a company may want, if they are fickle. Social dynamics theories like the 'tipping point' draw attention towards the importance of influential peers, connectors, mavens and salesmen in setting off trends.

i-Mode became a hit with two separate groups of innovators: fashion-conscious youth and traditional geeks. It went beyond these segments in volume and value by drawing on findings from the highly quantitative science of complexity and the psychology behind social epidemics, say the authors.

It promoted personal passions (by paying attention to social symbol aspects of cellphones) and adapted swiftly to customer preferences even if they seemed counter-intuitive or non-profitable at first (e.g., ringtones, wallpaper).

Mobile data services are about intimacy of ownership and constant contact, the ability to share experiences socially. They are about much more than purely utilitarian services, the authors advise.

i-Mode also effectively harnessed disparities like perceptions of being in second place, and gaps in information access like trade information. It defined its vision as making users' experiences more comfortable, convenient and supportive of their thinking lives. (The name DoCoMo itself is a play on the Japanese phrase 'doko demo', which means 'everywhere.')

The management also instilled a sense of urgency among employees and impatience with the order of things in order to focus heavily on the success of future data communications services. The message was to concentrate on 'emotional' and service satisfaction for users rather than just material satisfaction (as per Maslow's hierarchy). Unusual steps were resorted to, such as publishing an interview as a full-page advertisement in a leading daily—thus getting out the message and vision of mobile data services to key players in industry.

DoCoMo also had a fair share of luck and serendipity in the choice and availability of managers, evangelists, salesmen, intrapreneurs, creative heads, technology strengths and alliance partners. 'The energy bouncing around among these people created not just a new product but a radical new creative culture. It successfully transformed itself into the kind of organisation that makes its own luck,' the authors observe.

This helped the organisation make intuitive management leaps and set off chain reactions at crucial steps, like backing phones instead of PDAs or choosing C-HTML instead of WML.

'i-Mode's formula seems clear: luck and timing combined with innovative vision and quick execution,' according to the authors. Some local enabling cultural factors include the long commute times for most Japanese workers, and social conventions forbidding cellphone conversations in trains: hence the ripe market for mobile data services.

But some of these are spreading to other locations in the rest of the world (e.g., restaurants, schools), and some data services (e.g., based on Pokemon characters) can be sold everywhere in the world—thus the Japanese experience should not be dismissed as something unique and inapplicable to the rest of the world, the authors caution.

DoCoMo nurtured a sense of fun in the workplace as a way of encouraging adaptive behaviour, overcoming fear, creating spontaneity, connecting to others, providing a sense of perspective, and playing from the heart. Fun drives innovation; DoCoMo is certainly a leader in providing fun to customers and employees, the authors explain.

DoCoMo also draws on sources of inner strength, a 'kind, gentle strength,' including from its two key leaders, as it embarks on new journeys in multimedia space and global markets. It is confident of its experiences in location-based services, information distribution, remote monitoring services, settlement business, alliancing strategies, and media convergence.

DoCoMo has learnt fast and had the courage to admit its mistakes as well. It has certainly had its share of mistakes: the slow growth of its venture into Hong Kong, and writing down of massive investments in KPN Mobile in Europe. Other differences lie in infrastructure: Japan's 3G infrastructure is built from scratch, but in Europe the network will be built on the backbone of the existing GSM network, thus posing new learning challenges.

Book review by Madanmohan Rao

30. The Freedom Economy

Peter Keen and Ron Mackintosh, *The Freedom Economy: Gaining the m-Commerce Edge in the Era of the Wireless Internet*. New York: McGraw-Hill, 2001, 260 pp.

Looking for advice on how and where to harness the emerging wireless world for business gain? *The Freedom Economy* identifies three focused areas of opportunities for agile business in a mobile world: customer relationships, knowledge mobilisation, and logistics.

The material is clearly presented, well researched, and offers business managers concrete steps for assessing the m-commerce scenario and identifying growth and revenue opportunities. Ten chapters cover a wide range of material including international competitive scenarios, wireless security, cyberlaws, risk assessment, and wireless infrastructure.

Peter Keen is the author of *The eProcess Edge*, and Ron Mackintosh is the founder of e-business integrator *Differentis*.

The world's wireless technology base is comparable to where Internet infrastructure was in the early 1990s and PC technology in the early 1980s, and m-commerce applications are at the stage Internet-based commerce was in 1996, the authors begin. But m-commerce technology innovation is faster than Internet time and even more chaotic.

The bulk of the analysis in the book revolves around the following parameters of the wireless world:

1. Phases of evolution of the technology and applications: promise, chaos/clutter, and business mainstream.

2. Infrastructure categories: phone-centric (3G, 2.5G, WAP, DoCoMo), device-centric (PDAs, portable music players), enterprise-centric (WLANs, wearable computers, laptops), universal (GPS, Bluetooth, RF bar-coding).

3. Application categories (based on Braudel's Rule, drawn from Fernand Braudel's classic *Civilisation and Capitalism*): freedoms

(which change the limits of the possible in the structures of daily life), conveniences (attractive and easy-to-use services), features (new functional elements), and untargeted/unfocused (no specific application).

Analysing the business ecosystem from these parameters, the authors show how phone-centric m-commerce offers the largest long-term opportunity for building customer relationships in the global mass market—but faces challenges at the level of standardised devices, operating systems and applications. Its success thus rests on the planned new wireless infrastructures becoming implemented within planned timetables and planned capabilities.

'Device-centric m-commerce is often the simplest and most immediately available vehicle for introducing special-purpose applications. Enterprise-centric m-commerce provides the widest window into a company's total computing, communications and e-commerce resources,' the authors advise.

Location specificity, time sensitivity, and effective design of m-services are key for building m-commerce relationships in the mass market. Examples already implemented include m-payment services for soda vending machines, product search, car rescue, movie ticket payments, mobile ads, travel services, timetables, parking meter location and payment, downloadable music, m-banking, online auctions, and news alerts.

Companies implementing such m-services at 'the moment of value' include GM's OnStar service for automobile utilities, Vindigo's PDA guide for restaurants, Sweden's patient information systems in ambulances, and even Grameen Phone in Bangladesh which creates entirely new relationship spaces via phone call facilities reaching international migrant workers.

As for the early disappointments of WAP, the authors cite Johan Helm's book *Designing Wireless Services*, which points out that it is not the Web that mobile users need, but the information that is on the Web, displayed in the form most suited to their mobile devices.

At the level of logistics, m-commerce in organisations can help reduce inventories, cut purchase order costs, reduce the document flow, and speed up the supply chain. 'Clean processes and management by interface are the core sustainable advantage of the logistics leaders. They are the foundation of integrated commerce,' according to the authors.

Home Finish Builder's staff uses PDAs to coordinate work order reports about house repair. Adidas Canada uses handheld terminals for speeding up picking, checking and packing of goods. Sears plans to deploy 15,000 wireless handhelds for use in its receiving through replenishment cycles in 860 department stores. Wal-Mart is using Symbol Technologies' handhelds and WLANs for shortening check-out time. Weyerhauser uses handhelds to provide maps to loggers in timber tracts.

eDispatch.com uses smartphones for increasing dispatcher efficiency and performance. Agricultural machine manufacturer John Deere uses GPS for its crop identity tracking system. Drug wholesaler McKesson uses strap-on computers and WLANs in its 31 distribution centers, and reports an 80 per cent drop in incorrect items shipped. UK doctors use SMS messages to locate nurses at short notice.

The US Army uses Savi's RFID tags to track pallets in shipment. Singapore Airlines uses RFID tags to improve its cargo hub operations. UPS and FedEx provide mobile package tracking software for use on PDAs. Starbucks has installed WLANs for Internet access to make its coffeeshops more attractive for traveling office workers. Colorado schools are installing WLANs to make it easier for their students to access educational resources.

Mexican cement manufacturer Cemex has redesigned its business model so that its ready-mix trucks are on the road all day in a continuous flow, coordinated by dispatch software and mobile devices. Fisherman in Kerala are using mobile phones to dynamically reconfigure distribution chains depending on where the best prices for the catch can be identified.

'M-commerce adds greatly to the e-commerce contribution to the logistics revolution, which has in itself been truly astonishing,' according to the authors, in areas like fleet management, remote monitoring, and field communications.

M-commerce strategies focused on knowledge mobilisation can take knowledge management to an entirely new plane of performance and effectiveness, putting road warriors and field workers in the center of the information and communications world via mobile portals and on-demand expert services. Laptops have provided portability, but generally did not add mobility in this regard, according to the authors.

'The knowledge mobilisation opportunity is so huge that no company can afford not to grab it fast and hard,' the authors advise. This applies to external competitive/regulatory intelligence, structured internal knowledge, and especially informal internal knowledge.

Sun's field engineers are equipped with handhelds so that they have the collective knowledge of the company at easy disposal. Hotel giant Carlson's managers use iPaq handhelds with up-to-date information on room status and yield management. Boeing's engineers use laptops and WLANs to better access complex multimedia documents on the move. Bell Canada technicians use wearable computers and miniature cameras (Xybernaut Mobile Assistants) so that they can instantly tap office expertise even while they are on top of a pole.

'Mobile accessible information puts knowledge to work right at the demand points. Mobile technologies enhance communication, information and collaboration, the three cornerstones of knowledge building and usage,' the authors sum up.

One chapter covers the dark side of m-commerce, such as challenges to user privacy, lack of security in wireless communication, and poor reliability of m-infrastructure.

There is also a cacophony of marketing claims, along with spectacular flops like Iridium and uncertainty over the future investments of telecoms who have forked out billions in 3G license fees.

One chapter compares the mobile scenario in the US, Europe and Asia, in terms of penetration, hourly costs, level of usage of m-commerce, cultural attitudes, regulatory climates, and business innovation. Due to factors like inadequate roaming services and complexity of pricing plans, US businesses and consumers lag Europe in m-commerce—but the US has been leading in other areas like B2B and B2C e-commerce.

Companies which are pioneering in the technology and marketing of m-commerce like Nokia are well suited to meet the supply side in a demand-hungry region like Scandinavia—elsewhere, both the supply and demand must be built.

DoCoMo has been a huge success in Japan—but this may be an isolated case due to factors like higher wired Internet access costs, a more gadget crazy culture, and the ideographic nature of the language.

'In many ways, the key issue for businesses everywhere is to learn quickly from each type of approach. European firms need the US e-commerce pragmatism, and the US needs to think more in terms of architecture and infrastructure than hit-and-run quick fixes,' the authors advise.

'The absolute priority in m-commerce is to invent freedoms at the level of relationships, processes and knowledge, by exploiting the unique features of mobile devices linked by wireless to Web-based resources,' the authors urge.

Challenges for companies in establishing beachheads in the m-world arise due to the basic uncertainly of planning what they can't predict, in terms of technology maturation and consumer attitudes towards m-transactions. CIOs also run the risk of ending up as 'Chief Gadget Officers,' the authors joke.

Companies should therefore identify freedom opportunities (important services which just would not be possible without the m-channel), ensure their practicality, select the appropriate technology enablers, and then define the beachheads, the authors conclude.

Book review by Madanmohan Rao

31. Wireless Rules

Frederick Newell and Katherine Newell Lemon, *Wireless Rules: New Marketing Strategies for Customer Relationship Management Anytime, Anywhere*. New York: McGraw-Hill, 2001, 330 pp.

M-commerce and m-business are the hot new topics of the year, and dozens of books are already hitting the shelves—but *Wireless Rules* stands out with its strong emphasis on the opportunities and challenges facing marketers and practitioners of customer relationship management (CRM) in the wireless medium.

Frederick Newell is CEO of consulting firm Seklemian/Newell, and the author of *Loyalty.com* and *The New Rules of Marketing*. Katherine Newell Lemon is a marketing professor at Boston College, and co-author of *Driving Customer Equity*.

Thirty-one well-referenced chapters are divided into six sections, covering a wide range of topics like wireless content design, privacy, online news media, e-finance, telematics, and metrics; case studies are drawn from Asia, Europe and the US.

Useful online resources cited in the book include Wireless NetNow, Wireless Review, ComputerWorld.com, WAP.com, WAPforum.org, UseIT.com, WebMonkey, DestinationCRM.com, CRM-Forum.com, IdeaVirus.com, NAA.org, PervasiveWeekly.com, and CommNow.com.

The wireless channel will impact everyone in the marketing chain, from the CEO and CIO to sales forces and customer service reps, the authors begin. It will alter online and offline channels, move companies from 'tell and sell' to 'listen and learn' mode, and impact the 4Ps (product, price, promotion, place) as well as 4Cs (customer, company, competitors, collaborators) of marketing.

'Wireless communication is the biggest change to hit telecommunications since Alexander Graham Bell called Thomas Watson 125 years ago. It is the biggest change to hit marketing since the development of the Internet,' according to the authors.

An estimated 40 per cent of adults and 75 per cent of teenagers in the industrially advanced world will have always-on wearable devices by 2010.

In the wireless era of the 'Evernet,' customers care more than ever about accessibility, context, convenience, comparative shopping services, instant information, customisability, trust, privacy, loyalty rewards, consistent quality, simplified services, innovation, and fashion—all of which will have to be carefully incorporated in mobile marketing plans.

'Wireless and the Web together create new possibilities for multiway, multi-log communications,' according to the authors; this includes customer self-service as well as full service by the vendors.

The soul of CRM is finding ways to make customers' lives easier—to make it easier and more convenient, even more fun, for them to do business with your company. Wireless puts CRM 'on steroids' by creating better experiences and relationships. The big message in wireless is not m-commerce, but the chance to reinforce awareness and loyalty among customers.

As for m-advertising, the message must add value or convenience or both, for recipients. But there is also great opportunity for misuse and a backlash of over-regulation. The new paradigms for message delivery, measurement and regulation are being discussed and resolved by organisations like the Wireless Advertising Association (WAA) and Cellular Telecommunications and Internet Association (CTIA).

Responsible permission marketing tactics like confirmed opt-in should be the de facto standard to enable targeted, uniquely personal, compelling, and useful messages in the wireless advertising ecology, otherwise consumer trust will be lost. Users will need to be in control, otherwise they will resent such services as 'digital dog tags.' Marketers thus have to juggle a 'high wireless balancing act' with privacy, security and targeting options, the authors joke.

'You need to build in, not bolt on, security and privacy solutions for your wireless strategy. You are dealing with the integrity and viability of the channel,' they advise.

'You need to understand how to acceptably interrupt your customers—on their terms,' the authors advise. Permission has six steps—opt in, opt on, opt when, opt where, opt how, and opt now—covering location, time, and method for alerting and fulfillment.

'This is effectively a new economic order of the digital world. Think of the cellphone as the ultimate permission marketing tool,' the authors urge. Concepts of the mass marketing and database marketing era like share of market are being supplemented by share of customer and lifetime value; new dialogue skills and cross-functionality are called for, along with new technological and information processing agility.

As the automobile continues to become an ever more important part of the American lifestyle, so does the potential of telematics or in-vehicle data and voice services—estimated to reach a market size of $8 billion by 2005.

General Motors' OnStar (launched in 1996) and Ford's WingCast are key players to watch as they transform cars into 'mobile portals' or even 'Java browsers with tyres.' OnStar's offerings include emergency services, Virtual Advisor (news, e-mail and information along with GPS), and Infotainment (MP3 downloads).

Opportunities also open up here for satellite radio companies; challenges arise in ensuring driving safety along with access to these potentially distracting services.

On the content front, newspapers have a great chance to leverage local content for wireless users—but they may also blow it again and lose out to other players as they did to portals in the Internet era, observe media consultants Pete Zollman and Vin Crosbie.

News alerts, movie show times, politics, and classified ads can be successfully leveraged for wireless. This may require 'editorial scientists' and 'content engineers,' and can also tap photo services once bandwidth and devices improve.

Every new form of media redefines preceding forms, says Kevin Clark, director of IBM Mobile Computing, citing MIT's Ithiel deSola Pool.

In corporate space, wireless has lots of impacts on productivity and knowledge management. An estimated 30 per cent of the US workforce (or 36 million people) is out of the office traveling at any one time. Mobile technologies must be harnessed with care, taking into account integration and security issues for the enterprise data infrastructure.

Online banking and financial services via wireless channels are a sure winner as well, as evinced by Charles Schwab's ad campaign: 'Buy. Hold. Cell.' Mobile CRM is the second chance for those who were late in adopting the Web, the authors advise.

Success strategies for players in m-space include effective alliancing, partnering and collaborating. Many alliances have already been struck: between Yahoo and Telefonica Mooviles in Spain; US newspapers and SkyTel; AT&T and Net2Phone; PacketVideo and Universal Pictures; mySimon.com and ViaFone.

Return on investment metrics should revolve around share of customer access, airtime, and purchases. This includes frequency and length of interaction, extent of self-disclosure, level of opt in/out, and word of mouth.

The book offers numerous case studies of m-commerce and mobile CRM at work.

In the US, users can buy Amazon books via Sprint PCS phones. BestBuy's UsByAir.com service spans content sales and music purchases. TellMe Networks provides voice-activated navigation services for cellphone users using VoiceXML (Voice Extensible Markup Language).

Several companies in Los Angeles deliver air quality information to environmentally-conscious wireless users. Personalised traffic alerts to PDAs are delivered by TrafficStation.

Altavista offers restaurant locator information to WAP users; Vindigo provides city-specific information to handheld computers; and Windy1.com provides business information in Chicago.

GeePS.com and Itaris let retailers target their favourite customers in malls via localised messaging. AirClic offers barcode reader services for cellphones. Youth apparel company Alloy Online offers its wireless users free voicemail.

Rovent provides a bathroom locator service on the wireless portals Palm.net and YadaYada.com. Ebion in Ireland has introduced mobile betting, and iToke in Amsterdam even plans to deliver marijuana to cellphone users!

Galileo, MyTrip.com, Travelocity, and SmartRay.com bring personalised flight updates to the user's cellphone. Lufthansa lets frequent flyers choose their seating preferences via cellphones.

Yahoo already offers voice access to e-mail via 1-800-MY-YAHOO services in the US; Quest Communications offers services for voice-activated Web surfing. U-Inspire offers daily inspirational messages via PDAs and cellphones. Web gateways for SMS messaging are offered by sites like Web2Mobile.com.

Time Warner's mobile newsfeeds, launched in 1999, reach over 41 million subscribers. European golf fans subscribe to tournament updates on cellphones. US football teams use services like MobileFlash to deliver realtime information to fans.

Companies active in m-advertising services include DoubleClick, AvantGo, WindWire, SkyGo, Zing Wireless, InPhonic, Advertising.com, PhoneOnline.com, 365 Corporation, AdLive, Beyond Interactive (which coined the term 'WAPverts'), and Wireless Commerce (in Finland). Worldata and WebConnect offer permission marketing services.

Mobile workforce solutions are being offered by companies like Telispark; Landstar coordinates its freight fleet with wireless devices. FieldCentrix provides wireless access to inventory information for field technicians at companies like Mesa Energy Systems. SalesForce.com offers sales force automation services via wireless; Sales.com, a Siebel spinoff, is experimenting with such services as well.

In terms of international diffusion and innovation, the US wireless industry is 18 months behind Europe—which in turn is 18 months behind Japan, according to *The Economist*.

Widespread usage of cellphones among tens of millions of subscribers in Japan has led to the proliferation of 'thumb culture.' Chatting and personal services offered by companies like Bandai have boosted the revenues of NTT DoCoMo; even the police force circulates criminals' photographs via i-Mode cellphones.

The world's first WAP roaming service was initiated in July 2000 between Singapore and Hong Kong. eStreet in the town of Lulea in Sweden is well known as a real-life laboratory for testing mobile applications.

Further on down the road, new opportunities are arising via m-payment services, automated 'bot' services, personal area network technologies like Bluetooth, home networking, hotel networks, mobile barcode coupons, portable phone numbers across networks, and multimedia gaming.

Challenges arise in effective design, coping with a multiplicity of devices (beyond 'Swiss Army knife' cellphones), information architecture for archiving and accessing content on mobile devices (e.g., dialogue indexing), overcoming 'walled garden' syndrome, effective audience segmentation, and managing outsourced m-services.

'Consumers won't give up wireless devices easily. Once they are hooked, they are hooked,' the authors conclude. Make no mistake—wireless marketing will be here, and it will be dominant; companies must pay attention to frameworks and details while applying wireless rules, the authors conclude.

In sum, in the words of business guru Gary Hamel: 'Coolest of all is to invent an organisation that can win again and again in a permanently disruptive world.'

Book review by Madanmohan Rao

32. Constant Touch: A Global History of the Mobile Phone

John Agar, *Constant Touch: A Global History of the Mobile Phone.* Cambridge, UK: Icon Books, 2003, 172 pp.

'You can tell what a culture values by what it has in its bags and pockets,' begins popular science author John Agar in *Constant Touch: A Global History of the Mobile Phone*. By historical comparison, the pocket watch was a rarity in the 17th century, but become an expensive status symbol of business activity for industrial professionals in the 18th century—and is now owned by the majority of the population in the 21st century.

'Pocket watches provide the closest historical parallel to the remarkable rise of the mobile cellular phone in our own times,' observes Agar. Like pocket watches, cellphones too started off as expensive status symbols and cost as much as a car, but are now owned by over a billion people worldwide.

John Agar has taught in Manchester, London and Oxford. He was formerly director of the National Archive for the History of Computing. His previous books include *Science and Spectacle*, *Making Space for Science*, and *Turing and the Universal Machine*.

The 18 chapters in this slender pocketbook cover a wide range of topics: paths to mobility in different countries, regional diffusion patterns, the cultures of mobile users, and globalisation of the mobile phone industry.

Wireless technology has come a long way since the invention of wireless telegraphy in the late 19th century. The technological components were initially so heavy that they could be housed only in ships (Marconi's target audience) and cars (Ericsson's innovation). Miniaturisation of mobile communication devices has been spurred by innovations in portable electrical power supplies, liquid crystal displays, global sourcing of materials like tantalum (Australia, Congo) and nickel (Chile), manufacturing bases around the world, and the software industry.

The key regulatory challenges facing the industry have revolved around the scarcity of radio spectrum, coordinated by national and international bodies (e.g., FCC in the US, ITU globally). Engineers at Bell Labs pioneered the cellular model of managing mobile radio in 1947, consisting of massive fixed infrastructure of wires, switches and inter-carrier agreements to enable users to travel in 'cells' of coverage which re-used scarce frequencies.

Regional Diffusion

Car-bound radio communications were used in the US by the Detroit police force in the 1920s. Motorola developed the 'Handie-Talkie' and 'Walkie-Talkie' during World War II. These systems were still not mobile phones: the radio network remained separate from the telephone network.

Despite an early lead in mobile innovation, commercialisation of mobile telephony posed problems in the US due to regulatory obstacles, the sheer size of coverage required for the entire country, competing devices like pagers, and a socio-economic focus on the automobile as the symbol of mobility. The fragmented pattern of coverage began to consolidate as companies merged, leading to the rise of national players like McCaw (later bought by AT&T).

Paradoxically, the US lost the lead in digital mobile phones because its customers were generally satisfied with the first generation analogue cellular phones. It now has a more competitive position in emerging 3G markets due to San Diego-based Qualcomm's pioneering role in CDMA standards (which differs significantly from FDMA and TDMA).

Nordic countries like Sweden had strong diaspora connections to the US, a penchant for user-friendly industrial design, and consensus-based planning approaches. The Nordic Mobile Telephone (NMT) group was established in 1969 to develop a regional analogue cellular phone system with built-in features like roaming. Early use was largely by industry professionals, before becoming a household device.

In other parts of Europe, the systems were fragmented: for instance, Germany developed a new standard which used a personal identification card, the ancestor of the SIM card. Fortunately, a 'bureaucratic miracle' occurred in Europe in the inauguration of the digital GSM standard in 1982, which grew to enable a pan-European unified network.

With services officially launched in 1992, European coverage was nearly complete by 1995. 'Roaming was enthusiastically embraced as much by partying twenty-somethings in Ibiza as much as by European business executives,' says Agar. The mobile phone became less a business tool and much more an everyday object.

In the UK, early users of mobile radio included travel repair services, taxi firms, and large companies. In 1982, licences for analogue cellular services on the TACS standard were awarded to British Telecom (Cellnet) and a consortium including Racal and Millicom (Vodafone, which is now a major international player).

China launched TACS services in 1998, and overtook the US to become the world's largest cellphone subscriber base by 2002. In Japan, NTT DoCoMo created a smash hit with i-Mode in 1999, with voice as well as always-on data services more sophisticated than GSM's SMS or WAP. It appealed to a wide range of users: teenagers, single women, business professionals.

Mobile Culture

Some of the earliest socio-cultural impact areas of mobile phones revolved around private conversations conducted in public places like trains, leading to differing notions of etiquette and personal space balanced with the benefits of being in constant touch or resurrecting 'dead time' for professionals.

'City life is mobile. City life is fast. And in no region outside the Nordic countries have cellular phones become as culturally important as in the cities of the Pacific Rim,' says Agar. This includes the entrepreneurial culture of Hong Kong (where six licences were granted to service providers to keep up with huge demand), sociability of Australian city life, and even political revolution in the Philippines (where SMS was used to coordinate protests which overthrew the president in 2000).

In many countries of Asia where the extended family is important and where youth are drawn to cities, mobile phones are important tools of cohesion. In the Philippines, a group of subscribers formed an organisational called Txtpower, which has campaigned against proposals to tax txt. Decentralised technologies like the Internet and mobile communications thus enable shifts in governance towards networks rather than centralised hierarchical modes of organisation.

Interestingly enough, the power of texting was an accident—it was not considered to be important by telecom operators, but its power was discovered by users. In contrast, 'texting' on mobile phones was not as popular in the US due to incompatible phones, fewer perceptions of cost advantages of text over voice, and a prevalence of pagers and beepers.

While one aspect of the wireless environment has been mobility, another important aspect is the material culture. Products of companies like Nokia display superior design, elegance, style and branding, and are fast becoming fashion carriers. Cellphones have also been featured regularly in media products like films (*Wall Street*, *The Matrix*).

In family settings, cellphones are being used by parents to stay in touch with children out late at night, thus playing an important security and assurance role. Risks also arise, unfortunately, due to thefts of cellphones and attacks on their owners, as well as subscription fraud, loss of privacy, and the use of cellphones to coordinate crime. On the sensitivity frontier, some religious organisations in Europe have even banned the use of mobile transmission masts on churches. Youth around London have used mobile phones to coordinate illegal rave parties around the motorways.

Globalisation

Just as containerisation transformed the shipping industry and ports of the world in 1970s, the next generation of 3G mobile technology will transform existing cellular networks and devices at a heavy investment cost. The IMT-2000 is the umbrella for five different standards across Europe, the Americas and Asia.

'The success of third-generational mobile phones depends on the unknowable willingness of the public to buy them, and without good content—in the form of addictive entertainment or really useful services—a repetition of the WAP debacle is possible,' Agar concludes.

Book review by Madanmohan Rao

33. Mobile Usability

Christian Lindholm, Turkka Keinonen and Harri Kiljander, *Mobile Usability: How Nokia Changed The Face Of The Mobile Phone.* McGraw-Hill, 2003, 301 pp.

This is an excellent book about the complex process of user interface design and usability research in the world of mobile phones. It presents first-hand experiences and perspectives on phone/screen design philosophy, mobile interaction design, interface styles, task analysis of social bonding, field testing and usability engineering in Europe, Asia and the US.

All the authors have had considerable design experience with Nokia. The 13 chapters are divided into four sections: interface styles, cultures of usage, design methodology and evolving trends.

'In the relatively brief lifetime of the mobile phone, two major technological trends have taken hold: devices trend toward miniaturisation and applications, features and functions trend toward expansion. Taken together, they present us with an interesting paradox—squeezing more and more applications into smaller and smaller terminals is how we try to keep users satisfied, but it makes the devices harder to use,' the authors begin.

There is a trade-off between multiple rich features and usability of the device, which can be graphed along the 'usability knee' till the breakpoint of ease-of-use versus functionality. The authors trace the complex design issues around features which allow users to handle multiple phone calls (e.g., US users prefer to switch between calls, Europeans would rather prioritise the first call), phonebook management, time management, rich media, messaging and office applications.

'The logic of PC interfaces does not scale downward. Small interfaces are essentially different from big ones,' say the authors. The surface area of the terminal determines the size and number of buttons and screen. This difference in quantity leads to a difference in quality of interaction.

Unlike the PC industry, the cellular industry has no standard user interface. Applications have to be optimised for small screens on all layers of the interface: vocabulary layer, logic layer and functionality layer.

'Since the early 1980s, the mobile phone has undergone a transition from a business tool to an upscale accessory and ultimately to a ubiquitous consumer product,' the authors explain. They segment users into six categories: experiencers, impressors, controllers, maintainers, balancers and sharers. Further differences arise across cultures, languages, age groups (e.g., youthful 'screenagers') and occupational/professional sectors.

Users themselves have different preferences for intuitiveness, personalisation, simplicity, efficiency, backward compatibility, productivity, individuality, appearance, exclusivity and fashionability. Emerging technologies like digital imaging and Bluetooth throw in additional complexities for product design and functionality.

Phone and interface design complex span areas like snap-on covers, accessory connectors, interaction sequences, graphic design, menu layout, sounds/animations, localisation and downloadable software.

Nokia's approach has been to evolve two core styles: the two-softkey style (Series 30) and the one-softkey or Navi-key style (Series 20). The Navi-key has become one of the most valuable sub-brands of Nokia.

Based on these core styles, a 'manageable set' of user interface styles has been evolved. The experience of unsuccessful user interface (UI) features is also discussed, such as the Navi-roller for combined scrolling and clicking, or the one-row keyboard (due to lack of the intuitive feel for keys that a two-dimensional keyboard can provide).

'Application developers and service providers also want a reasonable degree of stability from the wireless device user interface platform, so that their services can run on as many handsets and handset generations as possible,' say the authors.

'Cellular mobile telephone development is becoming more software-focused. Product release cycles are gradually becoming shorter and shorter. Designing an ever-increasing number of handset features in an ever-shrinking development cycle can happen only where the underlying UI platform is reusable and scalable,' the authors caution. 'A solid UI backbone is essential for building good user interfaces for the mobile information society.'

Key menu design components include idle screen, application menu, hierarchical sub-menus and option lists. Users must also find the menus easy to use and remember shortcuts for frequent tasks. Sometimes, behaviours for mobile phones conflict with behaviours learned from landline phones (e.g., punching in the number before contacting the network).

Design teams and consumer device companies often need a product champion to get radical innovations into the market (as with Akiro Morita and the Sony walkman); Nokia experienced this with the invention of the Navi-key.

Introduced in 1991 in Europe, GSM networks (based on TDMA) now serve over a 100 countries except for the US (where analog networks still dominate). Useful design lessons are drawn in the book from other early interface design such as the Apollo guidance computer for astronauts. Mobile text input received a significant boost with the invention of predictive text method T9 by a US company called Tegic; the Nokia model 7110 with T9 was introduced in 1999.

Nokia conducts frequent cognitive walkthroughs, ethnography, usability tests, audits and design verification in its various markets, to determine how users relate to and operate their mobile phones (tools/toys). 'In order to feel, hear, smell and see the actual context for which you are designing, and to understand the daily life of your customers, you must experience the environment and meet the people,' the authors advise. At times, not even the questions are known when the research begins.

For instance, in India, environments can be noisy, humid and hot; SMS greetings are popular on religious and festive occasions; the gray market for mobile phones is strong; repairing devices is preferred to replacing them; phones are sometimes shared in families; street navigation is more via landmarks than maps; and language mixing (e.g., 'hinglish') is quite common.

In countries like Finland, TV-SMS interaction is quite common in programmes. As for driver interaction with mobile communication devices, experiments in understanding 'traffic psychology' were conducted via driver observations (e.g., number of glances at devices, duration of each glance) in cities like London, Munich and Salo.

Nokia also took part in the Mobile Media and Entertainment Services (MOMENTS) project, during 1995–1998, to study 3G multimedia services like street search, route selection, stock alerts, and animated weather front information. Accuracy of service, user response time, and even idle-time services (for 'spending time') were tested.

It is becoming important for interaction researchers and marketing researchers to collaborate in researching new communication patterns and emerging cultural phenomena.

An interesting phenomenon is that products change after their launch, as the meaning of the phone is renegotiated in the daily context of the users again and again. For instance, SMS has created a new culture, and users are learning behaviours that they would never have imagined before. 'Although mobile phones and other new digital communication tools can be seen to mediate human communication, they also shape it,' the authors observe.

Mobile phones are leading to more spontaneous activity planning, and the relation to time and space have become more flexible. 'It is possible to plan less and improvise more,' the authors observe.

From an implementation perspective, there are six different disciplines in Nokia's UI design teams: interaction design, localisation, graphic design, user interface platform, sound design and usability. These teams come together in contexts like understanding constraints and risks of mobile communication by automobile drivers.

From a workforce point of view, companies like Nokia believe in open-mindedness, high quality, good documentation, parallel task-processing, and constant rejuvenation. 'Reliable management tools that have been hardened in a crucible of daily use will remain essential,' the authors explain.

Products and designs have to exhibit consistency within and between applications and generations, which can be quite a challenge due to unpredictable technological development and chaotic interactions between emerging systems.

The authors also cite the work of human-computer interaction (HCI) usability guru Jakon Nielsen, who defines usability via learnability, memorability, efficiency, error rate and satisfaction. An expressive user interface must evolve with its user, say the authors. Mobile communication must also respect users' privacy.

The book has its humorous moments as well, with discussion of the 'ugly duckling' technology (growth of e-mail/SMS), 'spouse test' (a spouse would never give praise for the sake of praise), 'mom test' (least common denominator of user sophistication), 'generation trap' (waiting for current generation feedback before moving on the next technology generation), 'bimbo trap' (a feature so easy to use that it is insulting to the user), and 'WAP is crap' phase.

Book review by Madanmohan Rao

34. Smart Mobs: The Next Social Revolution

Howard Rheingold, *Smart Mobs: The Next Social Revolution—Transforming Cultures and Communities in the Age of Instant Access.* Perseus Books, 2003, 266 pp. $26.

Smart Mobs is an exciting and comprehensive exploration of the frontiers of the next major techno-cultural shift spawned by mobile social networks. Howard Rheingold is the author of the classics *Tools for Thought* (about the impact of PCs in the 1980s) and *The Virtual Community* (about the Internet in the 1990s) as well as former editor of *HotWired* and *The Whole Earth Review*.

The material is divided into eight chapters, and provides a highly readable mix of personal first-hand accounts as well as thoroughly referenced research (with 32 pages of references and URLs). The book also has an excellent online companion at www.smartmobs.com.

Topics covered range from wireless dating and political activism to reputation systems and privacy threats. Wireless technologies are proliferating in multiple forms: wireless Internet nodes, mobile phones, RFID tags, and embedded microchips (in 'smartifacts').

'Groups of people using these tools will gain new forms of social power, new ways to organise their interactions and exchanges just in time and just in place,' Rheingold begins. One should expect the unexpected when previously separate technologies meet—as with the marriage of the PC and telecom networks to create the Internet, which spawned possibilities that neither of the parent technologies exhibited in isolation.

'Just as existing notions of community were challenged by the emergence of social networks in cyberspace, traditional ideas about the nature of place are being challenged as computing and communication devices begin to saturate the environment,' says Rheingold.

'Mobile Internet, when it really arrives, will not just be a way to do old things while moving. It will be a way to do things that couldn't be done before,' Rheingold observes. The mobile phone, like the PC in the 1980s, is poised to make a jump from a useful toy adopted by a subculture to a disruptive technology that changes every aspect of society.

The Cutting Edge: Japan and the Nordic Countries

Numerous business and anthropological studies have focused on the pioneering creators and users of Japan's i-Mode wireless Internet service. Tokyo texters are sometimes called 'the thumb tribe,' and the *keitai* (mobile telephone) users have created their own private channels of communication amidst the neon and video of 21st century Japanese cities.

Mobile phones have triggered an inter-generational shift in Japan because they freed youth from the lack of privacy of landline communication at home, thus creating more possibilities for social networking via texting, according to Keio University professor Mizuko Ito. Meetings times and places can be fluidly negotiated (or 'softening' of time and 'blurring' of place). Keitai are viewed as fashion as well as technology.

The combination of data and voice services makes it possible for users to search for or look up information in real time, thus enabling a range of location-based services coupled with time-based promotional offers. The mobile phone can even be viewed as a 'remote control for your life.'

'Swarming'—the cyber-negotiated public behaviour of texting users—is a term frequently used in the other epicentre of pervasive mobile technology, Helsinki. Finland's notable contributions to the world of the Internet include IRC (invented by Jarkko Oikarinen in 1988) and Linux (by Linus Torvalds).

The unexpected success of SMS is a sign that people are appropriating a communication technology for social purposes. By 2000, Finns exchanged more than a billion SMS messages a year. The wired ecology of subcultures in Finland has created a new 'shared urban living space' combining physical location and mobile virtual communities.

Some groups of mobile users in Stockholm also call themselves the 'Mob,' indulging in games with bots and gangs. Stockholm is also home to the company. It's Alive, which created the world's first location-based mobile game.

The SMS subculture—the style, content and the very act of messaging itself—provides new raw material for identity formation, social bonding, individual personality and peer-to-peer networking.

Technologies of Cooperation

Many public goods ride on game theory models of coopetition, which balance competition, self-interest and possession with volunteerism, self-governance and principles of social equity. 'The Internet was deliberately designed by hackers to be an innovation commons, a laboratory for collaboratively creating better technologies,' Rheingold observes. Unfortunately, the overload of spam is a threat.

In the emerging and maturing social cyberspaces, people themselves are becoming autonomous communication nodes, and some are even becoming portals. Today's media networks display a combination of one-to-one, one-to-many and many-to-many communication.

An interesting featured of digital media infrastructure is that computers and electronic computers are among the rare technologies that grow more powerful and less expensive simultaneously, thanks to growth patterns characterised by Moore's Law (doubling of computing power every 18 months), Metcalfe's Law (value of a network grows with the square of the number of nodes), Sarnoff's Law (value of broadcast networks is proportionate to the number of viewers), and Reed's Law (value of a social network grows exponentially with the user base).

Computation Nations

Grid computing, peer-to-peer networking and distributed computation are new forms of technology-assisted cooperation. The Search for Extra-Terrestrial Intelligence (SETI) project leverages community computing resources via the SETI@home initiative. United Devices ropes in individuals who supply CPU cycles to corporate research initiatives.

'Napster was the killer app that awoke the world to the disruptive potential of P2P power,' Rheingold observes. Though Napster has been shut down, new forms like Gnutella have emerged. P2P networks are emerging for sharing not just files and computing cycles but decisions and judgements.

Ubicomp: Ubiquitous and Wearable Computers

Virtual and physical worlds are becoming increasingly enmeshed as computing devices. They are built into physical locations and wearable devices—creating smart rooms, digital cities, and global communication 'skins.'

Computers will eventually be so pervasive that they will disappear, as with the case of electricity which most urban citizens have taken for granted for over a century. Despite early years of hype about virtual reality, many of the capabilities and issues have stayed on. 'Sometimes a technological development appears to dead-end, when it is really in the process of side-stepping,' Rheingold rightly observes. Some even call its present incarnation 'artificial reality' or 'augmented reality.'

Researchers at companies like HP are looking at what happens to life in a city, home or office when the physical world becomes browsable and clickable via embedded devices, scannable surfaces and individual URLs—or when bits and atoms have been merged.

Online Reputation Systems

The Internet, now assisted by mobile technologies, has spawned new kinds of reputation systems (e.g., on e-Bay, Amazon, Slashdot, Epinions), leading to new kinds of power structures and networks for smart mobs. At the same time, unfortunately, there has been a rash of flamers, spammers, frauds and bigots.

'More recently, the phenomenon of weblogging, which enables thousands of Web surfers to publish and update their own lists of favourite Websites, has tipped online recommendation-sharing into an epidemic,' according to Rheingold.

Online 'webs of trust' have to be designed carefully to weed out low-quality contributions, optimise audience size, prevent moderator abuse, avoid ballot stuffing, and reward good community members. Reputation, altruism and punishment are key social components of such systems.

Wi-Fi: Wireless Quilts

The inter-connection of Wi-Fi networks or wireless LANs one neighbourhood at a time has a potential of 'unwiring' the world in a way far different from the proposed massive 3G networks. Anarchic as well as civic approaches to wireless networking are emerging, as with the hotspots provided in New York's parks by NYC Wireless. Precedents exist in the patchwork growth model of earlier networks like FidoNet and Usenet.

Neighbourhood area networks (NANs) are being built by the Bay Area Wireless Users Group (BAWUG) in California as well. In terms of hacker culture, new activities like 'whacking' (wireless hacking) and 'war driving' (roaming in search of open wireless networks) are emerging.

Regulatory challenges can arise in terms of spectrum allocation, but technologies like 'software configurable radio' can alleviate some of these. 'Imperfect but decentralised and free technologies are a critical way to induce innovation and grow the network,' advises cyberlaw guru Lawrence Lessig. This may be accompanied by 'creative destruction,' but that is part of the inevitable game of freewheeling innovation and capitalism.

Mobile Power

On January 20, 2001, president Joseph Estrada of the Philippines became the first head of state in history to lose power in a protest movement coordinated in part by smart mobs using SMS to broadcast and receive messages. Similar 'netwar' techniques have been used by demonstrators against the WTO in Seattle, gasoline price hike protestors in Britain and bicyle activists in San Francisco.

'The Philippines might furnish early indicators of the way mobile communications could affect other countries where it is more cost-effective to jump directly to a wireless infrastructure,' Rheingold observes.

The ease of use of SMS lends itself to forwarding jokes, rumours and chain letters, spawning a 'mania' in the Philippines—so much so that texting in many contexts is not seen as rude any more.

The proliferation of mobile devices can also spark new forms of P2P journalism and 'cultural reclamation' as more and more people become equipped and motivated to chronicle and influence events. The combination of peer-to-peer and face-to-face can lead to emergence of new forms of swarm intelligence and social minds.

Panopticon or Cooperation?

At the same time, Rheingold warns that the technology can be used to improve life and liberty and also degrade it, as both a weapon of social control and a means of resistance. 'Even the beneficial effects will have side effects,' he cautions.

'Wearable computing communities' will have to deal with the consequences of leaving behind a 'trail of electronic breadcrumbs' which will become a 'moving cloud,' leading to Orwellian concerns about a panopticon-like surveillance society.

'Detailed information about the minute-by-minute behaviours of entire populations will become cost-effective and increasingly accurate,' Rheingold cautions. The population itself could become a collective surveillant: Big Everybody.

Threats can also arise to quality of life as we become increasingly dependent on our digital tools and operate in a mode of perpetual interruption and hypercoordination. New forms of discipline and social protocols will be called for in this 'post-human era.'

Rheingold ends by calling for more interdisciplinary studies of cooperation systems, cognitive and social impacts of new media, reshaping of urban environments, and regulatory models.

Book review by Madanmohan Rao

35. Roam: Making Sense of the Wireless Internet

Bruno Giussani, *Roam: Making Sense of the Wireless Internet*. Random House Business Books, 2001, 320 pp.

This book is an informative and hype-free roadmap of the wireless Internet economy, spanning major markets in Europe, the US and Asia. The 12 chapters are divided into four sections: market evolution, mobile content, handheld devices and user issues.

Bruno Giussani (www.giussani.com) is a European media analyst and formerly Internet columnist for *The New York Times*. He has been involved with Internet services companies in Switzerland and is an advisor to non-profits like Netaid.

'I have no doubt that we have entered a long-term trend towards ubiquitous connectivity, and that wireless will redefine many corporations and recreate significant portions of our physical and social reality,' says Giussani.

According to media scholar Manuel Castells, the 'flow of power' of the industrial age has given way to the 'power of flows' of the Internet age.

Wireless technologies exhibit a wide range of manifestations: 3G, Bluetooth, GPRS, CDMA 2000, SMS, WLAN and DAB (digital audio broadcasting). The wireless network layers will mesh with the wired Internet to create a 'total area network' (TAN).

According to Martin Cooper, inventor of the portable cellular network in 1973 (when he was director of research at Motorola), the ability to use the spectrum has doubled every 30 months for the last 105 years ('Cooper's Law').

The increasingly intimate relationship to the mobile phone and the connectedness it allows has led to descriptors like 'handy' in Germany and 'the small hand' in Scandinavia.

The usage of mobile phones by victims during the tragic events of September 11 illustrated, among other things, the power of mobile phones 'to channel one person's consciousness into another's a few seconds before death.' Instant mobile communication even helped the passengers on the fourth hijacked aircraft to find out about events in New York and Washington, and act accordingly.

In terms of market evolution, telecom operators who missed the first wave of the Internet (well capitalised by companies like Amazon, e-Bay, Yahoo) did not want to replicate the mistake by missing the second—and maybe last—chance to get a bigger piece of the Internet cake, according to Giussani.

The year 2000, with the general IT and telecom downturn, marked a disastrous period for telecom operators in Europe who became intoxicated by their own marketing hype about the mobile information society. Fantasy projections and over-optimistic forecasts were the order of the day in 1999 and 2000 as telecom operators behaved like 'lemmings going over the edge of a cliff.' The British 3G spectrum auctions of April 2000 were equivalent to a 'financial timebomb' due to the steep licence fees (totalling 36.8 billion euros) and stringent service obligations involved.

Serious regulatory debates continue to revolve around the relevance of the auctioning model for telecom licences, revenue-sharing options, reconciling public-private interests, impact on long-term national economy/infrastructure, scheduling of licence fee payments, the constraints raised by the scarcity of available spectrum, and alternatives like a 'beauty contest' based purely on merits of bidding companies' proposals.

Finland, home of Nokia, was the first country to attribute UMTS licences (in March 1999). The licences were given away for free, thus avoiding burdening operators with unnecessary debt.

Many crucial pieces of the 3G utopia such as compelling content, affordable devices, back-end fine-tuning of billing, call handover, interconnectivity and user profiling are tricky to put in place quickly and smoothly.

As for the early disappointments with WAP, it sparked off numerous puns and digs at the acronym such as 'Waplash,' 'Wapathy,' 'Wait and Pay,' 'Wrong Approach to Portability,' 'Where Are the Phones?' and even 'WAP is crap.'

WAP in the late 1990s was a case of over-promise and under-delivery, according to Giussani. The user's expectation of usability, cost and perceived value were not met. As with the early 'walled gardens' of CompuServe and AOL in the Internet Age, mistakes were repeated as cellular operators in Europe also created walled gardens and stuck to a narrow revenue focus rather than create a new industry via revenue sharing with content players.

However, WAP has worked relatively well in smaller niche sectors, like the medical community. The British medical site doctors.net.uk provides registered practitioners with a WAP link to the Medline reference database. Packet-switched networks like GPRS will probably enhance usage in terms of speed and give WAP a second chance.

On the other hand, NTT DoCoMo's i-Mode succeeded much better in Japan thanks to its being marketed as a mobile lifestyle devoid of any major comparison with the Internet which might have created wrong customer expectations. It was launched right from the onset with services from big Japanese banks (thus lending an aura of credibility and trust), affordable messaging services, and billing solutions for preferred content partners.

'We have not been looking for killer apps, we have tried instead to create a killer environment,' explains i-Mode creator Takeshi Natsuno. The environment lets service providers innovate, thanks to a win-win revenue sharing scheme (as compared with fragmented and even confrontational approaches in European and US markets).

Due to the popularity of services during moments of commute or idle time, there is a vast space opening up for 'a whole economy centred on satisfying two-minute needs'—helping kill short periods of idle time by doing purposeful or fun things.

DoCoMo has also tied up with gaming services platforms like Playstation, and has Java phones (called I-Appli) for business applications like fleet management, reporting, sales inventory lookup and mobile workforce updates.

But Giussani has reservations about the potential of i-Mode's expansion to Europe, particularly due to higher cost of the handset and the looser grip on the European market by operators as compared with NTT DoCoMo in Japan.

On the other hand, SMS has been a runaway success in Europe and Asia. Introduced as a standard store-and-forward feature of the GSM communication system in 1991 (typically for operator messages to subscribers about voicemail), it allows creation of messages of 160 characters (in Latin alphabet, or 70 characters for Arabic or Chinese).

'SMS is an accidental sensation,' says Giussani. The world's first SMS message was sent in 1993 by Vodafone in Britain, and the world's first Web-SMS gateway was created in South Africa by Mobile Telephone Networks. The grassroot craze rapidly took most mobile industry executives by surprise.

The book is full of case studies of wireless applications in action. For instance, the World Economic Forum in Davos (where Giussani was director of Internet strategy) used Wi-Fi-enabled iPAQs loaded with the conference agenda and wireless e-mail for delegates. FedEx has been using handhelds to record deliveries for several years now. In the US, two-thirds of all mobile calls are made from cars.

Numerous wireless startups are profiled in the book, such as Elektrosmog, Seattle Wireless, Flytxt, Mobilocity, Airwave, MobileStar, Wayport, BlueKite, uB-Mobile, CellPoint, Omnisky, and WorldZap.

Wi-Fi startup SFLan's motto is 'radio locally, fibre globally.' Bluetooth startups are hoping to be able to connect 'piconets' into a 'scatternet.' ArrayComm plans to use smart antennae technology as in the human ear to filter out non-relevant sounds via adaptive array processing.

Jippii Group of Finland has seven in-house composers whose task is to write the next big 15-second hits delivered as ringtones, accounting for as much copyright fees as all Finnish radio stations in 2000.

In Sweden, a cocktail information service sends ingredient lists by SMS. Chocolates can be ordered via SMS in Finland thanks to a service called GSM Chocolate. Real-time auctions are conducted via SMS in Germany. Builders and plumbers in Britain get job notifications via SMS.

French banks send account alerts via SMS. Swiss smokers trying to quit get supportive messages via SMS; some Swiss residents also get SMS reminders for taking out garbage; and Swiss railway tickets can be delivered via SMS after Web payment.

Germany and the UK have even had SMS poetry competitions. Students in the UK have been contacted for blood donation campaigns coordinated via SMS. Rave parties in many European clubs are promoted via SMS to regulars. The CelebSightings service helps mobile users track celebrities in New York and Los Angeles.

'Bridget Jones' Diary' fans can receive SMS messages from 'Bridget' Motorola and CodeOnline developed 2,000 questions for SMS trivia competitions. nGames hopes to extend playing of the same game across multiple platforms and locations (called 'pervasive gaming'). iProx helps users track the movements of consenting friends who sign up for the LBS service.

Gambling is another promising stalwart in the mobile world, as is pornography (thanks to companies like Sexwaps, Erotigo and Palmstories).

Dating services heavily use SMS. And in the Philippines, SMS even helped topple president Joseph Estrada via coordinated popular protests.

On the flip side, MMS, the next big thing after SMS, is in danger of being overhyped, Giussani cautions. Serious challenges also arise with mobile spamming, and some countries are contemplating anti-spam legislation.

New opportunities are opening up for voice portals and SMS-based portals. A good wireless advertisement must be actionable, relevant and not affect service performance, according to Omnisky. Barcode scanning will become a viable option on many handheld devices. Queue-busting applications with handhelds are becoming used in an increasing number of stores.

In the workspace, mobile applications can help supply chain operations, knowledge gathering/dissemination, monitoring of machines, and customer satisfaction. The supply chain can be turned into the warehouse if wireless solutions are effectively harnessed.

The US army uses RFID tags to enable 'just-in-time' logistics. Bell Canada, California's Sempra Energy and Britain's Idea Boilers are using wireless data transmission solutions for gas installers, technicians and engineers.

On the device front, a plethora of appliances are emerging: PDAs, smartphones, special-purpose devices, tablet PCs and even digital paper. They fall into four categories: dedicated, integrated, modular, and federated. The stakes are high for players like Palm, Microsoft and Symbian on the operating system front. The telematics market—wireless in cars—is dominated by OnStar, Wingcast, Egery and Connect.

Challenges arise in grafting together devices which are 'genetically different.' Ideally, mobile devices should pass the '4-B' test: they should be usable on the beach, in a bus, in the bathroom and in bed.

On the user front, challenges arise in creating usable services in areas like security, which need to ensure authentication, access control, confidentiality, integrity and non-repudiation. A well-designed system may even ensure that the mobile phone becomes a single authentication device for a user's needs.

Legal challenges can arise in terms of defining and enforcing acceptable limits of radiation, signal jammers in sensitive locations and public spots, location of antennae, and even their design.

Further on down the road, emerging frontiers for wireless including extending connectivity to rural areas in developing countries. 'Connectivity is productivity,' says Iqbal Quadir, founder of Grameen Phone in Bangladesh.

Book review by Madanmohan Rao

36. The Wireless Internet Opportunity for Developing Nations

The Wireless Internet Institute (eds), *The Wireless Internet Opportunity for Developing Nations*. Wireless Internet Institute/infoDev/ UNICT Task Force, 2003, 127 pp.

Can networks based on Wi-Fi and Wi-Max help developing regions of the world catch up in terms of access to online information and communication services? What regulatory changes can help developing countries leverage unlicensed radio spectrum for online applications ranging from e-government to e-health? How can entrepreneurs and the not-for-profit sector form alliances with technology players and government agencies on the wireless Internet frontier?

Intriguing questions indeed, and a host of wireless Internet players have formed the Wireless Internet Institute (W2I: www.w2i.org) to address these concerns. This publication was prepared for a conference held at the UN in New York in 2003, and is co-published by W2I, the World Bank's infoDev programme and the UNICT Task Force.

The wireless Internet value chain includes equipment, applications, integrators and operators, listed as follows:

1. Equipment: chipsets (Intel, Cisco, Agere, TI, Broadcom), access points and client radios (Alvarion, Linksys, D-Link, Symbol, 3Com, Motorola, Lucent, Alcatel, Siemens), antennas and amplifiers (Buffalo, Hughes), wireless-enabled access devices (Dell, IBM, HP, Nokia, Palm, Sony)
2. Software and applications: security (BlueSocket, Elixar, Escalon), optimisation (Wild Packets, Mesh Networks), network management (Bridgewater, Airwave), accounting (iPass, GRIC, Nomadix)
3. Integrators and operators: integration (Nortel, IBM, Wipro, Alcatel, Siemens), private hotspots (iPass, T-Mobile, Wayport, GRIC, Boingo, hotels, airports, restaurants), public hotspots (NYC Wireless, BAWUG, Madrid Wireless) and industry organisations (Wi-Fi Alliance, ITU, ACM, IEEE, WECA).

Wireless technologies have a key role to play in helping developing countries leapfrog other platforms due to the considerable speed and relatively low costs of deployment, according to the UN's Kofi Annan.

Cellular wireless is already accelerating in many developing countries; they exceed wireline penetration in many countries, e.g., in those of Africa. Globally, cellular connections surpassed fixed lines by 2002. But information gaps are growing between developed and developing countries, and even among developing countries.

'The Internet's biggest advantage over other communication technologies is that it can provide an array of solutions and products,' say the authors; these include information and collaboration for e-governance, e-health, e-education and e-commerce.

New networks based on standards like Wi-Fi (IEEE 802.11, for local access) and Wi-Max (802.16, for backhaul) can help tap low-cost broadband Internet infrastructure and bridge the last-mile gap, thus accelerating technology transfer and adoption. Considerations of sustainability and profitability can be met with adequate planning and investment. Risks like theft of communication wires can reduced, as also costs of set-up.

The 802.11 capabilities are being built into more consumer devices like laptops and PDAs.

Intriguing possibilities open up via 'intelligent phones,' and mesh networks (which can help reduce the number of base stations). Newer devices can interact with multiple networks, thus enabling switching between services like WLAN VoIP and standard cellular mode. 'Software-defined radio' is another promising development on this front. Wireless local loops (WLL) are about a third of the corresponding landline costs, while packet-based broadband networks can cost as little as a ninth.

On the regulatory front, the World Radio Conference of 2003 allocated the 2.4 GHz and 5 GHz bands to licence-exempt applications (supported by 96 per cent of developed countries and 41 per cent of developing countries). Data standards are generally defined in IEEE, while cellular standards are coming from the ITU (GSM operates at 900 MHz).

Case Studies

The book also includes 12 case studies in five categories: remote regions, wireless ISPs, shared access, adaptive technologies, and rebuilding nations. The case studies are written up along the following parameters: goals, accomplishments, infrastructure, funding, regulatory environment and scalability.

The ISP WorldLink used Cisco Aironet products and Americom satellite links to provide Internet access at the base camp of Mount Everest. While useful to tourists, local demand for services is still feeble, and a local trainer will need to be found. The Amazon Association used solar energy, VSAT links from OnSat and a hotspot to provide services like e-health but faces serious challenges in financial viability.

ISP Africa Online uses Alvarion BreezeNet equipment to provide wireless access. Seventy per cent of users in Malawi are commercial clients and large farmers. Though not directly benefiting poor farmers, it has given donors and NGOs better access to manage their activities. But the market size is small, and costs of the Internet connection to the US are very high. Baja Wireless uses D-Link equipment to provide wireless Internet access in Mexico, but Telmex has prohibited it from offering VoIP services. E-Link Americas, funded by the Institute for Connectivity of the Americas in Ottawa, is planning on Internet access via DVB-RCS (digital video broadcasting return channel via satellite) and a network of local country-specific partnerships. Wireless ISP UniNet has had its WLAN equipment confiscated in South Africa by regulators.

The McCormack Institute plans to use wireless technologies to broaden Internet access from the universities of African countries like Kenya and Senegal, to a wider user base of citizens via community resource centres. SARI (Sustainable Access in Rural India) uses WLL technology from local company n-Logue to provide Internet access via kiosk operators for e-governance and agricultural consultations. DakNet in the south Indian state of Karnataka uses a government bus as a mobile access point for store-and-forward data services. An Indian-manufactured bicycle-pedalled generator was used by the Jhai Foundation to power Wi-Fi-enabled PCs in a remote village in Laos.

Internet Project Kosovo (IPKO) launched broadband wireless connectivity in Kosovo in 1999 in the weeks after the war, to UN agencies, NGOs, cybercafes and businesses. UNDP uses WLANs at telekiosks in the airport and post offices in Kabul.

'The inspiring mobile telephony success story of Grameen Phone in rural Bangladesh shows how entrepreneurial villagers can be recruited to adopt a technology that provides them greater connectivity and therefore economic opportunity on a whole new scale,' says the authors. Its network is accessible in more than 35,000 villages, providing telephone access to millions while fostering a generation of village micro-entrepreneurs.

The Road Ahead

Despite some of these early successes, the fact remains that hotspots and VoIP are still perceived as too disruptive and threatening to incumbent operators in some countries. Other political restrictions can also be slapped onto information access. Strategies ranging from direct confrontation and open demos to negotiation and external alliancing should be developed.

Issues need to be resolved in terms of interaction between Wi-Fi/Wi-Max and 3G (e.g., more nomadic applications with 802.16e, and mid-bandwidth data services with 3G). No-interference working plans and pro-active peer agreements will still need a lot of attention for players in the unlicensed spectrum bands.

Challenges will arise in ensuring acceptable levels of network performance (e.g., by reducing the number of hops), and creating affordable yet workable interconnections between various telecom service providers and ISPs.

Business models will be tough to work out in low-income areas, low-demand communities, and poorly financed districts. Building understanding and capacity in managing online information service centres will be a challenge; local ownership and a sense of pride must be promoted. In extremely poor areas, even basic power supply will remain an issue.

Recommendations

In conclusion, W2I has a number of recommendations for developing countries on the wireless Internet front:

1. Establish national consensus on the benefits of wireless Internet applications.
2. Support the use of unlicensed spectrum, and ICT standards like Wi-Fi and Wi-Max.
3. Create a competitive market for Internet services.
4. Foster cooperation among potential actors: operators, content players, local communities and financiers.
5. Support experimentation in new services, and take the lead in areas like e-government.
6. Share best practices at national and global levels.

Book review by Madanmohan Rao

Index